THE QUALITATIVE THEORY OF ORDINARY DIFFERENTIAL EQUATIONS

AN INTRODUCTION

Fred Brauer and John A. Nohel
University of Wisconsin

Jin Rong Liao
Aug. 1991

DOVER PUBLICATIONS, INC., NEW YORK

To our parents

Published in Canada by General Publishing Company, Ltd., 30 Lesmill Road, Don Mills, Toronto, Ontario.

Published in the United Kingdom by Constable and Company, Ltd., 10 Orange Street, London WC2H 7EG.

This Dover edition, first published in 1989, is an unabridged, corrected republication of the work first published by W. A. Benjamin, Inc., New York, 1969.

Manufactured in the United States of America
Dover Publications, Inc., 31 East 2nd Street, Mineola, N.Y. 11501

Library of Congress Cataloging-in-Publication Data

Brauer, Fred.
The qualitative theory of ordinary differential equations.

Reprint. Originally published: New York : W. A. Benjamin, 1969.
Bibliography: p.
Includes index.
1. Differential equations. 2. Stability. I. Nohel, John A. II. Title.
QA372.B823 1989 515.3′5 88-30943
ISBN 0-486-65846-5 (pbk.)

PREFACE

In this book we present a self-contained introduction to some important aspects of modern qualitative theory for ordinary differential equations. Since only a minimal background in techniques of solution of differential equations such as is normally acquired in an elementary undergraduate course is assumed, the book is accessible to any student of physical sciences, mathematics, or engineering who has a good knowledge of calculus and of the elements of linear algebra. In addition, algebraic results are stated as needed; the less familiar ones are proved either in the text or in appendixes.

Our principal objective is the study of stability theory in Chapters 4, 5 and the applications considered in Chapter 6. Here the reader will find a rigorous but elementary treatment of several topics which are of importance in modern engineering and physics. In effect, these methods provide the justification of many approximations ordinarily assumed to be valid. At the same time readers interested in mathematical theory will find here an introduction to several topics of current research interest. Chapters 1, 2, and 3 serve as preparation. Some of this preparatory material has already been presented (but in less detail) in [2]. The reader familiar with Chapters 6 and 7 of [2] can begin the present book with Chapter 4.

There are, of course, several interesting and important topics which we have omitted and for several reasons. For example, the Poincaré-Bendixson theory of plane autonomous systems and the theory of the index of a critical point are not included, because we can hardly improve upon the beautiful presentation by W. Hurewicz in [13]. Other topics, such as the use of fixed point theorems and implicit function theorems require a more sophisticated background than we wish to assume. For this reason we have also omitted the important methods of Poincaré and of averaging for establishing existence of and approximation to periodic solutions of almost linear perturbed systems. In a different direction, the reader interested in pursuing the study of boundary value problems is referred to Chapter 5 of [2] for an introduction and to Chapters 7 to 12 of [4] or to [26] for a more advanced treatment.

At the University of Wisconsin we have each taught one-semester courses to juniors and seniors (in mathematics, physical sciences, and engineering) based on the first five chapters. There is, however, ample material here for a one-semester course for those students who are prepared to begin with Chapter 4.

Throughout the text and at the end of several chapters, the reader will find numerous exercises, some routine, and some more difficult. These are designed to help follow the argument and to provide a better understanding of the subject and, thus, they form an essential part of the book. The interested reader will find many problems from the first three chapters solved in our book, *Problems and Solutions in Ordinary Differential Equations* (W. A. Benjamin, Inc. New York, 1968).

It is impossible to acknowledge all the help, direct and indirect, from which we have benefited in the preparation of this book. Certainly, the stimulating influences of Professors Norman Levinson and Earl Coddington, and of our colleagues Charles Conley, Jacob Levin, and Wolfgang Wasow, have been valuable. Professors H. A. Antosiewicz, J. K. Hale, A. Strauss, and A. D. Ziebur read portions of a preliminary version of the manuscript and we acknowledge with pleasure their many useful suggestions. Professors B. Berndt, D. Ferguson, and J. Williamson gallantly served as our assistants in teaching some of this material and we are grateful for their helpful comments. In preparing Chapter 6 we have found valuable ideas in some unpublished lecture notes by Professor Lawrence Markus and we are grateful to him for making them available to us. Finally, it is a pleasure to acknowledge the help of Mrs. Phyllis J. Rickli who has converted almost illegible handwriting into a clean manuscript, and the help of the staff of W. A. Benjamin, Inc., in the construction of this book. Naturally any errors that may remain in spite of all this assistance are our responsibility, and we would appreciate being advised of them.

Fred Brauer
John A. Nohel

Madison, Wisconsin
October 1968

CONTENTS

Chapter 1 | SYSTEMS OF DIFFERENTIAL EQUATIONS

Differential equations originated in Newton's efforts to explain the motion of particles. In modern science and technology, the mathematical description of complex physical processes often leads to systems of ordinary differential equations. However, we shall use only the very simple mass-spring systems to illustrate this and leave other examples to the exercises. In spite of their simplicity these examples are prototypes of mathematical models of other much more complicated physical systems. Further, these examples serve to motivate and illustrate much of the theory with which we will be concerned.

Before turning to the construction of these mathematical models, we recall some aspects of the Newtonian model for the motion of a system of particles. In this model, it is assumed that a body, called a particle, can be represented as a point having mass. (We shall assume knowledge of the rather difficult concept of mass; for practical purposes, mass can be measured by the weight of the body.) It is assumed that, in the absence of "forces," the motion of each particle is unaccelerated and therefore is straight-line motion with constant, perhaps zero, velocity (**Newton's first law**). The presence of acceleration is therefore to be interpreted as a sign of the presence of a **force**. This is a vector quantity* given by

Newton's second law:

If **F** is the force acting on a particle of mass m moving with a velocity

* Throughout the book vectors will be indicated by boldface type.

(vector) **v**, then

$$\mathbf{F} = \frac{d}{dt}(m\mathbf{v})$$

The vector quantity $m\mathbf{v}$ is called the momentum of the particle. If the mass is constant, Newton's second law may be written as

$$\mathbf{F} = m\frac{d\mathbf{v}}{dt} = m\mathbf{a},$$

where **a** is the acceleration vector of the particle. For a system consisting of several particles, Newton's laws are applicable to each particle.

In the Newtonian model, the gravitational force can be shown (experimentally) to be proportional to mass, so that problems involving gravitational forces on particles near the earth's surface can be handled conveniently by assuming that the acceleration g due to gravity is constant.

1.1 A Simple Mass-Spring System

A weight mass of m is suspended from a rigid horizontal support by means of a very light spring (see Figure 1.1). The weight is allowed to move only along a vertical line (no lateral motion in any direction is permitted). The spring has a natural (unstretched) length L when no weight is suspended from it.

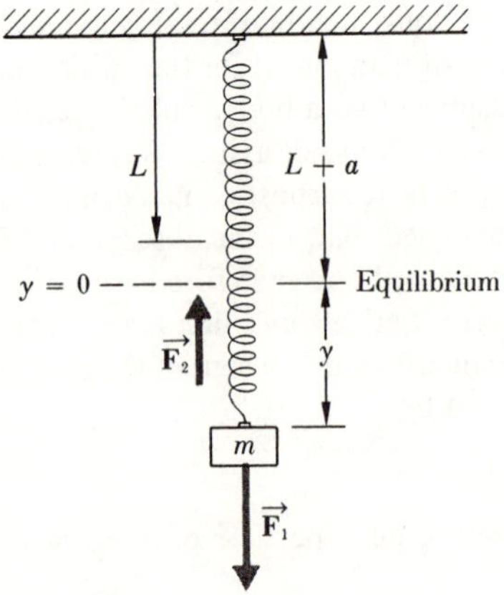

Figure 1.1

When the weight is attached, the system has an equilibrium position at which the spring is stretched to a length $L + a$, where a is a positive number. We may set the system in motion by displacing the weight from this equilibrium position by a prescribed amount and releasing it either from rest or with some prescribed initial velocity. Our task is to describe in mathematical terms the motion of the system.

Since the motion is restricted to a vertical line, the position of the weight can be described completely by the displacement y from the equilibrium position (see Figure 1.1). The mathematical equivalent of the motion of the mass-spring system will then be a function ϕ such that $y = \phi(t)$ describes the position of the weight for each value of $t \geq 0$, where $t = 0$ represents the starting time of the motion. In order to determine the motion, that is, to determine the function ϕ, we must impose additional restrictions on ϕ. For example, if we displace the weight a distance y_0 and then release it, we would require that $\phi(0) = y_0$. If we release it from rest at this position, we will also require that $\phi'(0) = 0$. Experience suggests that with these additional conditions, the motion is completely determined.

In order to obtain a mathematical model for this system, we must use appropriate physical principles and certain simplifying approximations. The basic tool is Newton's second law. We first give mathematical expressions for the forces acting on the weight using physical principles and approximations, and then using Newton's second law we obtain an equation that must be satisfied by the function ϕ. Our assumptions are as follows:

(a) The spring has zero mass.
(b) The weight can be treated as though it were a particle of mass m.
(c) The spring satisfies **Hooke's law**, which states that the spring exerts a restoring force on the weight toward the equilibrium position; the magnitude of this force is proportional to the amount by which the spring is stretched from its natural length. The constant of proportionality $k > 0$ is called the spring constant, and a spring obeying Hooke's law is called a **linear spring**.
(d) There is no air resistance and the only external force is a constant vertical gravitational attraction.

We stress the fact that a different set of physical assumptions would lead to a different mathematical model. Further, the accuracy of a particular mathematical model in predicting physical phenomena will depend primarily on the reasonableness of the physical assumptions.

Newton's second law is stated above in terms of vectors. Since in this problem the motion is restricted to a line, the vectors involved are one-dimensional, and vector notation is not needed. With reference to Figure 1.1, we shall measure the displacement y from equilibrium ($y = 0$), choosing the downward direction as positive. The force of gravity F_1 in Figure 1.1 is mg, and the restoring force of the spring F_2 is $-k(y + a)$ by Hooke's law. Observe that Figure 1.1 has been drawn with $y > 0$ so that F_2 is directed upward.

• EXERCISE

1. Sketch the analogue of Figure 1.1 with $y < 0$ and compute the forces F_1 and F_2 in this case.

The total force acting on the weight is

$$F_1 + F_2 = mg - k(y + a)$$

The equilibrium position occurs when this total force is zero. Therefore, at equilibrium, $mg - k(0 + a) = 0$, or $a = mg/k$. Thus we can rewrite the total force at any position y of the mass as

$$F_1 + F_2 = mg - k\left(y + \frac{mg}{k}\right) = -ky.$$

By Newton's second law,

$$F_1 + F_2 = -ky = \frac{d}{dt}(mv) = m\frac{d^2y}{dt^2}$$

Therefore the motion of the system is specified by the equation

$$\frac{d^2y}{dt^2} + \frac{k}{m}y = 0 \tag{1.1}$$

Equation (1.1) is the mathematical model for the mass-spring system under the assumptions (a), (b), (c), (d). It is a **differential equation** (of the second order). To obtain specific information about a particular motion, we must specify other information. For example, we have already remarked that if we release the weight from rest with an initial displacement y_0, we must impose the pair of **initial conditions**

$$\phi(0) = y_0 \qquad \phi'(0) = 0 \tag{1.2}$$

The mathematical problem is then to find a function ϕ defined for all $t \geq 0$ satisfying the differential equation (1.1), that is, $\phi''(t) + (k/m)\phi(t) = 0$ for

$t \geq 0$, and the initial condition (1.2). Such a function is called **a solution of the differential equation (1.1) obeying the initial conditions (1.2).** We may hope that this solution will give a good approximation to the actual motion of a real mass-spring system, and that we can use the solution to predict properties of the motion that can be measured experimentally.

We can modify the model in several ways by attempting to use physical laws that are closer to reality. For example, leaving assumptions (a), (b), (c) intact, we could replace assumption (d) by

(d′) There is air resistance proportional to the velocity in addition to the gravitational attraction.

In this case, the mathematical model for the mass-spring system will be the equation

$$\frac{d^2y}{dt^2} + \frac{b}{m}\frac{dy}{dt} + \frac{k}{m}y = 0 \tag{1.3}$$

together with the initial conditions (1.2), in place of (1.1) and (1.2). The term $(b/m)\,dy/dt$ is the appropriate mathematical translation of the resistance force of air, where b is a nonnegative constant.

• EXERCISE

2. Derive equation (1.3). [*Hint:* In Figure 1.1 there will now be a force F_3 arising from assumption (*d′*).]

Another possible model is obtained by replacing (c) by the assumption that there is a restoring spring force that is not necessarily linear and leaving assumptions (a), (b), (d) intact. In this case, we replace Equation (1.1) by

$$\frac{d^2y}{dt^2} + g(y) = 0 \tag{1.4}$$

where $g(y)$ is a so-called nonlinear spring term. To conform with reality, we might assume that $g(y)$ is positive when y is positive and negative when y is negative. The precise form of the function g depends on the physical law assumed in place of Hooke's law; we might have $g(y) = (k_1/m)y + k_2 y^3$.

Differential equations such as (1.1), (1.3), or (1.4) describe the "equation of motion" of particular systems. As we shall see, their solutions describe the nature of all the possible motions of the physical system as predicted by each mathematical model. When conditions such as (1.2) are added, we single out one or more special solutions to predict the behavior of the system if the motion starts from some particular state or configuration.

The reader may wish to acquire additional facility in the mathematical formulation of simple physical problems. The exercises below provide some practice in this, and give rise to differential equations that will be used frequently for illustrative purposes.

• EXERCISES

3. A pendulum is made by attaching a weight of mass m to a very light and rigid rod of length L mounted on a pivot so that the system can swing in a vertical plane (see Figure 1.2). The weight is displaced initially to an angle θ_0 from the vertical and released from rest.

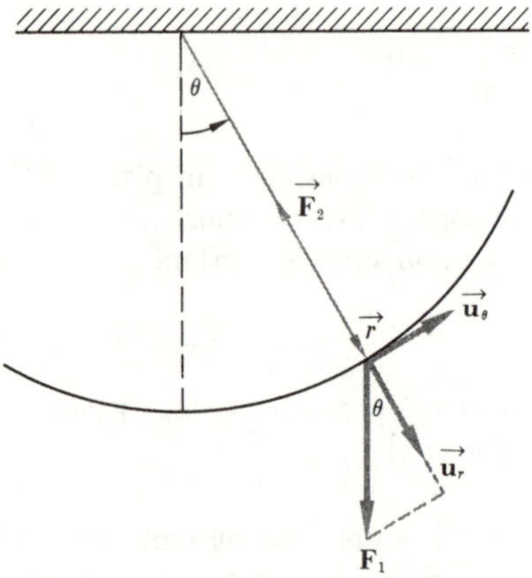

Figure 1.2

Derive, or look up (see, for example, [2]), the equations of motion of the pendulum under the following assumptions:

(i) The rod is rigid, of constant length L, and of zero mass.
(ii) The weight can be treated as though it were a particle of mass m.
(iii) There is no air resistance; the pivot is without friction; and the only external force present is a constant vertical attraction.

[*Hint:* At any time t, the gravitational force $\mathbf{F}_1$ has magnitude mg and is directed downward. There is also a force $\mathbf{F}_2$ of tension in the rod of magnitude T directed along the rod toward the pivot.]

Answer:

$$-mL\left(\frac{d\theta}{dt}\right)^2 = mg\cos\theta - T$$
$$mL\frac{d^2\theta}{dt^2} = -mg\sin\theta \tag{1.5}$$

where $\theta(0) = \theta_0$ and $\theta'(0) = 0$. Note that the first of these equations contains also the unknown quantity T. However, if the angle θ can be determined from the second equation, then the magnitude of the tension T can be found from the first equation; in fact, $T = mg\cos\theta + mL(d\theta/dt)^2$. Therefore, the motion of the pendulum is completely determined by the second equation, which may be written in the form

$$\frac{d^2\theta}{dt^2} + \frac{g}{L}\sin\theta = 0 \tag{1.6}$$

4. Suppose we replace assumption (iii) in Exercise 3 by:

(iv) The pendulum encounters resistance, due to the pivot and the surrounding air, which is proportional to the velocity vector;

and leave the remaining assumptions unchanged. Show that the equation of motion for this system is

$$\frac{d^2\theta}{dt^2} = -\frac{g}{L}\sin\theta - \frac{k}{m}\frac{d\theta}{dt} \tag{1.7}$$

in place of (1.6). The last term is the appropriate mathematical translation of the additional resistance force. Note that equation (1.7) reduces to (1.6) if $k = 0$.

5. What is the magnitude of the tension assuming that θ has been found from Equation (1.7)?

1.2 Coupled Mass-Spring Systems

Consider a mass m_1 suspended vertically from a rigid support by a weightless spring of natural length L_1, with a second mass m_2 suspended from the first by means of a second weightless spring of natural length L_2 as shown in Figure 1.3. We shall make the same assumptions here as we did for the single mass-spring system in Section 1.1. In particular, we assume that the masses m_1 and m_2 can be treated as point masses, and that both springs obey Hooke's law and have respectively the spring constants k_1, k_2. We let $y_1(t)$, $y_2(t)$ be

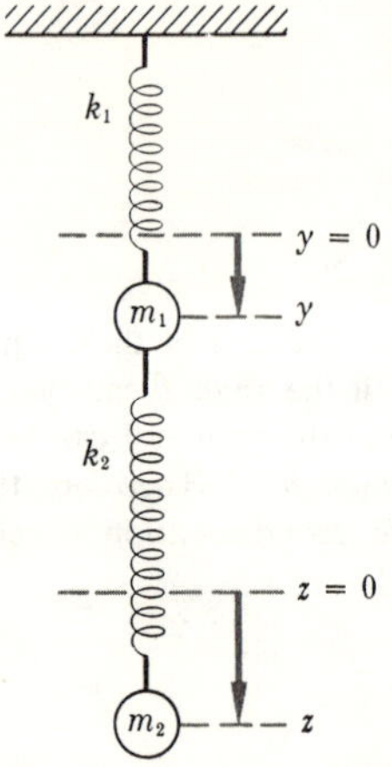

Figure 1.3

the respective displacements at time t of the masses m_1, m_2 from equilibrium (that is, the point at which the system remains at rest, before being set into motion). As in the simple case, the quantities y_1 and y_2 are vector functions of time. However, since the motion is along a straight line, no confusion will arise if vector notation is not employed. We also assume that air resistance is negligible and that no external forces other than gravity act on the system.

The description of the model is completed by specification of $y_1(0)$, $y_1'(0)$, $y_2(0)$, $y_2'(0)$, the initial displacement and initial velocity of each mass. To derive the equations for the motion under the present hypotheses, we apply the same technique as in the simple case in Section 1.1. It is easy to see that at any time t the net force acting on the mass m_2 is $-k_2[y_2(t) - y_1(t)]$, while that acting on the mass m_1 is $-k_1y_1(t) + k_2[y_2(t) - y_1(t)]$.

• EXERCISE

1. Show that at any time t the net force acting on m_1 is $-k_1y_1(t) + k_2[y_2(t) - y_1(t)]$. [*Hint:* Let $L_1 + a_1$ be the equilibrium position of the mass m_1 measured downward from the vertical support; let $L_1 + L_2 + a_2$ be the position of m_2 measured downward from the support, assuming that m_1 is zero. Now add the mass m_1 to get the equilibrium position of the system. Write down the sum of the forces acting on m_1 amd m_2, and then evaluate the constants a_1, a_2 as in Section 1.1. Note that the final expression for the net force is independent of m_1, m_2, L_1, L_2, a_1, a_2.]

Thus by Newton's second law we have immediately

$$\begin{aligned} m_1 y'' &= -k_1 y_1 + k_2(y_2 - y_1) \\ m_2 y_2'' &= -k_2(y_2 - y_1) \end{aligned} \tag{1.8}$$

as the differential equations describing the motion. As pointed out above, we also prescribe the initial values $y_1(0)$, $y_1'(0)$, $y_2(0)$, $y_2'(0)$. Thus our problem has led us to an initial value problem for **a system of two differential equations**, each of second order.

By a solution of this problem we mean a pair of functions ϕ_1, ϕ_2 defined for $t \geq 0$, twice differentiable, satisfying for each t the equation (1.8) and for $t = 0$ the given initial conditions. Naturally, the same questions about the accuracy of the present model and about the reasonableness of the various hypotheses can be asked, just as in the single mass-spring system considered in Section 1.1.

• EXERCISES

2. Derive the equations of motion of the system shown in Figure 1.3 if it is assumed that air resistance is proportional to velocity.

3. Consider three masses m_1, m_2, m_3 connected by means of three springs (obeying Hooke's law) with constants k_1, k_2, k_3 and moving on a frictionless horizontal table as shown in Figure 1.4, with the mass m_3 subjected to a given

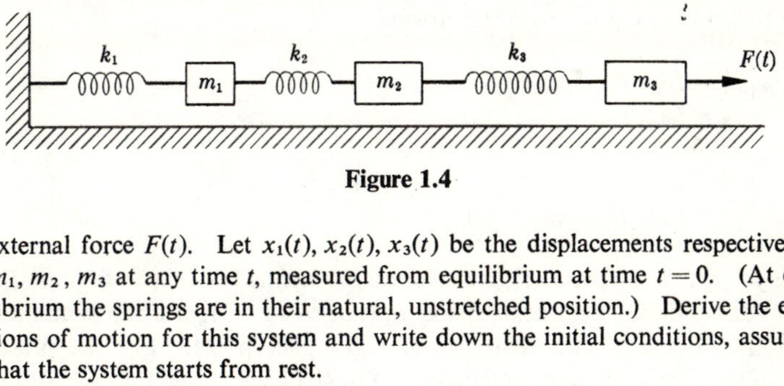

Figure 1.4

external force $F(t)$. Let $x_1(t)$, $x_2(t)$, $x_3(t)$ be the displacements respectively of m_1, m_2, m_3 at any time t, measured from equilibrium at time $t = 0$. (At equilibrium the springs are in their natural, unstretched position.) Derive the equations of motion for this system and write down the initial conditions, assuming that the system starts from rest.

It is clear that if n springs and n masses are used in the above problems, then the equations of motion would consist of n equations for the displacements of the masses, each equation being of second order.

• EXERCISE

4. Use Kirchoff's law (sum of voltage drops around a closed circuit equals zero) to write the differential equations satisfied by the currents i_1 and i_2 in the

idealized circuit shown in Figure 1.5, where L_1, L_2 are given constant inductances; R_1, R_2 are given constant resistances; and E is a given impressed voltage. (Recall that $Li'(t)$ is the voltage drop across an inductor of inductance L due to a current $i(t)$ and $Ri(t)$ is the voltage drop across a resistance R due to a current $i(t)$.)

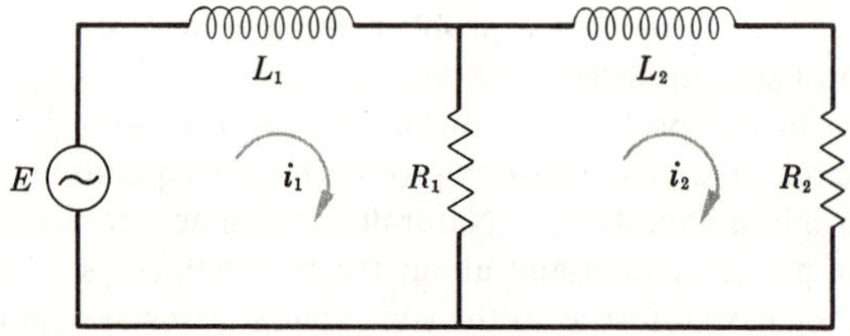

Figure 1.5

1.3 Systems of First-Order Equations

The examples in Section 1.2 cannot be conveniently expressed in terms of single differential equations. In this chapter we shall study systems of first-order differential equations of the form

$$\begin{cases} y_1' = f_1(t, y_1, y_2, \ldots, y_n) \\ y_2' = f_2(t, y_1, y_2, \ldots, y_n) \\ \vdots \\ y_n' = f_n(t, y_1, y_2, \ldots, y_n) \end{cases} \tag{1.9}$$

where $f_1, f_2, \ldots, f_n$ are n given functions defined in some region D of $(n+1)$-dimensional Euclidean space and $y_1, y_2, \ldots, y_n$ are the n unknown functions. For a precise definition of region, see p. 24. For the present, the intuitive notion is quite adequate. We shall see below that the systems considered in Sections 1.1 and 1.2 are special cases of the system (1.9). **To solve (1.9) means to find an interval I on the t axis and n functions $\phi_1, \ldots, \phi_n$ defined on I such that**

(i) $\phi_1'(t), \phi_2'(t), \ldots, \phi_n'(t)$ **exist for each** t **in** I,

(ii) **the point** $(t, \phi_1(t), \ldots, \phi_n(t))$ **remains in** D **for each** t **in** I,

(iii) $\phi_j'(t) = f_j(t, \phi_1(t), \phi_2(t), \ldots, \phi_n(t))$ for **each** t **in** I $(j = 1, \ldots, n)$.

Naturally, the functions f_j may be real or complex valued. We shall assume the real case unless otherwise stated. While the geometric interpretation (see, for example, [2, p. 15]) is no longer so immediate as in the case $n=1$, a solution of (1.9) (that is, a set of n functions $\phi_1, \ldots, \phi_n$ on an interval I) can be visualized as a curve in the $(n+1)$-dimensional region D, with each point p on the curve given by the coordinates $(t, \phi_1(t), \ldots, \phi_n(t))$ and with $\phi_i'(t)$ being the component of the tangent vector to the curve in the direction y_i. This interpretation reduces to the familiar one for $n=1$ and the curve in D defined by any solution of (1.9) can therefore again be called a **solution curve.** **The initial value problem** associated with a system such as (1.9) is the problem of finding a solution (in the sense defined above) passing through a given point P_0: $(t_0, \eta_1, \eta_2, \ldots, \eta_n)$ (we do not write $(t_0, y_{10}, \ldots, y_{n0})$ to avoid double subscripts) of D. In general, we cannot expect to be able to solve (1.9) except in very special cases. Nevertheless, it is desired to obtain as much information as possible about the behavior of solutions of systems. For this reason we shall develop a considerable amount of theory for systems of differential equations.

Example 1. Consider the differential equation $y' = y^2$. Here $n = 1$, the region D is the whole (t, y) space, $f(t, y) = y^2$ is defined everywhere, and

$$\phi(t) = \frac{\eta}{1 - \eta(t - t_0)}$$

is a solution on an interval I containing t_0 for which $\phi(t_0) = \eta$; for example, if $\eta > 0$, this solution exists for $-\infty < t < t_0 + 1/\eta$. To see this we verify that ϕ satisfies (i), (ii), (iii) of the definition.

• EXERCISES

1. Construct the above solution ϕ by the method of separation of variables (see, for example, [2, Ch. 2]) and sketch the graph.
2. What is the interval of validity of the above solution ϕ if $\eta < 0$?
3. Discuss the case $\eta = 0$.

Example 2. Consider the system

$$y_1' = y_1{}^2$$

$$y_2' = y_1 + y_2$$

Here $n = 2$, the region D is the whole (t, y_1, y_2) space, and

$$\phi_1(t) = \frac{\eta_1}{1 - \eta_1(t - t_0)}$$

$$\phi_2(t) = \eta_2 \exp(t - t_0) + \eta_1 \int_{t_0}^{t} \frac{\exp(t - s)}{1 - \eta_1(s - t_0)} ds$$

is a solution on an interval I containing t_0 for which $\phi_1(t_0) = \eta_1$, $\phi_2(t_0) = \eta_2$. For $\eta_1 > 0$ this solution exists for $-\infty < t < t_0 + (1/\eta_1)$.

• EXERCISES

4. Verify the statements made for the system in Example 2 above.
5. Construct the above solution by combining the method of separation of variables (Exercise 1 above) with the method of solving linear differential equations of first order (see, for example, [2, Ch. 2]).
6. Discuss the case $\eta_1 < 0$.
7. Discuss the case $\eta_1 = 0$
8. Verify that each of the following functions or sets of functions is a solution of the given differential equation or system satisfying the given initial conditions. Determine the interval of validity in each case.
   (a) $y' = ty^3$; $(t_0, \eta) = (0, 1)$; $\phi(t) = (1 - t^2)^{-1/2}$.
   (b) $y' = ty^3$; $(t_0, \eta) = (0, -1)$; $\phi(t) = -(1 - t^2)^{-1/2}$.
   (c) $y_1' = y_2$, $y_2' = y_1$; $(t_0, \eta_1, \eta_2) = (0, 1, 1)$; $\phi_1(t) = e^t$, $\phi_2(t) = e^t$.
   (d) $y_1' = y_2$, $y_2' = y_1$; $(t_0, \eta_1, \eta_2) = (0, 1, -1)$; $\phi_1(t) = e^{-t}$, $\phi_2(t) = -e^{-t}$
9. Consider the differential equation

$$y' = \begin{cases} 0 & (t \le 0; -\infty < y < \infty) \\ 2y^{1/2} & (t \ge 0; 0 \le y < \infty) \\ y^2 & (t \ge 0; -\infty < y < 0) \end{cases}$$

   (a) Determine whether

$$\phi(t) = \begin{cases} 1 & (t < 0) \\ (t + 1)^2 & (t \ge 0) \end{cases}$$

   is a solution on $-\infty < t < \infty$.
   (b) Is $\phi(t)$ continuous everywhere?
   (c) Is $\phi'(t)$ continuous everywhere?
10. Consider the differential equation $y' = 2/(t^2 - 1)$ with $f(t, y) = 2/(t^2 - 1)$ defined on each of the domains $D_1 = \{(t, y) \mid -\infty < t < -1, |y| < \infty\}$, $D_2 = \{(t, y) \mid -1 < t < 1, |y| < \infty\}$, and $D_3 = \{(t, y) \mid 1 < t < \infty, |y| < \infty\}$. Verify that

$$\phi(t) = \log\left|\frac{t - 1}{t + 1}\right|$$

is a solution of this equation on each of the intervals $-\infty < t < -1$, $-1 < t < 1$, and $1 < t < \infty$. (A graph of ϕ will show why $y = \phi(t)$ is not a solution on an interval such as $-2 < t < 2$.)

11. Consider the differential equation $y' = (y^2 - 1)/2$. Verify that

$$\phi(t) = \frac{1 + ce^t}{1 - ce^t}$$

is a solution of this equation on an appropriate interval I, for any choice of the constant c. [*Hint:* Try this first for specific values of c such as $c = 0$, $c = 1$, $c = -1$.] Draw graphs for ϕ for each choice of c.

We observe that (1.8) and the systems derived from physical considerations in the exercises in Sections 1.1 and 1.2 are systems of second-order equations, while (1.9) is a system of first-order equations. We shall show that the system (1.9) of first-order equations is sufficiently general to include all such problems, and in particular all single nth-order equations are included as a special case in (1.9). We shall also see that the theory of nth-order equations is a special case of the corresponding theory for systems of first-order equations.

Example 3. Consider the second-order equation

$$y'' = g(t, y, y') \tag{1.10}$$

where g is a given function. Put $y = y_1$, $y' = y_2$; we then have $y_1' = y' = y_2$ and from (1.10), $y'' = y_2' = g(t, y_1, y_2)$. Thus (1.10) can be described by the system of two first-order equations

$$\begin{aligned} y_1' &= y_2 \\ y_2' &= g(t, y_1, y_2) \end{aligned} \tag{1.11}$$

which is a special case of (1.9) with $n = 2$, $f_1(t, y_1, y_2) = y_2$, $f_2(t, y_1\, y_2) = g(t, y_1, y_2)$. To prove that (1.10) and (1.11) are equivalent, let ϕ be a solution of (1.10) on some interval I; then $y_1 = \phi(t)$, $y_2 = \phi'(t)$ is a solution of (1.11) on I since $y_1' = \phi' = y_2$ and $y_2' = \phi'' = g(t, \phi(t), \phi'(t)) = g(t, y_1, y_2)$. Conversely, let ϕ_1, ϕ_2 be a solution of (1.11) on I, then $y = \phi_1(t)$ (that is, the first component) is a solution of (1.10) on I since $y'' = \phi_1'' = (\phi_1')' = \phi_2' = g(t, \phi_1, \phi_2) = g(t, \phi_1, \phi_1') = g(t, y, y')$.

• EXERCISES

12. Write a system of two first-order differential equations equivalent to the second-order equation

$$\theta'' + \frac{g}{L}\sin\theta = 0$$

with initial conditions $\theta(0) = \theta_0$, $\theta'(0) = 0$, which determines the motion of a simple pendulum (Section 1.1).

13. Show that the equation $y''' + 3y'' - 4y' + 2y = 0$ is equivalent to the system of three first-order equations

$$\begin{cases} y_1' = y_2 \\ y_2' = y_3 \\ y_3' = -2y_1 + 4y_2 - 3y_3 \end{cases}$$

Example 3. The scalar equation of nth order

$$y^{(n)} = g(t, y, y', \ldots, y^{(n-1)}) \tag{1.12}$$

can be reduced to a system of n first-order equations by the change of variable $y_1 = y$, $y_2 = y'$, $\ldots$, $y_n = y^{(n-1)}$. Then (1.12) is seen to be equivalent to the system

$$\begin{cases} y_1' = y_2 \\ y_2' = y_3 \\ \quad\vdots \\ y_{n-1}' = y_n \\ y_n' = g(t, y_1, y_2, \ldots, y_n) \end{cases} \tag{1.13}$$

which is another special case of (1.9). The proof of the equivalence of (1.12) and (1.13) is only a slight generalization of the proof in Example 3.

• **EXERCISE**

14. Establish the equivalence of (1.12) and (1.13).

Example 5. Returning to the system (1.8) of two second-order equations governing the motion of the system of two masses in Figure 1.3, we let $y_1 = y$, $y_2 = y'$, $y_3 = z$, $y_4 = z'$, and we obtain that (1.8) is equivalent to the system of four first-order equations

$$\begin{aligned} y_1' &= y_2 \\ y_2' &= -\left(\frac{k_1}{m_1} + \frac{k_2}{m_1}\right) y_1 + \frac{k_2}{m_1} y_3 \\ y_3' &= y_4 \\ y_4' &= -\frac{k_2}{m_2}(y_3 - y_1) \end{aligned} \tag{1.14}$$

which is another special case of (1.9).

• **EXERCISES**

15. Establish the equivalence of the systems (1.9) and (1.14).

16. Write the systems of second-order equations derived in Exercises 2, 3, and 4, Section 1.2, as equivalent systems of first-order equations.

17. Reduce the system

$$y_1' + y_2' = y_1^2 + y_2^2$$
$$2y_1' + 3y_2' = 2y_1 y_2$$

to the form (1.9). [*Hint:* Solve for y_1' and y_2'.]

1.4 Vector-Matrix Notation for Systems

In Example 5, Section 1.3, we obtained the system (1.14), of four first-order equations. We notice that we can describe this system completely by giving the matrix of coefficients

$$A = \begin{pmatrix} 0 & 1 & 0 & 0 \\ -\dfrac{k_1 + k_2}{m_1} & 0 & \dfrac{k_2}{m_1} & 0 \\ 0 & 0 & 0 & 1 \\ \dfrac{k_2}{m_2} & 0 & -\dfrac{k_2}{m_2} & 0 \end{pmatrix}$$

If, in addition, we define the column vectors

$$\mathbf{y} = \begin{pmatrix} y_1 \\ y_2 \\ y_3 \\ y_4 \end{pmatrix} \qquad \mathbf{y}' = \begin{pmatrix} y_1' \\ y_2' \\ y_3' \\ y_4' \end{pmatrix}$$

then we may write the system (1.14) in the compact form

$$\mathbf{y}' = A\mathbf{y}$$

where the right-hand side is the usual matrix-vector product. We shall see that we can always represent a system of first-order differential equations as a single first-order vector differential equation.

We define $\mathbf{y}$ to be a point in n-dimensional Euclidean space, E_n, with coordinates $(y_1, y_2, \ldots, y_n)$*. Unless otherwise indicated, E_n will represent **real** n-dimensional Euclidean space; that is, the coordinates $(y_1, \ldots, y_n)$ of the vector $\mathbf{y}$ are real numbers. However, the entire theory developed here carries over to the complex case with only minor changes, which will be indicated where necessary. We next define functions

$$\hat{f}_j(t, \mathbf{y}) = f_j(t, y_1, \ldots, y_n) \qquad (j = 1, \ldots, n)$$

and thus the system (1.9) can be written in the form

$$\begin{cases} y_1' = \hat{f}_1(t, \mathbf{y}) \\ y_2' = \hat{f}_2(t, \mathbf{y}) \\ \vdots \\ y_n' = \hat{f}_n(t, \mathbf{y}) \end{cases} \tag{1.15}$$

Proceeding heuristically (we will be more precise below), we next observe that the functions $\hat{f}_1, \ldots, \hat{f}_n$ can be regarded as n components of the vector-valued function $\mathbf{f}$ defined by

$$\mathbf{f}(t, \mathbf{y}) = (\hat{f}_1(t, \mathbf{y}), \ldots, \hat{f}_n(t, \mathbf{y}))$$

We also define

$$\mathbf{y}' = (y_1', \ldots, y_n')$$

Thus the system of n first-order equations (1.9), and all the systems that arose earlier in this section (see also (1.15)), can be written in the very compact form

$$\mathbf{y}' = \mathbf{f}(t, \mathbf{y}) \tag{1.16}$$

The system (1.16) resembles the familiar single first-order equation $y' = f(t, y)$, with y, f replaced by the vectors $\mathbf{y}$, $\mathbf{f}$, respectively.

Example 1. We may write the system (1.11),

$$y_1' = y_2$$
$$y_2' = g(t, y_1, y_2)$$

* The distinction between $\mathbf{y}$ as a column vector and $\mathbf{y}$ as a point in E_n will be clear from the context, and should cause no confusion.

as $\mathbf{y}' = \mathbf{f}(t, \mathbf{y})$ with $\mathbf{y} = (y_1, y_2)$ and

$$\hat{f}_1(t, \mathbf{y}) = f_1(t, y_1, y_2) = y_2$$
$$\hat{f}_2(t, \mathbf{y}) = f_2(t, y_1, y_2) = g(t, y_1, y_2)$$

so that

$$\mathbf{f}(t, \mathbf{y}) = (y_2, g(t, y_1, y_2)).$$

• EXERCISE

1. Write the systems (1.13), (1.14), and the systems in Exercises 13, 16, and 17 of Section 1.3, each in the form $\mathbf{y}' = \mathbf{f}(t, \mathbf{y})$ and in each case determine the vector $\mathbf{f}$.

We shall assume that the reader is familiar with the elements of vector algebra. However, we recall certain basic well-known facts, as well as cite some that may be new, in order to proceed with a systematic study of (1.16). We define the zero vector $\mathbf{0}$ (the origin of E_n) by $\mathbf{0} = (0, \ldots, 0)$ and for any point $\mathbf{y} \in E_n$ we define $c\mathbf{y}$, where c is any real or complex number, by the relation

$$c\mathbf{y} = (cy_1, cy_2, \ldots, cy_n)$$

If $\mathbf{y}$ and $\mathbf{z}$ are two vectors in E_n, we define their sum $\mathbf{y} + \mathbf{z}$ to be the vector

$$\mathbf{y} + \mathbf{z} = (y_1 + z_1, y_2 + z_2, \ldots, y_n + z_n)$$

and, of course, $\mathbf{y} - \mathbf{z} = \mathbf{y} + (-\mathbf{z})$. Two vectors $\mathbf{y}$ and $\mathbf{z}$ are equal if and only if $y_i = z_i$ $(i = 1, \ldots, n)$. The Euclidean length of the vector $\mathbf{y}$ is defined by the relation

$$\|\mathbf{y}\| = [|y_1|^2 + \cdots + |y_n|^2]^{1/2} = \left[\sum_{i=1}^{n} |y_i|^2\right]^{1/2}$$

Notice that $|y_i|$ is well-defined for y_i complex and thus $\|\mathbf{y}\|$ is also defined for a complex vector $\mathbf{y}$. We need the notion of length in order to measure distances between solutions of systems. However, for the purpose of dealing with systems of differential equations such as (1.16), it turns out to be more convenient to define a different quantity for the length (or norm) of a vector $\mathbf{y}$ than the familiar Euclidean length, namely

$$|\mathbf{y}| = |y_1| + |y_2| + \cdots + |y_n| = \sum_{i=1}^{n} |y_i|$$

Again, $|\mathbf{y}|$ is well defined for either real or complex vectors $\mathbf{y}$. No confusion need arise from using the absolute value sign for different purposes; on the left-hand side $|\mathbf{y}|$ is the notation for length of the vector $\mathbf{y}$; on the right-hand side we sum the absolute values of the components of $\mathbf{y}$. Observe, for example, that if $\mathbf{y} = (3 + i, 3 - i)$, then $\|\mathbf{y}\| = [|3 + i|^2 + |3 - i|^2]^{1/2} = (10 + 10)^{1/2} = (20)^{1/2}$ and $|\mathbf{y}| = |3 + i| + |3 - i| = |\mathbf{y}| = \sqrt{2}\|\mathbf{y}\|$. In general, the quantities $\|\mathbf{y}\|$ and $|\mathbf{y}|$ are related, as follows.

• **EXERCISE**

2. If $\mathbf{y} \in E_n$, show that

$$\|\mathbf{y}\| \leq |\mathbf{y}| \leq \sqrt{n}\|\mathbf{y}\|$$

[*Hint:* Use the inequality $2|uv| \leq |u|^2 + |v|^2$ and show that $\|\mathbf{y}\|^2 \leq |y|^2 \leq n\|\mathbf{y}\|^2$.]

The important point about this inequality is that $|\mathbf{y}|$ is small if and only if $\|\mathbf{y}\|$ is small.

The length function $|\mathbf{y}|$ has the following important properties:

(i) $|\mathbf{y}| \geq 0$ and $|\mathbf{y}| = 0$ if and only if $\mathbf{y} = \mathbf{0}$.
(ii) If c is any complex number, $|c\mathbf{y}| = |c|\,|\mathbf{y}|$.
(iii) For all $\mathbf{y}$ and $\mathbf{z}$, $|\mathbf{y} + \mathbf{z}| \leq |\mathbf{y}| + |\mathbf{z}|$.

The proofs are immediate from well-known properties of complex numbers. For example, to prove (ii) we have

$$|c\mathbf{y}| = \sum_{j=1}^{n} |cy_j| = \sum_{j=1}^{n} |c|\,|y_j| = |c| \sum_{j=1}^{n} |y_j| = |c|\,|\mathbf{y}|$$

Similarly, for (iii) we use the inequality $|u + v| \leq |u| + |v|$ valid for any complex numbers u and v.

• **EXERCISE**

3. Show that the Euclidean length $\|\mathbf{y}\|$ of a vector $\mathbf{y}$ also satisfies the properties (i), (ii), (iii) above. [*Hint:* To prove (iii) you will need to apply the Schwarz inequality for sums, that is,

$$\left|\sum_{i=1}^{n} a_i b_i\right|^2 \leq \sum_{i=1}^{n} |a_i|^2 \sum_{i=1}^{n} |b_i|^2$$

to the term $2(\sum_{i=1}^{n} y_i \bar{z}_i)$ which arises in the expansion of

$$\|\mathbf{y} + \mathbf{z}\|^2 = \sum_{i=1}^{n} |y_i + z_i|^2 = \sum_{i=1}^{n} (y_i + z_i)\overline{(y_i + z_i)}$$

$$= \sum_{i=1}^{n} |y_i|^2 + 2\left(\sum_{i=1}^{n} y_i \bar{z}_i\right) + \sum_{i=1}^{n} |z_i|^2.]$$

Using the length function, we define **the distance between two vectors y and z**, $d(\mathbf{y}, \mathbf{z})$ by the relation

$$d(\mathbf{y}, \mathbf{z}) = |\mathbf{y} - \mathbf{z}|$$

The distance function $d(\mathbf{y}, \mathbf{z})$ has the following important properties for arbitrary vectors $\mathbf{y}$, $\mathbf{z}$, $\mathbf{v}$:

(i) $d(\mathbf{y}, \mathbf{z}) \geq 0$ and $d(\mathbf{y}, \mathbf{z}) = 0$ if and only if $\mathbf{y} = \mathbf{z}$.
(ii) $d(\mathbf{y}, \mathbf{z}) = d(\mathbf{z}, \mathbf{y})$.
(iii) $d(\mathbf{y}, \mathbf{z}) \leq d(\mathbf{y}, \mathbf{v}) + d(\mathbf{v}, \mathbf{z})$ (triangle inequality).

The proofs of these properties follow immediately from the corresponding properties (i), (ii), (iii) of the length function. For example, to prove (iii) we have $d(\mathbf{y}, \mathbf{z}) = |\mathbf{y} - \mathbf{z}| = |(\mathbf{y} - \mathbf{v}) + (\mathbf{v} - \mathbf{z})| \leq |\mathbf{y} - \mathbf{v}| + |\mathbf{v} - \mathbf{z}| = d(\mathbf{y}, \mathbf{v}) + d(\mathbf{v}, \mathbf{z})$.

Any function satisfying the properties (i), (ii), (iii) is called a distance function. For example, $\rho(\mathbf{y}, \mathbf{z}) = \|\mathbf{y} - \mathbf{z}\|$ for any vectors $\mathbf{y}$, $\mathbf{z}$ is such a function, and represents the Euclidean distance between the points $\mathbf{y}$ and $\mathbf{z}$ in E_n.

• **EXERCISE**

4. Show that $\rho(\mathbf{y}, \mathbf{z}) = \|\mathbf{y} - \mathbf{z}\|$ also satisfies the properties of a distance function. [*Note:* The proof of (iii) is harder than for the distance function d. You will need to use the Schwarz inequality, as in Exercise 3.]

To define continuity, differentiability, and integrability of vector functions, we need the notion of limit for vectors. We first use the distance function d to define convergence. A sequence $\{\mathbf{y}^{(k)}\}$ of vectors in E_n is said to converge to the vector $\mathbf{y}$ if and only if $d(\mathbf{y}^{(k)}, \mathbf{y}) = |\mathbf{y}^{(k)} - \mathbf{y}| \to 0$ as $k \to \infty$ and in this case we write $\lim_{k\to\infty} \mathbf{y}^{(k)} = \mathbf{y}$ or $\{\mathbf{y}^{(k)}\} \to \mathbf{y}$. Since

$$|\mathbf{y}^{(k)} - \mathbf{y}| = |y_1^{(k)} - y_1| + |y_2^{(k)} - y_2| + \cdots + |y_n^{(k)} - y_n|$$

where $\mathbf{y}^{(k)} = (y_1^{(k)}, \ldots, y_n^{(k)})$, $\mathbf{y} = (y_1, \ldots, y_n)$, the above definition says that the sequence of vectors $\{\mathbf{y}^{(k)}\} \to \mathbf{y}$ as $k \to \infty$ if and only if each component of $\mathbf{y}^{(k)}$ tends to the corresponding component of the vector $\mathbf{y}$ (the components form sequences of real or complex numbers). It is clear that all properties of limits of sequences of complex numbers may now be assumed to hold for sequences of vectors without further explanation.

If we use the Euclidean distance function $\rho(\mathbf{y}, \mathbf{z}) = \|\mathbf{y} - \mathbf{z}\|$, we say that the sequence $\{\mathbf{y}^{(k)}\}$ converges to the vector $\mathbf{y}$ if and only if $\rho(\mathbf{y}^{(k)}, \mathbf{y}) = \|\mathbf{y}^{(k)} - \mathbf{y}\| \to 0$ as $k \to \infty$. It seems clear that the concept of convergence should not depend on the particular distance function used. We establish this for the distance functions $d(\mathbf{y}, \mathbf{z})$ and $\rho(\mathbf{y}, \mathbf{z})$ in Exercise 5.

• EXERCISE

5. Let $\{\mathbf{y}^{(k)}\}$ be a sequence of vectors. Show that $|\mathbf{y}^{(k)} - \mathbf{y}| \to 0$ as $k \to \infty$ if and only if $\|\mathbf{y}^{(k)} - \mathbf{y}\| \to 0$ as $k \to \infty$. [*Hint:* Use Exercise 2.]

A vector-valued function $\mathbf{g} = \mathbf{g}(t)$ *is a correspondence that assigns to each number t in an interval I one and only one vector* $\mathbf{g}(t)$*; we write*

$$\mathbf{g}(t) = (g_1(t), \ldots, g_n(t))$$

and we call g_k *the* kth *component* (real- or complex-valued scalar function) *of the vector function* $\mathbf{g}$. Because of our definition of convergence we shall now define $\mathbf{g}$ to be continuous, differentiable, or integrable on I if and only if each component of $\mathbf{g}$ has this property.

If $\mathbf{g}$ is differentiable on I we denote its derivative by $\mathbf{g}'$ and we define

$$\mathbf{g}' = (g_1', g_2', \ldots, g_n')$$

Similarly, if $\mathbf{g}$ is continuous on I we denote its integral from a to b (a and b on I) by $\int_a^b \mathbf{g}(s)\,ds$ and we define

$$\int_a^b \mathbf{g}(s)\,ds = \left(\int_a^b g_1(s)\,ds, \int_a^b g_2(s)\,ds, \ldots, \int_a^b g_n(s)\,ds\right)$$

We take note of the very important inequality

$$\left|\int_a^b \mathbf{g}(s)\,ds\right| \le \int_a^b |\mathbf{g}(s)|\,ds \qquad (a < b) \tag{1.17}$$

To prove (1.17) we have, successively

$$\left|\int_a^b \mathbf{g}(s)\,ds\right| = \left|\int_a^b g_1(s)\,ds\right| + \cdots + \left|\int_a^b g_n(s)\,ds\right|$$
$$\le \int_a^b |g_1(s)|\,ds + \cdots + \int_a^b |g_n(s)|\,ds = \int_a^b |\mathbf{g}(s)|\,ds$$

• EXERCISE

6. Justify each step in the proof of inequality (1.17). Note that in the middle step you have ordinary absolute values.

It is also true that

$$\left\| \int_a^b \mathbf{g}(s)\, ds \right\| \le \int_a^b \|\mathbf{g}(s)\|\, ds$$

for any continuous vector $\mathbf{g}$, but the proof is more difficult than the one for (1.17).

We can now return to the system

$$\mathbf{y}' = \mathbf{f}(t, \mathbf{y}) \tag{1.16}$$

where the vector-valued function $\mathbf{f}$ is defined in some $(n+1)$-dimensional region D in $(t, y_1, y_2, \ldots, y_n)$ space. To find a solution of (1.16) (compare Section 1.3) means to find a real interval I and a vector function $\boldsymbol{\phi}$ defined on I such that:

(i) $\boldsymbol{\phi}'(t)$ exists for each t on I;
(ii) the point $(t, \boldsymbol{\phi}(t))$ lies in D for each t on I;
(iii) $\boldsymbol{\phi}'(t) = \mathbf{f}(t, \boldsymbol{\phi}(t))$ for every t on I.

Thus the analogy between (1.16) and a single scalar equation of first order is complete. Just as for the scalar equation, to solve an initial value problem for the system (1.16) with the initial condition $\boldsymbol{\phi}(t_0) = \boldsymbol{\eta}$, where $(t_0, \boldsymbol{\eta})$ is a point in D, means to find a solution $\boldsymbol{\phi}$ of (1.16) in the above sense passing through the point $(t_0, \boldsymbol{\eta})$ of D, that is, satisfying $\boldsymbol{\phi}(t_0) = \boldsymbol{\eta}$. While it is not in general possible to solve (1.16) explicitly, we can illustrate the concepts with some simple problems.

Example 2. The system

$$y_1' = y_2$$
$$y_2' = y_1$$

is of the form (1.16) with $\mathbf{y} = (y_1, y_2)$, $\mathbf{f}(t, \mathbf{y}) = (y_2, y_1)$. We may also write it as $\mathbf{y}' = A\mathbf{y}$, where $A = \begin{pmatrix} 0 & 1 \\ 1 & 0 \end{pmatrix}$. Then also $\mathbf{f}(t, \mathbf{y}) = A\mathbf{y}$. Clearly, $A\mathbf{y}$ is well-defined on D, all of (t, y_1, y_2) space, and $\boldsymbol{\phi}(t) = (e^t, e^t)$ is a solution valid for $-\infty < t < \infty$, since (i), (ii), (iii) of the definition are satisfied. Note that $c\boldsymbol{\phi}$, c a constant, is also a solution.

• EXERCISES

7. Can you find (guess) another solution $\boldsymbol{\psi}(t)$ of the system in Example 2 on $-\infty < t < \infty$ that is not of the form $c\boldsymbol{\phi}(t)$?

8. Find a solution $\boldsymbol{\phi}$ of the system

$$y_1' = -y_1$$
$$y_2' = y_1 + y_2$$

that satisfies the initial condition $\boldsymbol{\phi}(0) = (2, 1)$. [*Hint:* Solve the first equation and substitute in the second equation. The reader may wish to refer to [2, Section 2.2] for this method of solution. What is the interval I of validity?]

9. Find a solution $\boldsymbol{\phi}$ of the system

$$y_1' = -y_1$$
$$y_2' = y_1 + t y_2$$

satisfying the initial condition $\boldsymbol{\phi}(0) = (2, 1)$.

10. Describe a method for solving the "triangular system"

$$\begin{cases} y_1' &= a_n y_1 + a_{12} y_2 + \cdots & + a_{1n} y_n \\ y_2' &= \quad a_{22} y_2 + \cdots & + a_{2n} y_n \\ \vdots & \\ y_{n-1}' &= & a_{n-1,n-1} y_{n-1} + a_{n-1,n} y_n \\ y_n' &= & a_{nn} y_n \end{cases}$$

where a_{ij} $(j \geq i)$ are constants; note that a_{ij} with $j < i$ are zero.

11. Find a solution $\boldsymbol{\phi}$ of the system

$$y_1' = y_1 + y_2 + f(t)$$
$$y_2' = y_1 + y_2$$

where $f(t)$ is a continuous function, satisfying the initial condition $\phi(0) = (0, 0)$. [*Hint:* Define $v(t) = y_1(t) + y_2(t)$.]

12. In Exercise 8 compute another solution $\boldsymbol{\psi}$ satisfying the initial condition $\boldsymbol{\psi}(0) = (2, 2)$. Then compute $|\boldsymbol{\phi}(t) - \boldsymbol{\psi}(t)|$, where $\boldsymbol{\phi}$ is the solution in Exercise 8.

The reader will notice that all the examples in the exercises above are of the form (1.16), but at the same time they are: (1) linear in the components of $\mathbf{y}$; and (2) of a very special "triangular" form, which makes it possible for us to solve them explicitly. We shall have much more to say about general linear systems in Chapter 2.

1.5 The Need for a Theory

The task of formulating a mathematical model for the motion of a physical system such as the mass-spring system or the simple pendulum leads to a differential equation; different physical approximations of the same system lead to different models (that is, different differential equations). Let us look at just one of these, namely, the one for the simple pendulum starting from an initial angle θ_0 at rest,

$$\frac{d^2\theta}{dt^2} + \frac{g}{L}\sin\theta = 0 \tag{1.6}$$

Suppose we intend to use this model to determine the motion; we would first have to find a solution $\theta = \phi(t)$ of the equation that satisfies (1.6), and the initial conditions

$$\phi(0) = \theta_0 \qquad \phi'(0) = 0 \tag{1.18}$$

Unfortunately, no solution expressible in terms of elementary functions can be found. It would be disconcerting if there were no solution! This would seem to mean that there would have to be an internal inconsistency in the set of assumptions that were made about the physical approximation. Since an actual pendulum certainly moves, this would mean that our model is quite useless, and we would have to construct a new model. Therefore, in order for a mathematical model to be useful it must at least have solutions; a very important aspect of mathematical theory has to do with proving that certain classes of differential equations *have* solutions. For brevity we describe this as the **existence problem**. Indeed, as far as our specific problem is concerned, we can prove that the differential equation (1.6) has a solution satisfying (1.18) (see Theorem 1.2, Theorem 3.3, p. 123, or Exercise 2, Section 3.1, p. 110, and Exercise 12, p. 118).

This is not the only requirement that is desirable for a useful model. Suppose we displace a pendulum to an angle θ_0 and release it and watch the resulting motion of the pendulum. Experience suggests that if we could repeat the experiment exactly, we would get **exactly** the same motion. Described differently, this is the hypothesis of determinism; a particular set of initial conditions must result in exactly one motion. Applied to a differential equation, this means that there should be **exactly one solution for a given set of initial conditions**. For brevity, we refer to this as the **uniqueness problem**. For example, if we start an experiment with the pendulum at rest with zero displacement [$\theta_0 = 0$], we know from experience that it will stay at rest. This motion is described by saying that $\theta \equiv 0$ for all $t \geq 0$. This means that we would want to be sure that the only solution $\phi(t)$ of the equation (1.6) that obeys $\phi(0) = 0$, $\phi'(0) = 0$, is $\phi(t) \equiv 0$. This is indeed the case (see Theorem 1.2, or Theorem 3.4, p. 125, or Exercise 1, Section 3.3, p. 126).

There is a third property that experience suggests as a requirement for a satisfactory model. Experiments cannot in fact be repeated in **exactly** the same way. However, if all of the initial conditions are **almost** exactly the same, we expect the outcomes to be almost the same. We therefore desire that the solutions of our mathematical model should also have this property. Stating this in mathematical language, we say that the solutions of a differential equation ought to depend continuously on the values of the initial conditions. We refer to this property as **continuity of the solution with respect to initial conditions**. (See Theorem 3.7, p. 135.)

Thus, a mathematical model of a physical process should have the following three properties.

(a) A solution satisfying the given initial conditions exists.
(b) Each set of initial conditions leads to a unique solution (that is, two solutions that satisfy the same initial conditions are identical).
(c) The solutions depend continuously on the initial conditions.

Mathematicians have shown that wide classes of differential equations obey the requirements (a), (b), (c), even for equations for which there is no possible method for finding the solutions explicitly. We will state some of these general results without proofs in Section 1.6; they will guarantee, for example, that the various mass-spring systems and pendulum equations satisfy requirements (a), (b), (c).

• **EXERCISE**

1. Prove existence, uniqueness, and continuity of the solution ϕ of the equation $y'' = f(t)$, where f is continuous for t in an interval I, such that $\phi(t_0) = y_0$, $\phi'(t_0) = z_0$ for some t_0 in I. [*Hint:* Use the fundamental theorem of the calculus.] Note that the solution is a continuous function of (t, t_0, y_0, z_0).

1.6 Existence, Uniqueness, and Continuity

In what follows we let D represent a region in $(n + 1)$-dimensional space; this is a set of points with the property that given any point $(t_0, \boldsymbol{\eta})$ in D, the interior of the $(n + 1)$-dimensional "box"

$$B = \{(t, y) \,\big|\, |t_0 - t_0| < a, |\mathbf{y} - \boldsymbol{\eta}| < b\}$$

will, for $a, b > 0$ and sufficiently small, lie entirely in D. (We note that if we use the Euclidean norm $\|\mathbf{y} - \boldsymbol{\eta}\| < b$, then the set

$$C = \{(t, y) \,\big|\, |t - t_0| < a, \|\mathbf{y} - \boldsymbol{\eta}\| < b\}$$

would specify a "cylinder" whose cross section by a hyperplane $t =$ constant would be an n-dimensional sphere.) The most important special cases: the whole space, a half space $\{(t, \mathbf{y}) \,\big|\, 0 < t < \infty, |y| \geq 0\}$, and "infinite strips" (for example, $\{(t, \mathbf{y}) \big| |t - t_0| < \infty, |y| < 2\}$), have the above property of a region.

The reader should observe that if $(t_0, \mathbf{y}_0)$ is on the "boundary" of D, every box centered at $(t_0, \mathbf{y}_0)$ contains points that are not in D. (In fact, this statement can be used to give a precise definition of the boundary of D.) In other words, boundary points are not included in D. If $(t_0, \mathbf{y}_0)$ is a point of D that lies "close" to the boundary, then the numbers a and b in the definition of region must necessarily be chosen small.

In what follows, whenever we say that a certain function $\mathbf{f}$ is continuous on D we are assuming that it is continuous at all points of D, but we make no assumption concerning its behavior on the boundary. If $\mathbf{f}$ is continuous at some boundary points of D as well as in D, then $\mathbf{f}$ is still continuous on the region D. In many problems that arise, this is precisely the situation, but the continuity of $\mathbf{f}$ at some boundary points is not essential.

• EXERCISES

1. Consider the region $D = \{(t, y) | (t-1)^2 + (y+2)^2 < 4\}$ in E_2.
(a) Construct at least two rectangles centered at the following points, which lie entirely in D: $(1, -2)$, $(0, -1)$, $(2, -3)$, $(1, -\frac{1}{10})$.
(b) Show that the points $(1, 0)$ and $(-1, -2)$ are boundary points.
(c) Consider the collection of points $\{(t, y) | (t-1)^2 + (y+2)^2 \le 4\}$. Is this a region?

2. Which of the following collections of points are regions in E_2?
(a) $\{(t, y) | t > 0, -\infty < y < \infty\}$.
(b) $\{(t, y) | -\infty < t < \infty, -2 \le y < \infty\}$.
(c) $\{(t, y) | 0 < t < 1, -\infty < y < \infty\}$.
(d) $\{(t, y) | -\infty < t < \infty, -\infty < y < \infty\}$.

3. Show that the collection of points $\{(t, y) | -1 < t < 1, y = t^2\}$ in E_2 contains only boundary points.

4. Consider the region $D = \{(t, y) | t^2 + y^2 < 1\}$ in E_2. Are the following functions continuous on D?

(a) $f(t, y) = \dfrac{1}{1 + t^2 + y^2}$.

(b) $g(t, y) = \dfrac{1}{1 - t^2 - y^2}$.

5. Which of the following collections of points are regions? Sketch, if possible. (Unless implied otherwise, $\mathbf{y}$ is in E_n.)
(a) $\{(t, \mathbf{y}) | 0 < t < 1, 1 < |\mathbf{y}| < 2\}$.
(b) $\{(t, \mathbf{y}) | 0 < t < 1, 1 \le \|\mathbf{y}\| \le 2\}$.
(c) $\{(t, \mathbf{y}) | (t-1)^2 + \|\mathbf{y}\|^2 < 7\}$.
(d) $\{(t, y_1, y_2) | t > 0, y_1^2 + y_2^2 < 2t^2\}$.

6. Consider the region $D = \{(t, y_1, y_2) | t^2 > y_1^2 + y_2^2\}$ in E_3. Are the following functions continuous on D?

(a) $f(t, y_1, y_2) = \dfrac{1}{y_1^2 + y_2^2 - t^2}$.

(b) $g(t, y_1, y_2) = \dfrac{y_1}{t}$.

(c) $\dfrac{\partial f}{\partial y_1}(t, y_1, y_2)$, where f is as in part (a).

(d) $\dfrac{\partial g}{\partial y_2}(t, y_1, y_2)$, where g is as in part (b).

We are now ready to state our first important result.

Theorem 1.1. *Let* $\mathbf{f}$ *be a vector function* (with n components) *defined in a region* D *of* $(n+1)$*-dimensional Euclidean space. Let the vectors* $\mathbf{f}$, $\partial\mathbf{f}/\partial y_k$ $(k = 1, \ldots, n)$ *be continuous in* D. *Then given any point* $(t_0, \boldsymbol{\eta})$ *in* D *there exists a unique solution* $\boldsymbol{\phi}$ *of the system*

$$\mathbf{y}' = \mathbf{f}(t, \mathbf{y}) \tag{1.16}$$

satisfying the initial condition $\boldsymbol{\phi}(t_0) = \boldsymbol{\eta}$. *The solution* $\boldsymbol{\phi}$ *exists on any interval* I *containing* t_0 *for which the points* $(t, \boldsymbol{\phi}(t))$, *with* t *in* I, *lie in* D. *Furthermore, the solution* $\boldsymbol{\phi}$ *is a continuous function of the "triple"* $(t, t_0, \boldsymbol{\eta})$.

If the region D is the entire $(t, \mathbf{y})$ space, then it follows from Theorem 1.1 that every solution exists as long as its norm remains finite. This obvious remark will be quite useful in showing whether a solution exists for all t. An example will help clarify this point.

Example 1. Consider the scalar differential equation $y' = \alpha y$ (α constant). Here $f(t, y) = \alpha y$ and $\partial f/\partial y\,(t, y) = \alpha$. Both f and $\partial f/\partial y$ are continuous in the whole (t, y) plane. Theorem 1.1 shows that there is a unique solution ϕ of $y' = \alpha y$ through every point (t_0, y_0) in the plane. It is easily verified that

$$\phi(t) = y_0 \exp[\alpha(t - t_0)]$$

is a solution of this initial value problem. Therefore $\phi(t) = y_0 \exp[\alpha(t - t_0)]$ is the **only** solution. Also, since $|\phi(t)| = |y_0| \exp[\alpha(t - t_0)]$ is finite whenever t is finite, all points $(t, \phi(t))$, $-\infty < t < \infty$, lie in D and therefore this solution ϕ exists for $-\infty < t < \infty$.

• EXERCISES

7. Find a solution of $y' = \alpha y$ through $(1, 0)$. Is this the only solution through $(1, 0)$?

8. (a) Show that $-1/t$ is a solution of $y' = y^2$ passing through $(-1, 1)$.
   (b) Show that $\phi(t) = -1/t$ is the only solution of $y' = y^2$ passing through $(-1, 1)$. Be sure to determine an appropriate region D before applying Theorem 1.1.
   (c) What is the largest interval on which $\phi(t) = -1/t$ is a solution of $y' = y^2$ through the point $(-1, 1)$? The reader should observe that Exercise 8 shows that a solution of $y' = f(t, y)$ does not necessarily exist for all t even though f and $\partial f/\partial y$ are continuous in the whole plane.

Example 2. Discuss the problem of existence and uniqueness of solutions of the initial value problem for the system

$$y_1' = ty_2 + y_3$$

$$y_2' = (\cos t)y_1 + t^2 y_3$$

$$y_3' = y_1 - y_2$$

This system is of the form $\mathbf{y}' = \mathbf{f}(t, \mathbf{y})$ with $\mathbf{y} = (y_1, y_2, y_3)$, $\mathbf{f}(t, \mathbf{y}) = (ty_2 + y_3, (\cos t)y_1 + t^2 y_3, y_1 - y_2)$; hence $\mathbf{f}(t, \mathbf{y})$ is continuous for $|t| < \infty$, $|\mathbf{y}| < \infty$. Moreover, $\partial \mathbf{f}/\partial y_1 = (0, \cos t, 1)$, $\partial \mathbf{f}/\partial y_2 = (t, 0, -1)$, $\partial \mathbf{f}/\partial y_3 = (1, t^2, 0)$, which are also continuous for $|t| < \infty$, $|\mathbf{y}| < \infty$. Thus D is all of four-dimensional (t, y_1, y_2, y_3) space, and by Theorem 1.1, through any point $(t_0, \boldsymbol{\eta})$ there passes a unique solution $\boldsymbol{\phi}$ existing on some interval containing t_0. It can be shown, see the corollary to Theorem 2.1, p. 39, that the solution $\boldsymbol{\phi}$ actually exists on the interval $-\infty < t < \infty$.

• EXERCISES

9. Verify that the systems in Exercises 8, 9, 10, and 11 of Section 1.4, satisfy the hypothesis of Theorem 1.1 in the domain $D: -\infty < t < \infty$, $|\mathbf{y}| < \infty$.

10. Discuss the existence and uniqueness of solutions of the system

$$y_1' = y_1^2$$
$$y_2' = y_1^2 + y_2$$

11. Find a solution $\boldsymbol{\phi} = (\phi_1, \phi_2)$ of the system in Exercise 10 that satisfies the initial condition $\phi_1(-1) = 1$, $\phi_2(-1) = 0$. Discuss the interval I on which the solution $\boldsymbol{\phi}$ exists.

We now consider solutions $\phi(t)$ of the **scalar second-order differential equation**

$$y'' = g(t, y, y') \tag{1.10}$$

with initial conditions $\phi(t_0) = y_0$, $\phi'(t_0) = z_0$, where $g(t, y, z)$ is defined in a region D of (t, y, z) space. As we have seen in Section 1.1, several applications lead to problems of this form. In Example 3, Section 1.3, we showed that (1.10) is equivalent to the system

$$\begin{aligned} y_1' &= y_2 \\ y_2' &= g(t, y_1, y_2) \end{aligned} \tag{1.11}$$

by using the substitution $y = y_1$, $y' = y_2$. Thus if ϕ is a solution of (1.10) satisfying the initial conditions $\phi(t_0) = y_0$, $\phi'(t_0) = z_0$, then $\boldsymbol{\psi}(t) = (\psi_1(t), \psi_2(t)) = (\phi(t), \phi'(t))$ is the corresponding solution of (1.11) satisfying the initial condition

$$\boldsymbol{\psi}(t_0) = (\phi(t_0), \quad \phi'(t_0)) = (y_0, z_0)$$

In addition, as shown in Example 1, Section 1.4, the system (1.11) is precisely of the form of Equation (1.16) with

$$\mathbf{f}(t, \mathbf{y}) = (y_2, g(t, y_1, y_2))$$

to which Theorem 1.1 is applicable. Therefore, as an immediate corollary of this theorem, we have the following result on existence, uniqueness, and continuity of solutions of **the initial value problem** for scalar second-order differential equations.

Theorem 1.2. *Let g, $\partial g/\partial y$, and $\partial g/\partial z$ be continuous in a given region D. Let (t_0, y_0, z_0) be a given point of D. Then there exists an interval containing t_0 and exactly one solution ϕ, defined on this interval, of the differential equation $y'' = g(t, y, y')$ that passes through (t_0, y_0, z_0) (that is, the solution ϕ satisfies the initial conditions $\phi(t_0) = y_0$, $\phi'(t_0) = z_0$). The solution exists for those values of t for which the points $(t, \phi(t), \phi'(t))$ lie in D. Further, the solution ϕ is a continuous function, not only of t, but of t_0, y_0, z_0 as well (in fact, of the quadruple (t, t_0, y_0, z_0)).*

If the region D is the entire (t, y, z) space, then it follows from Theorem 1.2 that every solution exists for those values of t for which the solution and its derivative remain finite.

Example 3. Consider the scalar equation $y'' + k_1 y' + k_2 \sin y = 0$, where k_1 and k_2 are constants. This equation is a model of the motion of a damped simple pendulum with no external forces (see Equation (1.7)). It can be written in the form $y'' = g(t, y, y')$ with $g(t, y, z) = -k_1 z - k_2 \sin y$, $\partial g/\partial y = -k_2 \cos y$, $\partial g/\partial z = -k_1$. Clearly, g, $\partial g/\partial y$, and $\partial g/\partial z$ are continuous for all

(t, y, z), no matter what the constants k_1 and k_2 are. Thus Theorem 1.2 shows that given any triple (t_0, y_0, z_0), the differential equation $y'' + k_1 y' + k_2 \sin y = 0$ has a unique solution ϕ with $\phi(t_0) = y_0$, $\phi'(t_0) = z_0$. Note that here the region D can be taken to be the whole three-dimensional space. It can also be shown that $|\phi(t)|$ and $|\phi'(t)|$ are finite, and that this implies that the solution ϕ exists on $-\infty < t < \infty$. (This will be done in Chapter 5.) It follows, in particular, that the various models for the simple pendulum discussed in Section 1.1 have these properties.

• EXERCISES

12. Discuss the existence and uniqueness of solutions ϕ of $y'' + ky = 0$ (k constant) with initial conditions $\phi(t_0) = y_0$, $\phi'(t_0) = z_0$.

13. Discuss the existence and uniqueness of solutions ϕ of $y'' + p(t)y' + q(t)y = f(t)$, with initial conditions $\phi(t_0) = y_0$, $\phi'(t_0) = z_0$, where p, q, and f are given functions continuous on some interval $a < t < b$, where $a < t_0 < b$.

14. (a) Show that $\phi(t) \equiv 0$ is the only solution of $y'' + p(t)y' + q(t)y = 0$ satisfying the initial condition $\phi(0) = \phi'(0) = 0$, if p and q are continuous on some interval containing 0 in its interior.
    (b) Show that if $\psi(t)$ is a solution of $y'' + p(t)y' + q(t)y = 0$ that is tangent to the t axis at some point $(t_1, 0, 0)$, then $\psi(t) \equiv 0$.

15. It is easily verified that $c_1 \cos 2t + c_2 \sin 2t$ is a solution of $y'' + 4y = 0$ on $-\infty < t < \infty$ for every choice of the constants c_1 and c_2.
    (a) Determine c_1 and c_2 so that this solution satisfies the initial conditions

$$\phi\left(\frac{\pi}{4}\right) = 1, \qquad \phi'\left(\frac{\pi}{4}\right) = 2$$

    (b) Write down the solution ϕ satisfying the initial conditions

$$\phi\left(\frac{\pi}{4}\right) = 1, \qquad \phi'\left(\frac{\pi}{4}\right) = 2$$

    and prove that this is the only solution satisfying these conditions.

16. Consider the differential equation

$$y'' = \begin{cases} y & (t \geq 0, -\infty < y < \infty) \\ 0 & (t < 0, -\infty < y < \infty) \end{cases}$$

    (a) Is the function

$$\phi(t) = \begin{cases} 1 & (t < 0) \\ e^t & (t \geq 0) \end{cases}$$

    a solution on $-\infty < t < \infty$?
    (b) Is $\phi(t)$ continuous everywhere?
    (c) Is $\phi'(t)$ continuous everywhere?
    (d) Can you apply Theorem 1.2 to obtain the existence of a unique solution ψ such that $\psi(0) = 1$, $\psi'(0) = 1$? Explain fully.

17. Show that the only solution of

$$y'' + ty' + (1 + t^2)y^2 = 0$$

that touches the t axis at some point $(t_0, 0)$ is the identically zero solution.

Finally, there are problems where higher-order scalar differential equations are encountered. For example, in elasticity theory, equations of the fourth order arise naturally. We are interested in equations of order n, of the form $y^{(n)} = h(t, y, y', \ldots, y^{(n-1)})$. We observe that the cases when $n = 1$ and $n = 2$ are the ones we have already discussed. Here h is a function defined in some region D in the $(n + 1)$-dimensional $(t, y_1, y_2, \ldots, y_n)$ space. Combining Example 4, Section 1.3 with Theorem 1.1, we obtain, exactly as in the second-order scalar case, the following result on existence, uniqueness, and continuity of solutions of the initial value problem for scalar differential equations of order n.

Theorem 1.3. *Let h, $\partial h/\partial y_1, \ldots, \partial h/\partial y_n$ be continuous in a given region D. Let $(t_0, \eta_1, \ldots, \eta_n)$ be a given point of D. Then there exists an interval containing t_0 and exactly one solution ϕ, defined on this interval, of the differential equation $y^{(n)} = h(t, y, y', \ldots, y^{(n-1)})$ that passes through $(t_0, \eta_1, \ldots, \eta_n)$, [that is, the solution ϕ satisfies the initial conditions $\phi(t_0) = \eta_1, \phi'(t_0) = \eta_2, \ldots, \phi^{(n-1)}(t_0) = \eta_n$]. The solution exists for those values of t for which the points $(t, \phi(t), \phi'(t), \ldots, \phi^{(n-1)}(t))$ lie in D. Further, the solution ϕ is a continuous function of the $(n + 2)$ variables $t, t_0, \eta_1, \ldots, \eta_n$.*

The proofs of all the results stated in this section may be found in Chapter 3.

- **EXERCISES**

18. (a) Show that the differential equation $y^{(4)} + 2y'' + 3y = 0$ has a unique solution ϕ satisfying the initial conditions $\phi(1) = 1$, $\phi'(1) = 0$, $\phi''(1) = -1$, $\phi'''(1) = 2$.

(b) Show that $\psi(t) \equiv 0$ is the unique solution of this equation satisfying the initial conditions $\psi(-1) = \psi'(-1) = \psi''(-1) = \psi'''(-1) = 0$.

19. Show that $\psi(t) \equiv 5$ is the unique solution of $y''' + (y - 5)^2 = 0$ satisfying the initial conditions $\phi(t_0) = 5$, $\phi'(t_0) = \phi''(t_0) = 0$ for any t_0, $-\infty < t_0 < \infty$.

20. Use the result of Example 5 (Section 1.3) to deduce the existence and uniqueness of solutions of the coupled mass-spring system, Equation (1.8), p. 9.

1.7 The Gronwall Inequality

In the systematic study of systems of differential equations we shall often need to make use of an important inequality, which we now digress to state and prove. This inequality, known as the Gronwall inequality, will be applied frequently in what follows.

Theorem 1.4. (Gronwall Inequality.) *Let K be a nonnegative constant and let f and g be continuous nonnegative functions on some interval* $\alpha \le t \le \beta$ *satisfying the inequality*

$$f(t) \le K + \int_\alpha^t f(s)g(s)\,ds$$

for $\alpha \le t \le \beta$. *Then*

$$f(t) \le K \exp\left(\int_\alpha^t g(s)\,ds\right)$$

for $\alpha \le t \le \beta$.

Proof. Let $U(t) = K + \int_\alpha^t f(s)g(s)\,ds$, and observe that $U(\alpha) = K$. Then $f(t) \le U(t)$ by hypothesis, and, by the fundamental theorem of integral calculus and because $g(t) \ge 0$, we obtain

$$U'(t) = f(t)g(t) \le U(t)g(t) \qquad (\alpha \le t \le \beta)$$

We multiply this inequality by $\exp(-\int_\alpha^t g(s)\,ds)$ and apply the identity

$$U'(t) \exp\left(-\int_\alpha^t g(s)\,ds\right) - U(t)g(t) \exp\left(-\int_\alpha^t g(s)\,ds\right) = \left[U(t) \exp\left(-\int_\alpha^t g(s)\,ds\right)\right]'$$

to obtain

$$\frac{d}{dt}\left[U(t) \exp\left(-\int_\alpha^t g(s)\,ds\right)\right] \le 0$$

Integration from α to t gives

$$U(t) \exp\left(-\int_\alpha^t g(s)\,ds\right) - U(\alpha) \le 0$$

or, since $f(t) \le U(t)$ and $U(\alpha) = K$,

$$f(t) \le U(t) \le K \exp\left(\int_\alpha^t g(s)\,ds\right) \qquad (\alpha \le t \le \beta)$$

which is the desired inequality. ∎

• EXERCISES

1. Let K_1 and K_2 be positive constants and let f be a continuous nonnegative function on an interval $\alpha \leq t \leq \beta$ satisfying the inequality

$$f(t) \leq K_1 + K_2 \int_\alpha^t f(s)\,ds$$

Show that

$$f(t) \leq K_1 \exp\,[K_2(t-\alpha)]$$

2. Find all continuous nonnegative functions f on $0 \leq t \leq 1$ such that

$$f(t) \leq \int_0^t f(s)\,ds, \qquad 0 \leq t \leq 1$$

3. Let $f(t)$ be a nonnegative function satisfying the inequality

$$f(t) \leq K_1 + \varepsilon\,(t-\alpha) + K_2 \int_\alpha^t f(s)\,ds,$$

on an interval $\alpha \leq tp \leq \beta$, where ε, K_1, K_2 are given positive constants. Show that

$$f(t) \leq K_1 \exp\,[K_2(t-\alpha)] + \frac{\varepsilon}{K_2}(\exp[K_2(t-\alpha)] - 1)$$

[*Hint:* Consider

$$U(t) = K_1 + \varepsilon(t-\alpha) + K_2 \int_\alpha^t f(s)\,ds]$$

4. Find all continuous (not necessarily differentiable) functions $f(t)$ such that

$$[f(t)]^2 = \int_0^t f(s)\,ds \qquad t(\geq 0)$$

Chapter 2 | LINEAR SYSTEMS, WITH AN INTRODUCTION TO PHASE SPACE ANALYSIS

2.1 Introduction

We shall study linear systems of differential equations (that is, the system $\mathbf{y}' = \mathbf{f}(t, \mathbf{y})$ in which $\mathbf{f}(t, \mathbf{y})$ is linear in the components of $\mathbf{y}$), in considerable detail.

Example 1. Consider the system

$$\begin{aligned} y_1' &= y_1 - ty_2 + e^t \\ y_2' &= t^2 y_1 - y_3 \\ y_3' &= y_1 + y_2 - y_3 + 2e^{-t} \end{aligned} \tag{2.1}$$

which is linear in y_1, y_2, y_3 and of the form $\mathbf{y}' = \mathbf{f}(t, \mathbf{y})$ with $\mathbf{y} = (y_1, y_2, y_3)$, $\mathbf{f}(t, \mathbf{y}) = (y_1 - ty_2, t^2y_1 - y_3, y_1 + y_2 - y_3) + (e^t, 0, 2e^{-t})$. We observe that the vector $(y_1 - ty_2, t^2y_1 - y_3, y_1 + y_2 - y_3)$ can be represented as the matrix vector product $A(t)\mathbf{y}$ with

$$A(t) = \begin{pmatrix} 1 & -t & 0 \\ t^2 & 0 & -1 \\ 1 & 1 & -1 \end{pmatrix}$$

and with $\mathbf{y}$ regarded as a **column vector**. Thus the system (2.1) can be written as $\mathbf{y}' = A(t)\mathbf{y} + \mathbf{g}(t)$, where $\mathbf{g}(t)$ is the given vector $(e^t, 0, 2e^{-t})$, also regarded as a column vector.

• **EXERCISE**

1. Represent each of the systems in Example 2 (p. 21) and in Exercises 8, 9, 10, 11, (p. 22), Section 1.4, in the form $\mathbf{y}' = A(t)\mathbf{y} + \mathbf{g}(t)$. Identify the matrix $A(t)$ and the vector $\mathbf{g}(t)$.

More generally we see that the system $\mathbf{y}' = \mathbf{f}(t, \mathbf{y})$, with $\mathbf{f}(t, \mathbf{y})$ linear in the components of $\mathbf{y}$, has the form

$$\begin{cases} y_1' = a_{11}(t)y_1 + a_{12}(t)y_2 + \cdots + a_{1n}(t)y_n + g_1(t) \\ y_2' = a_{21}(t)y_1 + a_{22}(t)y_2 + \cdots + a_{2n}(t)y_n + g_2(t) \\ \vdots \quad\quad \vdots \quad\quad\quad\quad\quad\quad\quad\quad\quad \vdots \quad\quad\quad \vdots \\ y_n' = a_{n1}(t)y_1 + a_{n2}(t)y_2 + \cdots + a_{nn}(t)y_n + g_n(t) \end{cases} \tag{2.2}$$

and can be represented as

$$\mathbf{y}' = A(t)\mathbf{y} + \mathbf{g}(t) \tag{2.3}$$

where

$$A(t) = \begin{pmatrix} a_{11}(t)a_{12}(t) \cdots a_{1n}(t) \\ a_{21}(t)a_{22}(t) \cdots a_{2n}(t) \\ \vdots \\ a_{n1}(t)a_{n2}(t) \cdots a_{nn}(t) \end{pmatrix} \quad \text{and} \quad \mathbf{g}(t) = \begin{pmatrix} g_1(t) \\ g_2(t) \\ \vdots \\ g_n(t) \end{pmatrix}$$

and where $\mathbf{y}$ is the column vector with components $y_1, \ldots, y_n$.

• **EXERCISE**

2. Write the scalar linear equation $y^{(n)} + a_1(t)y^{(n-1)} + \cdots + a_{n-1}(t)y' + a_n(t)y = b(t)$ as a system $\mathbf{y}' = A(t)\mathbf{y} + \mathbf{g}(t)$. Determine the matrix $A(t)$ and the vector $\mathbf{g}(t)$.

We assume that the reader is familiar with the elementary matrix operations of addition and multiplication, and with the properties of determinants. We will deal for the most part with n-by-n matrices and with vectors, either column vectors (n-by-1 matrices) or occasionally also row vectors (1-by-n matrices). We shall denote square matrices by capital letters and vectors by small boldface letters. We denote by 0 the n-by-n matrix with all elements zero and by E the n-by-n identity matrix, that is, the matrix with each diagonal element 1 and all other elements zero. We have $AE = EA = A$ for every n-by-n matrix A; also, $E\mathbf{b} = \mathbf{b}$ for any column vector $\mathbf{b}$. We recall that the n-by-n matrices

A and B are said to commute if and only if $AB = BA$. Unless otherwise stated, all matrices will be n-by-n.

If A is a matrix, we let $\det A$ denote the determinant of A. We have the following basic properties, which are assumed to be familiar.

(i) For matrices A and B, $\det(AB) = \det A \cdot \det B$.
(ii) If $\det A \neq 0$, A is called nonsingular and has an inverse A^{-1} such that

$$AA^{-1} = A^{-1}A = E$$

Moreover,

$$A^{-1} = \frac{\tilde{A}}{\det A}$$

where $\tilde{A}$ is the n-by-n matrix whose elements $\tilde{a}_{ij}$ are the cofactors of a_{ji} in A $(i, j = 1, \ldots, n)$.

• **EXERCISE**

3. Use the result of (i) and (ii) to show that $\det(A^{-1}) = 1/\det A$.

(iii) Consider the system of linear algebraic equations

$$A\mathbf{x} = \mathbf{b} \tag{2.4}$$

where A is a given matrix, $\mathbf{b}$ is a given (column) vector, and $\mathbf{x}$ is the unknown vector. The system (2.4) has a unique solution if and only if $\det A \neq 0$. This solution is given by $\mathbf{x} = A^{-1}\mathbf{b}$; in particular, if $\mathbf{b} = \mathbf{0}$, then $\mathbf{x} = \mathbf{0}$. If $\mathbf{b} = 0$, the system has a **nontrivial** solution (that is, a solution $\mathbf{x} \neq 0$) if and only if $\det A = 0$.

(iv) If $\det A \neq 0$, the n columns of A considered as vectors are linearly independent, and conversely, if the n columns of A are linearly independent, then $\det A \neq 0$.

• **EXERCISES**

4. Prove property (iv).
5. State (look up) the theorem concerning the system (2.4) in the case $\det A = 0$ but $\mathbf{b} \neq 0$.

We define the **norm** (length) **of a matrix** A, denoted by $|A|$, by

$$|A| = \sum_{i,j=1}^{n} |a_{ij}| \tag{2.5}$$

that is, as the sum of the absolute values of all the elements. Notice that if A is n-by-1 or 1-by-n, that is, a vector, then (2.5) reduces to our previous definition of the length (norm) of a vector. We readily verify that the matrix norm satisfies the following properties:

(i) $|A + B| \leq |A| + |B|$

(ii) $|AB| \leq |A| \cdot |B|$

(iii) $|A\mathbf{b}| \leq |A| \cdot |\mathbf{b}|$

for matrices A, B of complex numbers and column vectors $\mathbf{b}$ with n components. The above norm is convenient for our purposes; other matrix norms satisfying the properties (i), (ii), (iii) are possible.

• EXERCISE

6. Prove the properties (i), (ii), (iii) of the norm $|A|$.

In Section 1.4, (p. 19) we defined the concept of convergence of a sequence of vectors in terms of a vector norm and used it to discuss continuity, differentiability, and integrability of vector functions. We now use the matrix norm (2.5) to do the same for matrices.

Definition. *The sequence of matrices* $\{A^{(k)}\}$ *converges to the matrix* A *if and only if the sequence of real numbers* $\{|A - A^{(k)}|\}$ *has limit zero, and in this case we write*

$$\{A^{(k)}\} \to A \quad \text{or} \quad \lim_{k \to \infty} A^{(k)} = A$$

Clearly, because of the definition of the norm, this means that $\{A^{(k)}\} \to A$ if and only if the sequence $\{a_{ij}^{(k)}\}$ of complex numbers, representing the element in the ith row and jth column in the matrices $\{A^{(k)}\}$, converges to the element a_{ij} of the matrix A as $k \to \infty$ for each of the n^2 elements $(i, j = 1, \ldots, n)$. **A matrix function $A(t)$ is a correspondence that assigns to each point t of an interval I one and only one n-by-n matrix $A(t)$.** Using the remark following the definition of convergence of a sequence of matrices, we see that it is consistent to say that a matrix function $A(t)$ is continuous, differentiable, or integrable on an interval I if and only if each of its n^2 elements $a_{ij}(t)$ is continuous, differentiable, or integrable respectively on I.

We shall often need to use the important inequality

$$\left| \int_c^d A(t)\mathbf{b}(t)\, dt \right| \leq \int_c^d |A(t)|\, |\mathbf{b}(t)|\, dt \tag{2.6}$$

for $c < d$, assuming, for example, that $A(t)$ and $\mathbf{b}(t)$ are continuous on $c \le t \le d$.

• EXERCISES

7. Prove the inequality (2.6). [*Hint:* Use (1.17), p. 20, and property (iii), p. 36.]

8. Let $\Phi(t)$ be a nonsingular matrix, differentiable on a real t interval I. Prove that $\Phi^{-1}(t)$ is differentiable and find a formula for $(\Phi^{-1})'(t)$. [*Hint:* For the second part use $\Phi(t)\Phi^{-1}(t) = E$ and differentiate.]

2.2 Existence and Uniqueness for Linear Systems

We now return to the linear system

$$\mathbf{y}' = A(t)\mathbf{y} + \mathbf{g}(t) \tag{2.3}$$

where we assume that the matrix $A(t)$ and the vector $\mathbf{g}(t)$ are continuous on an interval I. Then the vector function $\mathbf{f}(t, \mathbf{y}) = A(t)\mathbf{y} + \mathbf{g}(t)$ of Theorem 1.1 (p. 26) is continuous for $(t, \mathbf{y})$ in D, where D is the strip $\{(t, \mathbf{y}) \mid t \in I, |\mathbf{y}| < \infty\}$, and $\partial\mathbf{f}/\partial y_k = \operatorname{col}(a_{1k}(t), a_{2k}(t), \ldots, a_{nk}(t))$ $(k = 1, \ldots, n)$ where col stands for column; hence $\partial\mathbf{f}/\partial y_k$ are also continuous in D for $k = 1, \ldots, n$. Thus by Theorem 1.1, (2.3) has a unique solution $\boldsymbol{\phi}(t)$ passing through any given point $(t_0, \boldsymbol{\eta})$ with t_0 in I; and this solution exists on some interval containing the point t_0 in its interior. Theorem 1.1 also says that the solution $\boldsymbol{\phi}$ exists on any interval J containing the point t_0 and contained in the interval I for which the points $(t, \boldsymbol{\phi}(t))$ with t in J lie in D. For the present case of D this means that the solution exists on the whole interval I (finite or infinite) provided it can be proved that $|\boldsymbol{\phi}(t)|$, the norm of the solution $\boldsymbol{\phi}$, is bounded by a constant independent of t (such a bound is called an **a priori bound**). This is indeed always possible if I is a closed bounded interval as we now prove.

Theorem 2.1. *If $A(t)$, $\mathbf{g}(t)$ are continuous on some interval $a \le t \le b$, if $a \le t_0 \le b$, and if $|\boldsymbol{\eta}| < \infty$, then the system* (2.3) *has a unique solution $\boldsymbol{\phi}(t)$ satisfying the initial condition $\boldsymbol{\phi}(t_0) = \boldsymbol{\eta}$ and existing on the interval $a \le t \le b$.*

Proof. Let $\boldsymbol{\phi}(t)$ be the unique solution satisfying $\boldsymbol{\phi}(t_0) = \boldsymbol{\eta}$, existing for t on an interval J. To show that this solution exists on the whole interval $a \le t \le b$, it suffices, by the above remarks, to show that $|\boldsymbol{\phi}(t)|$ is bounded by a constant independent of t. For t in J, substitution of $\boldsymbol{\phi}$ into (2.3) gives

$$\boldsymbol{\phi}'(t) = A(t)\boldsymbol{\phi}(t) + \mathbf{g}(t)$$

Integration gives

$$\phi(t) - \phi(t_0) = \int_{t_0}^{t} A(s)\phi(s)\, ds + \int_{t_0}^{t} \mathbf{g}(s)\, ds \qquad (t \text{ in } J)$$

from which, using the initial condition and taking norms, we obtain

$$|\phi(t)| \leq |\eta| + \left|\int_{t_0}^{t} A(s)\phi(s)\, ds\right| + \left|\int_{t_0}^{t} \mathbf{g}(s)\, ds\right|$$

We continue with the argument for $t \geq t_0$; using properties of the norm and the inequality (2.6), we have

$$|\phi(t)| \leq |\eta| + \int_{t_0}^{t} |A(s)|\,|\phi(s)|\, ds + \int_{t_0}^{t} |\mathbf{g}(s)|\, ds \qquad (t \text{ in } J)$$

Since

$$\begin{aligned} |\eta| + \int_{t_0}^{t} |\mathbf{g}(s)|\, ds &\leq |\eta| + \max_{a \leq t \leq b} |\mathbf{g}(t)|(t - t_0) \\ &\leq |\eta| + \max_{a \leq t \leq b} |\mathbf{g}(t)|(b - a) = K_1 \end{aligned}$$

and, letting

$$K_2 = \max_{a \leq t \leq b} |A(t)|,$$

$$\int_{t_0}^{t} |A(s)|\,|\phi(s)|\, ds \leq \max_{a \leq t \leq b} |A(t)| \left(\int_{t_0}^{t} |\phi(s)|\, ds\right) = K_2 \int_{t_0}^{t} |\phi(s)|\, ds$$

this inequality can be written as

$$|\phi(t)| \leq K_1 + K_2 \int_{t_0}^{t} |\phi(s)|\, ds \qquad (t \text{ in } J)$$

where K_1 and K_2 are constants. Note that the constants K_1 and K_2 are nonnegative and independent of t, but that they do depend on $A(t)$, $\mathbf{g}(t)$, a,b.

• EXERCISE

1. Show that for $t \leq t_0$ we have $|\phi(t)| \leq K_1 + K_2 \int_t^{t_0} |\phi(s)|\, ds$.

Then by the Gronwall inequality (Section 1.7, p. 31), we obtain, for both $t \leq t_0$ and $t \geq t_0$,

$$|\boldsymbol{\phi}(t)| \le K_1 \exp(K_2|t - t_0|) \le K_1 \exp[(b-a)K_2] \qquad (t \text{ in } J)$$

This shows that $|\boldsymbol{\phi}(t)|$ is bounded by a constant. By Theorem 1.1 (p. 26), therefore, $\boldsymbol{\phi}(t)$ exists on the entire interval $a \le t \le b$, and this completes the proof. ∎

We remark that, interpreted geometrically, this proof shows that the solution remains inside an $(n+1)$-dimensional "box" of "base" $a \le t \le b$ and "height" $2|\mathbf{y}|$, where $|\mathbf{y}| \le K_1 \exp[K_2(b-a)]$.

We note also that if $A(t)$ and $\mathbf{g}(t)$ in (2.4) are continuous on $-\infty < t < \infty$, the above arguments apply to every **finite** subinterval; of course, in this case the solution $\boldsymbol{\phi}(t)$ need not remain bounded as $t \to \pm\infty$. The same remark applies if $A(t)$ and $\mathbf{g}(t)$ are continuous on $a < t < b$, but not necessarily on $a \le t \le b$. This leads to the following consequence of Theorem 2.1.

Corollary to Theorem 2.1. If $A(t)$, $\mathbf{g}(t)$ are continuous on an interval I, closed or open, finite or infinite, and if $t_0 \in I$, $|\boldsymbol{\eta}| < \infty$, then the equation (2.3) *has a unique solution $\boldsymbol{\phi}(t)$ satisfying the initial condition $\boldsymbol{\phi}(t_0) = \boldsymbol{\eta}$ and existing on I.*

• EXERCISES

2. Prove the above corollary.
3. Suppose $A(t)$ and $\mathbf{g}(t)$ are continuous for $-\infty < t < \infty$ and that

$$\int_{-\infty}^{\infty} |A(t)|\, dt < \infty$$

and

$$\int_{-\infty}^{\infty} |\mathbf{g}(t)|\, dt < \infty$$

Show that the solution $\boldsymbol{\phi}(t)$ of $\mathbf{y}' = A(t)\mathbf{y} + \mathbf{g}(t)$ exists for $-\infty < t < \infty$ and compute a bound for $|\boldsymbol{\phi}(t)|$ valid for $-\infty < t < \infty$.

4. State the analogue of Theorem 2.1 and its corollary for the scalar equation

$$a_0(t)y^{(n)} + a_1(t)y^{(n-1)} + \cdots + a_{n-1}(t)y' + a_n(t)y = b(t)$$

where $a_0, a_1, a_2, \ldots, a_n, b$ are continuous functions and $a_0 \ne 0$ on a bounded interval $a \le t \le b$.

2.3 Linear Homogeneous Systems

We are now ready to discuss the structure of solutions of the linear system (2.3), and we begin with the linear homogeneous system

$$\mathbf{y}' = A(t)\mathbf{y} \tag{2.7}$$

We assume that the n-by-n matrix $A(t)$ is continuous on an interval I and then by Theorem 2.1 and its corollary, we see immediately that given any point $(t_0, \boldsymbol{\eta})$, t_0 in I, there exists one and only one solution $\boldsymbol{\phi}$ of (2.7) such that $\boldsymbol{\phi}(t_0) = \boldsymbol{\eta}$. In particular, and this is most important for what follows, given the point $(t_0, \mathbf{0})$, t_0 any point of I, (2.7) has **the unique solution** $\boldsymbol{\phi} = \mathbf{0}$ on I, satisfying the initial condition $\boldsymbol{\phi}(t_0) = \mathbf{0}$; this is because by inspection $\mathbf{0}$ is always a solution of (2.7), and by Theorem 2.1 this is the only solution through $(t_0, \mathbf{0})$.

To obtain a result about the structure of solutions of the linear homogeneous system (2.7) we first observe that if $\boldsymbol{\phi}_1$ and $\boldsymbol{\phi}_2$ are any solutions of (2.7) on an interval I, and c_1 and c_2 are any (real or complex) constants, then the linearity of (2.7) tells us that

$$(c_1\boldsymbol{\phi}_1 + c_2\boldsymbol{\phi}_2)' = c_1\boldsymbol{\phi}_1' + c_2\boldsymbol{\phi}_2' = c_1 A\boldsymbol{\phi}_1 + c_2 A\boldsymbol{\phi}_2 = A(c_1\boldsymbol{\phi}_1 + c_2\boldsymbol{\phi}_2)$$

that is, $c_1\boldsymbol{\phi}_1 + c_2\boldsymbol{\phi}_2$ is again a solution of (2.7) on I. In the language of linear algebra this shows that the solutions of (2.7) form a vector space over the complex numbers. We denote this vector space by V.

We remind the reader that an abstract vector space over the real (or complex) numbers is a set of elements for which operations of addition and multiplication by scalars satisfying certain well-known properties are defined.

• EXERCISES

1. Look up the axioms for an abstract vector space.
2. Verify that the set of real (or complex) functions continuous on an interval I forms a vector space over the real (or complex) numbers, with addition and multiplication by scalars defined in the usual way.
3. Verify that the set of real (or complex) vector functions

$$\mathbf{f}(t) = (f_1(t), \ldots, f_n(t))$$

continuous on an interval I forms a vector space over the real (or complex) numbers, with addition and multiplication by scalars defined in the usual way.

A subset S of a vector space is called a **subspace** if it is closed under the formation of sums and products by scalars. It is easy to prove directly from the definition of a vector space that any subspace of a vector space is itself a vector space with the same operations.

• EXERCISE

4. Prove that a subspace of a vector space is a vector space.

Combining the results of Exercises 3 and 4, we see that the set of solutions V of (2.7) is a subspace of the vector space of continuous vector functions over the real (or complex) numbers, and is therefore a vector space. It is now natural to ask, What is the dimension of the vector space V? In order to discuss this problem, and indeed to define the dimension of a vector space, we must recall the definitions of linear dependence and independence of sets of vectors.

Definition. *A set of vectors* $\mathbf{v}_1, \mathbf{v}_2, \ldots, \mathbf{v}_k$ *is linearly dependent if there exist scalars* $c_1, c_2, \ldots, c_k$*, not all zero, such that the linear combination*

$$c_1\mathbf{v}_1 + c_2\mathbf{v}_2 + \cdots + c_k\mathbf{v}_k = \mathbf{0}$$

Definition. *A set of vectors* $\mathbf{v}_1, \mathbf{v}_2, \ldots, \mathbf{v}_k$ *is linearly independent if it is not linearly dependent.*

These definitions contain the definitions of linear dependence and independence of functions as very special cases (see, for example, Section 3.3, p. 69, of [2]). There the underlying vector space is the vector space of continuous functions on an interval I; see Exercise 2 above.

• EXERCISES

5. Formulate the definitions of linear dependence and independence of a set of k vector functions $\mathbf{f}_1(t), \ldots, \mathbf{f}_k(t)$ continuous on an interval I. (See Exercise 3 above for the underlying vector space.)
6. Show that the vectors

$$\mathbf{v}_1 = \begin{pmatrix} 1 \\ 0 \\ 0 \end{pmatrix}, \qquad \mathbf{v}_2 = \begin{pmatrix} 0 \\ 1 \\ 0 \end{pmatrix}, \qquad \mathbf{v}_3 = \begin{pmatrix} 0 \\ 1 \\ 1 \end{pmatrix}$$

are linearly independent in E_3. [*Hint:* Suppose they are linearly dependent and obtain a contradiction.]

7. Show that the vectors

$$\mathbf{v}_1 = \begin{pmatrix} 1 \\ 2 \\ 3 \end{pmatrix}, \qquad \mathbf{v}_2 = \begin{pmatrix} 1 \\ 3 \\ 5 \end{pmatrix}, \qquad \mathbf{v}_3 = \begin{pmatrix} 1 \\ 10 \\ -5 \end{pmatrix}, \qquad \mathbf{v}_4 = \begin{pmatrix} 0 \\ -1 \\ 17 \end{pmatrix},$$

are linearly dependent in E_3.

8. Show that if $r_1 \neq r_2$, the vectors (functions)

$$\mathbf{v}_1 = \exp(r_1 t), \qquad \mathbf{v}_2 = \exp(r_2 t)$$

are linearly independent in the space of continuous functions on $-\infty < t < \infty$.

9. Show that the vectors

$$\mathbf{v}_1 = \begin{pmatrix} \exp(r_1 t) \\ r_1 \exp(r_1 t) \end{pmatrix}, \qquad \mathbf{v}_2 = \begin{pmatrix} \exp(r_2 t) \\ r_2 \exp(r_2 t) \end{pmatrix}$$

are linearly independent in the space of continuous vector functions with two components on $-\infty < t < \infty$ provided $r_1 \neq r_2$.

10. Repeat Exercise 9 for the vectors

$$\mathbf{v}_1 = \begin{pmatrix} \cos t \\ -\sin t \end{pmatrix}, \qquad \mathbf{v}_2 = \begin{pmatrix} \sin t \\ \cos t \end{pmatrix}$$

A set S of vectors is said to form a **basis** of a vector space V if it is linearly independent and if every vector in V can be expressed as a linear combination of vectors in S.

• EXERCISES

11. Show that if S is a basis of a vector space V, then the expression of every vector in V as a linear combination of vectors in S is unique. [*Hint:* Suppose a vector $\mathbf{v}$ may be expressed as $\mathbf{v} = c_1\mathbf{v}_1 + c_2\mathbf{v}_2 + \cdots + c_k\mathbf{v}_k = d_1\mathbf{v}_1 + d_2\mathbf{v}_2 + \cdots + d_k\mathbf{v}_k$, where S consists of the vectors $\mathbf{v}_1, \mathbf{v}_2, \ldots, \mathbf{v}_k$. Show that $c_1 = d_1, c_2 = d_2, \ldots, c_k = d_k$.]

12. Show that the vectors $\mathbf{v}_1, \mathbf{v}_2, \mathbf{v}_3$ in Exercise 6 above form a basis of E_3.

13. Show that the vectors

$$\mathbf{w} = \begin{pmatrix} 1 \\ 2 \\ 3 \end{pmatrix}, \qquad \mathbf{w}_2 = \begin{pmatrix} 0 \\ 1 \\ 5 \end{pmatrix}, \qquad \mathbf{w}_3 = \begin{pmatrix} 0 \\ 0 \\ 4 \end{pmatrix}$$

also form a basis of E_3.

14. Show that any linearly independent set of three vectors in E_3 is a basis of E_3.

We can now define the dimension of a particular vector space V to be the number of elements in any basis of V. A vector space is called finite-dimensional if it has a finite basis. Thus, for example, E_n has dimension n. It can be shown that every basis of a finite-dimensional vector space has the same number of elements. We note that the space C of continuous functions on a finite interval is not finite-dimensional because the infinite set $1, t, t^2, \ldots, t^n, \ldots,$ is linearly independent on any interval and is contained in C.

We now return to the problem of finding the dimension of the vector space V of solutions of (2.7). We have the answer in the following basic results.

Theorem 2.2. *If the complex* n-by-n *matrix* $A(t)$ *is continuous on an interval* I, *then the solutions of the system*

$$\mathbf{y}' = A(t)\mathbf{y} \tag{2.7}$$

on I *form a vector space of dimension* n *over the complex numbers.*

Proof. We have already established that the solutions form a vector space V over the complex numbers. To establish that the dimension of V is n, we need to construct a basis for V consisting of n linearly independent vectors in V, that is, of n linearly independent solutions of (2.7) on I. We proceed as follows. Let t_0 be any point of I and let $\boldsymbol{\sigma}_1, \boldsymbol{\sigma}_2, \ldots, \boldsymbol{\sigma}_n$ be any n linearly independent points (vectors) in Euclidean n-space E_n. For example

$$\boldsymbol{\sigma}_j = \begin{pmatrix} 0 \\ \vdots \\ 0 \\ 1 \\ 0 \\ \vdots \\ 0 \end{pmatrix} \begin{matrix} \\ \\ \\ \leftarrow j\text{th row} \\ \\ \\ \\ \end{matrix} \qquad (j = 1, 2, \ldots, n)$$

are obviously n such vectors. By Theorem 2.1 and its corollary the system (2.7) possesses n solutions $\boldsymbol{\phi}_1, \boldsymbol{\phi}_2, \ldots, \boldsymbol{\phi}_n$, each of which exists on the entire interval I, and each solution $\boldsymbol{\phi}_j$ satisfies the initial condition

$$\boldsymbol{\phi}_j(t_0) = \boldsymbol{\sigma}_j \qquad (j = 1, 2, \ldots, n) \tag{2.8}$$

We first show that the solutions $\boldsymbol{\phi}_1, \boldsymbol{\phi}_2, \ldots, \boldsymbol{\phi}_n$ are linearly independent on I. Suppose they are not. Then there exist complex constants $a_1, a_2, \ldots, a_n$, **not all zero**, such that

$$a_1\boldsymbol{\phi}_1(t) + a_2\boldsymbol{\phi}_2(t) + \cdots + a_n\boldsymbol{\phi}_n(t) = 0 \qquad \text{for every } t \text{ on } I$$

In particular, putting $t = t_0$, and using the initial conditions (2.8), we have

$$a_1\boldsymbol{\sigma}_1 + a_2\boldsymbol{\sigma}_2 + \cdots + a_n\boldsymbol{\sigma}_n = 0$$

But this is impossible (**unless** $a_1, a_2, \ldots, a_n$ are all zero) because it contradicts the assumed linear independence of $\boldsymbol{\sigma}_1, \boldsymbol{\sigma}_2, \ldots, \boldsymbol{\sigma}_n$.

To complete the proof we must show that these n linearly independent solutions of (2.7) span V, that is, they have the property that every solution $\boldsymbol{\psi}(t)$ of (2.7) can be expressed as a linear combination of the solutions $\boldsymbol{\phi}_1, \boldsymbol{\phi}_2, \ldots, \boldsymbol{\phi}_n$. We proceed as follows. Compute the value of the solution $\boldsymbol{\psi}$ at t_0 and let $\boldsymbol{\psi}(t_0) = \boldsymbol{\sigma}$. Since the constant vectors $\boldsymbol{\sigma}_1, \boldsymbol{\sigma}_2, \ldots, \boldsymbol{\sigma}_n$ are linearly independent in Euclidean n-space E_n, they form a basis for E_n and there exist unique constants $c_1, c_2, \ldots, c_n$ such that the constant vector $\boldsymbol{\sigma}$ can be represented as

$$\boldsymbol{\sigma} = c_1\boldsymbol{\sigma}_1 + c_2\boldsymbol{\sigma}_2 + \cdots + c_n\boldsymbol{\sigma}_n$$

(see Exercise 12, for the case $n = 3$). Now consider the vector

$$\boldsymbol{\phi}(t) = c_1\boldsymbol{\phi}_1(t) + c_2\boldsymbol{\phi}_2(t) + \cdots + c_n\boldsymbol{\phi}_n(t)$$

Clearly, $\boldsymbol{\phi}(t)$ is a solution of (2.7) on I. (Why? Prove this.) Moreover, the initial value of $\boldsymbol{\phi}$ is (using (2.8))

$$\boldsymbol{\phi}(t_0) = c_1\boldsymbol{\sigma}_1 + c_2\boldsymbol{\sigma}_2 + \cdots + c_n\boldsymbol{\sigma}_n = \boldsymbol{\sigma}$$

Therefore $\boldsymbol{\phi}(t)$ and $\boldsymbol{\psi}(t)$ are both solutions of (2.7) on I with $\boldsymbol{\phi}(t_0) = \boldsymbol{\psi}(t_0) = \boldsymbol{\sigma}$. Therefore, by the uniqueness part of Theorem 2.1, $\boldsymbol{\phi}(t) = \boldsymbol{\psi}(t)$ for every t on I and the solution $\boldsymbol{\psi}(t)$ is expressed as the unique linear combination

$$\boldsymbol{\psi}(t) = c_1\boldsymbol{\phi}_1(t) + c_2\boldsymbol{\phi}_2(t) + \cdots + c_n\boldsymbol{\phi}_n(t) \qquad \text{for every } t \text{ on } I \qquad (2.9)$$

• **EXERCISE**

15. Show that this expression of $\boldsymbol{\psi}(t)$ as a linear combination of $\boldsymbol{\phi}_1(t), \ldots, \boldsymbol{\phi}_n(t)$ is unique. [*Hint:* See Exercise 11.]

Thus we have shown that the solutions $\boldsymbol{\phi}_1, \boldsymbol{\phi}_2, \ldots, \boldsymbol{\phi}_n$ of (2.7) span the vector space V. Since they are also linearly independent, they form a basis for the solution space V, and the dimension of V is n. This completes the proof of Theorem 2.2. ▌

We often say that the linearly independent solutions $\boldsymbol{\phi}_1, \ldots, \boldsymbol{\phi}_n$ form a **fundamental set of solutions**. There are clearly infinitely many different fundamental sets of solutions of (2.7). (Why?)

• **EXERCISES**

16. Prove the following analogue of Theorem 2.2 for systems with real coefficients If the real n-by-n matrix $A(t)$ is continuous on an interval I, then the real solutions of (2.7) on I form a vector space of dimension n over the real numbers.

17. Write the linear homogeneous scalar equation

$$a_0(t)y^{(n)} + a_1(t)y^{(n-1)} + \cdots + a_{n-1}(t)y' + a_n(t)y = 0$$

where $a_0, a_1, \ldots, a_n$ are continuous on I and $a_0(t) \neq 0$, as a system and interpret Theorem 2.2 for this equation.

We can interpret Theorem 2.2 in a different and useful way. A matrix of n rows whose columns are solutions of (2.7) is called a **solution matrix**. Now if we form an n-by-n matrix using the above n linearly independent solutions as columns, we will have a solution matrix on I, but also its columns will be linearly independent solutions of (2.7) on I. A solution matrix whose columns are linearly independent on I is called a **fundamental matrix** for (2.7) on I. Let us denote the fundamental matrix formed from the solutions $\boldsymbol{\phi}_1, \boldsymbol{\phi}_2, \ldots, \boldsymbol{\phi}_n$ as columns by $\boldsymbol{\Phi}$. Then the statement that every solution $\boldsymbol{\psi}$ is the linear combination (2.9) for some unique choice of the constants $c_1, \ldots, c_n$ is simply that

$$\boldsymbol{\psi}(t) = \boldsymbol{\Phi}(t)\mathbf{c} \tag{2.10}$$

where $\boldsymbol{\Phi}$ is the fundamental matrix constructed above and $\mathbf{c}$ is the column vector components $c_1, \ldots, c_n$. It is clear that if $\tilde{\boldsymbol{\Phi}}(t)$ is any other fundamental matrix of (2.7) in I, then the above solution $\boldsymbol{\psi}$ can be expressed as

$$\boldsymbol{\psi}(t) = \tilde{\boldsymbol{\Phi}}(t)\tilde{\mathbf{c}} \qquad \text{for every } t \text{ on } I$$

for a suitably chosen constant vector $\tilde{\mathbf{c}}$. Clearly every solution of (2.7) on I can be expressed in this form by using any fundamental matrix.

• **EXERCISE**

18. Given that $\boldsymbol{\psi}(t_0) = \boldsymbol{\sigma}$, determine the vector $\tilde{\mathbf{c}}$.

We see from the discussion above that to find any solution of (2.7) we need to find n linearly independent solutions on I, or equivalently, we need to find a fundamental matrix. A natural question, then, is the following: Suppose we have found a solution matrix of (2.7) on some interval I. Can we test in some simple way whether this solution matrix is a fundamental matrix? We shall see (Theorem 2.4) that any solution matrix $\Phi(t)$, not necessarily the special solution matrix used in (2.10), is a fundamental matrix of (2.7) if and only if $\det \Phi(t) \neq 0$ on I. However, it is convenient to establish an important auxiliary result first.

Theorem 2.3. (Abel's Formula.) *If Φ is a solution matrix of* (2.7) *on I and if t_0 is any point of I, then*

$$\det \Phi(t) = \det \Phi(t_0) \exp \left[\int_{t_0}^{t} \sum_{j=1}^{n} a_{jj}(s)\, ds \right] \quad \textit{for every } t \textit{ in } I \tag{2.11}$$

It follows immediately from Theorem 2.3 (since t_0 is arbitrary) that either $\det \Phi(t) \neq 0$ for each t in I or $\det \Phi(t) = 0$ for every t in I.

Proof of Theorem 2.3. Let us denote the columns of Φ by $\boldsymbol{\phi}_1, \boldsymbol{\phi}_2, \ldots, \boldsymbol{\phi}_n$; let $\boldsymbol{\phi}_j$ have components $(\phi_{1j}, \phi_{2j}, \ldots, \phi_{nj})$. Then the statement that $\boldsymbol{\phi}_j$ is a solution of (2.7) on I can be written in terms of components as

$$\phi'_{ij} = \sum_{k=1}^{n} a_{ik}\phi_{kj} \qquad (i, j = 1, \ldots, n) \tag{2.12}$$

It is now necessary to recall that the derivative of $\det \Phi$ is a sum of n determinants:

$$(\det \Phi)' = \begin{vmatrix} \phi'_{11} & \phi'_{12} & \cdots & \phi'_{1j} & \cdots & \phi'_{1n} \\ \phi_{21} & \phi_{22} & \cdots & \phi_{2j} & \cdots & \phi_{2n} \\ \vdots & \vdots & & \vdots & & \vdots \\ \phi_{n1} & \phi_{n2} & & \phi_{nj} & \cdots & \phi_{nn} \end{vmatrix} + \begin{vmatrix} \phi_{11} & \phi_{12} & \cdots & \phi_{1j} & \cdots & \phi_{1n} \\ \phi'_{21} & \phi'_{22} & \cdots & \phi'_{2j} & \cdots & \phi'_{2n} \\ \phi_{31} & \phi_{32} & \cdots & \phi_{3j} & \cdots & \phi_{3n} \\ \vdots & & & & & \\ \phi_{n1} & \phi_{n2} & \cdots & \phi_{nj} & \cdots & \phi_{nn} \end{vmatrix} + \cdots$$

$$+ \begin{vmatrix} \phi_{11} & \phi_{12} & \cdots & \phi_{1j} & \cdots & \phi_{1n} \\ \vdots & \vdots & & \vdots & & \vdots \\ \phi_{n-1,1} & \phi_{n-1,2} & & \phi_{n-1,j} & \cdots & \phi_{n-1,n} \\ \phi'_{n1} & \phi'_{n2} & & \phi'_{nj} & & \phi'_{nn} \end{vmatrix}$$

This fact is easily proved by induction.

Using (2.12) we obtain

$$(\det \Phi)' = \begin{vmatrix} \sum_{k=1}^{n} a_{1k}\phi_{k1} & \sum_{k=1}^{n} a_{1k}\phi_{k2} & \cdots & \sum_{k=1}^{n} a_{1k}\phi_{kn} \\ \phi_{21} & \phi_{22} & \cdots & \phi_{2n} \\ \vdots & \vdots & & \vdots \\ \phi_{n1} & \phi_{n2} & \cdots & \phi_{nn} \end{vmatrix}$$

$$+ \begin{vmatrix} \phi_{11} & \phi_{12} & \cdots & \phi_{1n} \\ \sum_{k=1}^{n} a_{2k}\phi_{k1} & \sum_{k=1}^{n} a_{2k}\phi_{k2} & \cdots & \sum_{k=1}^{n} a_{2k}\phi_{kn} \\ \phi_{31} & \phi_{32} & \cdots & \phi_{3n} \\ \vdots & \vdots & & \vdots \\ \phi_{n1} & \phi_{n2} & & \phi_{nn} \end{vmatrix} + \cdots$$

$$+ \begin{vmatrix} \phi_{11} & \phi_{12} & \cdots & \phi_{1n} \\ \vdots & \vdots & & \vdots \\ \phi_{n-1,1} & \phi_{n-1,2} & \cdots & \phi_{n-1,n} \\ \sum_{k=1}^{n} a_{nk}\phi_{k1} & \sum_{k=1}^{n} a_{nk}\phi_{k2} & \cdots & \sum_{k=1}^{n} a_{nk}\phi_{kn} \end{vmatrix}$$

Using elementary row operations, we can evaluate each determinant. For example, in the first determinant we multiply the second row by a_{12}, the third by $a_{13}, \ldots$, the nth by a_{1n}, add these $(n-1)$ rows, and then subtract the result from the first row. This leaves a_{11} as a factor of the resulting first row. Proceeding similarly with the other determinants we obtain

$$(\det \Phi)' = a_{11} \det \Phi + a_{22} \det \Phi + \cdots + a_{nn} \det \Phi$$

for every t on I, or equivalently,

$$(\det \Phi)' = \left(\sum_{k=1}^{n} a_{kk}(t) \right) \det \Phi \tag{2.13}$$

which is a first-order **scalar equation** for $\det \Phi$. Its solution is seen to be (2.11) without difficulty. ▮

• EXERCISES

19. Show that (2.11) is the unique solution of the scalar equation

$$y' = \left(\sum_{k=1}^{n} a_{kk}(t)\right) y$$

satisfying the initial condition $y(t_0) = \det \Phi(t_0)$.

20. Show that the jth determinant in the expression for $(\det \Phi')$ in the proof of Theorem 2.3 is $a_{jj} \det \Phi$.

21. Write the scalar equation $a_0(t)y^{(n)} + \cdots + a_{n-1}(t)y' + a_n(t)y = 0$ where $a_0, a_1, \ldots, a_n$ are continuous on I and $a_0(t) \neq 0$ on I as a system. Then show that for this specific system $\det \Phi$ is precisely the Wronskian (see Section 3.3, p. 75 of [2]). Theorem 2.3 for this scalar equation is the same as Theorem 3.4, p. 78 of [2], if $n = 2$.

We now use Theorem 2.3 to establish the main result.

Theorem 2.4. *A solution matrix* Φ *of* (2.7) *on an interval* I *is a fundamental matrix of* (2.7) *on* I *if and only if* $\det \Phi(t) \neq 0$ *for every* t *on* I.

Proof of Theorem 2.4. If $\det \Phi(t) \neq 0$ for every t on I, the columns of the solution matrix are obviously linearly independent on I, and therefore Φ is a fundamental matrix of (2.7) on I.

Conversely, if Φ is a fundamental matrix, then every solution $\boldsymbol{\phi}$ of (2.7) on I has the form $\boldsymbol{\phi}(t) = \Phi(t)\mathbf{c}$ for some constant vector $\mathbf{c}$. For each fixed t_0 in I and any vector $\boldsymbol{\phi}(t_0)$, because $\Phi(t)$ is a fundamental matrix, the system of algebraic equations $\boldsymbol{\phi}(t_0) = \Phi(t_0)\mathbf{c}$ has a unique solution for $\mathbf{c}$. Therefore (see property (iii) of determinants given in Section 2.1), $\det \Phi(t_0) \neq 0$. Now Abel's formula (Theorem 2.3) gives $\det \Phi(t) \neq 0$ for every t on I. This completes the proof of Theorem 2.4. ∎

The reader is warned that a matrix may have its determinant identically zero on some interval, although its columns are linearly independent. Indeed, let

$$\Phi(t) = \begin{pmatrix} 1 & t & t^2 \\ 0 & 2 & t \\ 0 & 0 & 0 \end{pmatrix}$$

Then clearly $\det \Phi(t) = 0$, $-\infty < t < \infty$, and yet the columns are linearly independent. This, according to Theorem 2.4, cannot happen for solutions of (2.7).

• **EXERCISE**

22. State Theorem 2.4 for the scalar equation in Exercise 21 (p. 48) in terms of the Wronskian determinant. (For $n = 2$, this is Theorem 3.3, p. 76 of [2].)

It is important from the practical point of view to remark that Theorems 2.3 and 2.4 together imply that to test whether a solution matrix of (2.7) is a fundamental matrix, it suffices to evaluate its determinant at one point! This point can frequently be chosen to make the calculation a simple one.

• **EXERCISES**

23. Show, with the aid of Theorem 2.3, that

$$\begin{pmatrix} \cos t & \sin t \\ -\sin t & \cos t \end{pmatrix}$$

is a fundamental matrix for the system $y' = Ay$ where

$$A = \begin{pmatrix} 0 & 1 \\ -1 & 0 \end{pmatrix}$$

✓24. Show, with the aid of Theorems 2.3, 2.4 that

$$\begin{pmatrix} \exp(r_1 t) & \exp(r_2 t) \\ r_1 \exp(r_1 t) & r_2 \exp(r_2 t) \end{pmatrix}$$

is a fundamental matrix for the system $y' = Ay$ where

$$A = \begin{pmatrix} 0 & 1 \\ -a_2 & -a_1 \end{pmatrix}$$

and r_1, r_2 are the distinct roots of the quadratic equation $z^2 + a_1 z + a_2 = 0$. (We shall learn in Section 2.5 how to construct this fundamental matrix.)

If Φ is a fundamental matrix of (2.7) on I, and C is a nonsingular constant (complex) matrix, then ΦC is clearly a solution matrix (prove this), and since $\det \Phi C = \det \Phi \cdot \det C$ it follows from Theorem 2.4 that ΦC is also a fundamental matrix on I.

• **EXERCISE**

25. Show that $C\Phi$, where C is a constant matrix and Φ is a fundamental matrix, need not be a solution matrix of (2.7).

Now suppose Φ and Ψ are two fundamental matrices of (2.7) on I. Letting ψ_j be the jth column of Ψ, it follows from (2.10) that $\psi_j = \Phi \mathbf{c}_j$, $j = 1, \ldots, n$, where $\mathbf{c}_j$ are suitable constant vectors. Therefore if we define C as the constant matrix whose columns are the vectors $\mathbf{c}_j, j = 1, \ldots, n$, we have at once that $\Psi(t) = \Phi(t)C$ for every t on I. Since $\det \Phi$ and $\det \Psi$ are both different from zero on I (why?), we also have $\det C \neq 0$ so that C is a nonsingular constant matrix. These remarks put together give the following relation between different fundamental matrices.

Theorem 2.5. *If Φ is a fundamental matrix for $\mathbf{y}' = A(t)\mathbf{y}$ on I and C is a nonsingular constant matrix, then ΦC is also a fundamental matrix for $\mathbf{y}' = A(t)\mathbf{y}$ on I. Every fundamental matrix of* (2.7) *is of this form for some nonsingular matrix C.*

- **EXERCISE**

26. Consider the system $\mathbf{y}' = A(t)\mathbf{y}$ where $A(t)$ is continuous for $-\infty < t < \infty$ and $A(t)$ is periodic with period 2π, that is, $A(t + 2\pi) = A(t)$. Show that if $\Phi(t)$ is a fundamental matrix on $-\infty < t < \infty$, then so is $\Phi(t + 2\pi)$. [*Hint:* Substitute and apply the theory of this section.] Thus, prove that

$$\Phi(t + 2\pi) = \Phi(t)C$$

for some nonsingular matrix C.

The result established in Exercise 26 is fundamental to the theory of linear systems with periodic coefficients. To proceed with this theory would require the result that the nonsingular constant matrix C has a logarithm, treated in Appendix 3. The interested reader may return to this topic in Section 2.9 after he has completed Section 2.6.

- **EXERCISES**

27. (a) Show that

$$\Phi(t) = \begin{pmatrix} t^2 & t \\ 2t & 1 \end{pmatrix}$$

is a fundamental matrix for the system $\mathbf{y}' = A(t)\mathbf{y}$ where

$$A(t) = \begin{pmatrix} 0 & 1 \\ -2/t^2 & 2/t \end{pmatrix}$$

on any interval I not including the origin.

(b) Does the fact that $\det \Phi(0) = 0$ contradict Theorem 2.4?

28. Show that if a real homogeneous system of two first-order equations has a fundamental matrix

$$\begin{pmatrix} e^{it} & e^{-it} \\ ie^{it} & -ie^{-it} \end{pmatrix}$$

then

$$\begin{pmatrix} \cos t & \sin t \\ -\sin t & \cos t \end{pmatrix}$$

is also a fundamental matrix. Can you find another real fundamental matrix?

The reader who is acquainted with the theory of scalar linear equations of order n should have noticed that we have now obtained here as a very special case essentially all of that theory, and that this has been accomplished with no additional effort.

2.4 Linear Nonhomogeneous Systems

We now use the theory developed in Sections 2.2 and 2.3 to discuss the form of solutions of the nonhomogeneous system

$$\mathbf{y}' = A(t)\mathbf{y} + \mathbf{g}(t) \tag{2.14}$$

where $A(t)$ is a given continuous matrix and $\mathbf{g}(t)$ is a given continuous vector on an interval I. The entire development rests on the assumption that we can find a fundamental matrix of the corresponding homogeneous system $\mathbf{y}' = A(t)\mathbf{y}$. The vector $\mathbf{g}(t)$ is usually referred to as a forcing term because if (2.14) describes a physical system, $\mathbf{g}(t)$ represents an external force. By Theorem 2.1 and its corollary, we know that given any point $(t_0, \boldsymbol{\eta})$, t_0 in I, there is a unique solution $\boldsymbol{\phi}$ of (2.14) existing in all of I such that $\boldsymbol{\phi}(t_0) = \boldsymbol{\eta}$.

To construct solutions of (2.14) we let $\Phi(t)$ be a fundamental matrix of the homogeneous system $\mathbf{y}' = A(t)\mathbf{y}$ on I; Φ exists as a consequence of Theorem 2.2 (see also remarks immediately following its proof). Suppose $\boldsymbol{\phi}_1$ and $\boldsymbol{\phi}_2$ are any two solutions of (2.14) on I. Then $\boldsymbol{\phi}_1 - \boldsymbol{\phi}_2$ is a solution of the homogeneous system on I.

• **EXERCISE**

1. Verify this fact.

By Theorem 2.2, and the remarks immediately following its proof, there exists a constant vector $\mathbf{c}$ such that

$$\boldsymbol{\phi}_1 - \boldsymbol{\phi}_2 = \Phi\mathbf{c} \tag{2.15}$$

• **EXERCISE**

2. Assuming uniqueness for the solution of the initial value problem for the homogeneous system, use the above argument to establish uniqueness of solutions of the initial value problem for (2.14).

Formula (2.15) tells us that to find any solution of (2.14), we need only know one solution of (2.14). (Every other solution differs from the known one by some solution of the homogeneous system.) There is a simple method, known as **variation of constants,** to determine a solution of (2.14) **provided we know a fundamental matrix for the homogeneous** system $\mathbf{y}' = A(t)\mathbf{y}$. Let Φ be such a fundamental matrix on I. We attempt to find a solution $\boldsymbol{\psi}$ of (2.14) of the form

$$\boldsymbol{\psi}(t) = \Phi(t)\mathbf{v}(t) \tag{2.16}$$

where $\mathbf{v}$ is a vector to be determined. (Note that if $\mathbf{v}$ is a constant vector, then $\boldsymbol{\psi}$ satisfies the homogeneous system and thus for the present purpose $\mathbf{v}(t) \equiv \mathbf{c}$ is ruled out.) Suppose such a solution exists. Then substituting (2.16) into (2.14), we find for all t on I

$$\boldsymbol{\psi}'(t) = \Phi'(t)\mathbf{v}(t) + \Phi(t)\mathbf{v}'(t) = A(t)\Phi(t)\mathbf{v}(t) + \mathbf{g}(t)$$

Since Φ is a fundamental matrix of the homogeneous system, $\Phi'(t) = A(t)\Phi(t)$, and the terms involving $A(t)\Phi(t)\mathbf{v}(t)$ cancel. Therefore if $\boldsymbol{\psi}(t) = \Phi(t)\mathbf{v}(t)$ is a solution of (2.14), we must determine $\mathbf{v}(t)$ from the relation

$$\Phi(t)\mathbf{v}'(t) = \mathbf{g}(t)$$

Since $\Phi(t)$ is nonsingular on I we can premultiply by $\Phi^{-1}(t)$ and we have, on integrating,

$$\mathbf{v}(t) = \int_{t_0}^{t} \Phi^{-1}(s)\mathbf{g}(s)\,ds \qquad (t_0,\ t \text{ on } I)$$

and therefore (2.16) becomes

$$\boldsymbol{\psi}(t) = \Phi(t)\int_{t_0}^{t} \Phi^{-1}(s)\mathbf{g}(s)\,ds \qquad (t_0,\ t \text{ on } I) \tag{2.17}$$

Thus if (2.14) has a solution $\boldsymbol{\psi}$ of the form (2.16), then $\boldsymbol{\psi}$ is given by (2.17). Conversely, define $\boldsymbol{\psi}$ by (2.17), where Φ is a fundamental matrix of the homogeneous system on I. Then, differentiating (2.17) and using the fundamental theorem of calculus, we have

$$\psi'(t) = \Phi'(t) \int_{t_0}^{t} \Phi^{-1}(s)\mathbf{g}(s)\, ds + \Phi(t)\Phi^{-1}(t)\mathbf{g}(t)$$

$$= A(t)\Phi(t) \int_{t_0}^{t} \Phi^{-1}(s)\mathbf{g}(s)\, ds + \mathbf{g}(t)$$

and using (2.17) again,

$$\psi'(t) = A(t)\psi(t) + \mathbf{g}(t)$$

for every t on I. Obviously $\psi(t_0) = \mathbf{0}$. Thus we have proved the **variation of constants formula**:

Theorem 2.6. *If Φ is a fundamental matrix of $\mathbf{y}' = A(t)\mathbf{y}$ on I, then the function*

$$\psi(t) = \Phi(t) \int_{t_0}^{t} \Phi^{-1}(s)\mathbf{g}(s)\, ds$$

is the (unique) *solution of* (2.14) *satisfying the initial condition*

$$\psi(t_0) = \mathbf{0} \qquad \textit{and valid on } I$$

Combining Theorem 2.6 with the remarks made at the beginning of this section, we see that every solution ϕ of (2.14) on I has the form

$$\phi(t) = \phi_h(t) + \psi(t) \tag{2.18}$$

where ψ is the solution of Equation (2.14) satisfying the initial condition $\psi(t_0) = \mathbf{0}$, and ϕ_h is that solution of the homogeneous system satisfying the same initial condition at t_0 as ϕ, for example, $\phi_h(t_0) = \eta$.

• EXERCISES

3. Consider the system $\mathbf{y}' = A\mathbf{y} + \mathbf{g}(t)$ where $A = \begin{pmatrix} 2 & 1 \\ 0 & 2 \end{pmatrix}$, $\mathbf{y} = \begin{pmatrix} y_1 \\ y_2 \end{pmatrix}$, $\mathbf{g}(t) = \begin{pmatrix} \sin t \\ \cos t \end{pmatrix}$. Verify that $\Phi(t) = \begin{pmatrix} e^{2t} & te^{2t} \\ 0 & e^{2t} \end{pmatrix}$ is a fundamental matrix of $\mathbf{y}' = A\mathbf{y}$. Find that solution ϕ of the nonhomogeneous system for which $\phi(0) = \begin{pmatrix} 1 \\ -1 \end{pmatrix}$.

4. Find the solution $\boldsymbol{\phi}$ of the system $\mathbf{y}' = A\mathbf{y} + \mathbf{g}(t)$ with A the same as in Exercise 3 above and with $\mathbf{g}(t) = \begin{pmatrix} 0 \\ e^{2t} \end{pmatrix}$, satisfying the initial condition $\boldsymbol{\phi}(0) = \begin{pmatrix} 1 \\ -1 \end{pmatrix}$.

5. Consider the system $\mathbf{y}' = A(t)\mathbf{y} + \mathbf{g}(t)$, where

$$A(t) = \begin{pmatrix} 0 & 1 \\ -2/t^2 & 2/t \end{pmatrix} \qquad \mathbf{g}(t) = \begin{pmatrix} t^4 \\ t^3 \end{pmatrix}$$

Find the solution $\boldsymbol{\phi}$ satisfying the initial condition $\boldsymbol{\phi}(2) = \begin{pmatrix} 1 \\ 4 \end{pmatrix}$ and determine the interval of validity of this solution. [*Hint:* Use the fundamental matrix given in Exercise 27, Section 2.3.].

6. Consider the second-order scalar equation

$$y'' + p(t)y' + q(t)y = f(t) \tag{2.19}$$

with p, q, f continuous on an interval I. Let ϕ_1, ϕ_2 be linearly independent (scalar) solutions of the homogeneous equation associated with (2.19).

(a) Write (2.19) as an equivalent system of two first-order equations and show that

$$\Phi(t) = \begin{pmatrix} \phi_1 & \phi_2 \\ \phi_1' & \phi_2' \end{pmatrix}$$

is a fundamental matrix of the associated homogeneous system on I.

(b) Use Theorem 2.6 to find the solution $\boldsymbol{\psi}$ of the inhomogeneous system in part (a) for which $\boldsymbol{\psi}(t_0) = \mathbf{0}$, t_0 on I. (Or obtain this solution directly.)

(c) Writing $\boldsymbol{\psi} = \begin{pmatrix} \psi_1 \\ \psi_2 \end{pmatrix}$, show that

$$\psi_1(t) = \int_{t_0}^{t} \frac{\phi_2(t)\phi_1(s) - \phi_1(t)\phi_2(s)}{W(s)} f(s)\, ds$$

where $W(t) = \det \Phi(t)$, is that solution of (2.19) for which $\psi_1(t_0) = 0$, $\psi_1'(t_0) = 0$. (Compare with formula (3.34), p. 105 of [2]; where $a_0(t) \equiv 1$, $a_1(t) = p(t)$, $a_2(t) = q(t)$.)

7. Consider the scalar equation of order n,

$$L_n(y) = a_0(t)y^{(n)} + a_1(t)y^{(n-1)} + \cdots + a_n(t)y = f(t)$$

where $a_0, a_1, \ldots, a_n$, and f are continuous on an interval I and $a_0(t) \neq 0$ on I. Let $\phi_1, \ldots, \phi_n$ be n linearly independent solutions of the homogeneous equation $L_n(y) = 0$.

(a) Write the equation $L_n(y) = f(t)$ as an equivalent system of n first-order equations and find a fundamental matrix of the corresponding homogeneous system.

(b) Use Theorem 2.6 to find a particular solution of the inhomogeneous system and thus deduce the general solution of $L_n(y) = f(t)$. (See also pp. 107–108 of [2].)

Exercises 6 and 7 above show that the entire theory of the linear nonhomogeneous scalar equation is contained in the development of this section.

2.5 Linear Systems with Constant Coefficients

The results of Section 2.4 show that in order to solve any linear system we need to find a fundamental matrix of the corresponding homogeneous system. If the homogeneous system has constant coefficients, that is, $A(t) \equiv A$, a constant matrix, we can obtain explicitly a fundamental matrix of the linear homogeneous system

$$\mathbf{y}' = A\mathbf{y} \tag{2.20}$$

and use this fundamental matrix to solve the inhomogeneous system by Theorem 2.6. We must first define the **exponential of a matrix**, e^M or $\exp M$, where M is an n-by-n matrix. **We say** (see also the discussion of convergence of a sequence of matrices using the matrix norm (2.5)) **that a series $\sum_{k=0}^{\infty} U_k$ of matrices converges if and only if the sequence $\{\sum_{k=0}^{n} U_k\}$ of partial sums converges,** where convergence of a sequence of matrices is defined in Section 2.1. The limit of this sequence of partial sums is called the **sum** of the series.

Combining the definition of convergence of a sequence of matrices with the Cauchy criterion for sequences of real or complex numbers, we can establish the following result:

Lemma 2.1. *A sequence $\{A_k\}$ of matrices converges if and only if given a number $\varepsilon > 0$, there exists an integer $N = N(\varepsilon) > 0$ such that $|A_m - A_p| < \varepsilon$ whenever $m, p > N$.*

• **EXERCISE**

1. Interpret Lemma 2.1 to obtain a similar criterion for the convergence of a series of matrices $\Sigma_{k=0}^{\infty} U_k$.

We now define $\exp M$ to be the sum of the series

$$\exp M = E + M + \frac{M^2}{2!} + \frac{M^3}{3!} + \cdots + \frac{M^k}{k!} + \cdots \tag{2.21}$$

(E is the n-by-n identity matrix.) We have a right to do this only if we can first show that this series converges for every complex n-by-n matrix M. To see whether it does, we define the partial sums

$$S_k = E + M + \frac{M^2}{2!} + \cdots + \frac{M^k}{k!} \tag{2.22}$$

and we use the Cauchy criterion for sequences of matrices (Lemma 2.1 above). Thus using the matrix norm, we have for $m > p$

$$|S_m - S_p| = \left| \sum_{k=p+1}^{m} \frac{M^k}{k!} \right| \leq \sum_{k=p+1}^{m} \frac{|M|^k}{k!}$$

• **EXERCISE**

2. Use properties of the matrix norm to justify the above calculation; note that this calculation is possible because the sums in (2.22) are finite.

From elementary calculus we know that for any matrix M, $e^{|M|} = \Sigma_{k=0}^{\infty} (|M|^k/k!)$ **(note that $|M|$ is a real number)**. Hence the sum $\Sigma_{k=0}^{m}(|M|^k/k!)$ is the partial sum of a series of positive numbers that is known to converge. Therefore, by the Cauchy criterion for convergence of series of real numbers, we see that given $\varepsilon > 0$, there exists an integer $N > 0$ such that

$$|S_m - S_p| < \varepsilon \qquad \text{for } m, p > N$$

This proves the convergence of the series on the right side of (2.21) and thus establishes the validity of (2.21) for every matrix M.

Noting that $|E| = n$, we have immediately

$$\begin{aligned}|\exp M| &\leq (n-1) + 1 + |M| + \frac{|M|^2}{2!} + \cdots + \frac{|M|^k}{k!} + \cdots \\ &= (n-1) + e^{|M|}\end{aligned}$$

It can be shown that if P is another n-by-n matrix, we have

$$\exp M \cdot \exp P = \exp(M + P) \tag{2.23}$$

if M and P commute ($MP = PM$).

• **EXERCISE**

3. Prove (2.23).

A useful property is that if T is a nonsingular n-by-n matrix,

$$T^{-1}(\exp M)T = \exp(T^{-1}MT). \tag{2.24}$$

• **EXERCISE**

4. Verify (2.24). [*Hint:* Use (2.21).]

We now establish the result for linear systems with constant coefficients

$$\mathbf{y}' = A\mathbf{y} \tag{2.20}$$

Theorem 2.7. *The matrix*

$$\Phi(t) = \exp(At) \tag{2.25}$$

is the fundamental matrix of (2.20) *with* $\Phi(0) = E$ *on* $-\infty < t < \infty$.

Proof of Theorem 2.7. That $\Phi(0) = E$ is obvious from (2.21). Using (2.21) with $M = At$ (well defined for $-\infty < t < \infty$ and every n-by-n matrix A), we have by differentiation of the power series (which is easily justified)

$$(\exp At)' = A + \frac{A^2 t}{1!} + \frac{A^3 t^2}{2!} + \cdots + \frac{A^k t^{k-1}}{(k-1)!} + \cdots = A \exp At$$

$-\infty < t < \infty$. Therefore $\exp At$ is a solution matrix of (2.20) (its columns are solutions of (2.20)). Since $\det \Phi(0) = \det E = 1$, Theorem 2.3 (Abel's formula) with $t_0 = 0$ gives

$$\det(\exp At) = \exp\left(\sum_{k=0}^{n} a_{kk}\right) t \neq 0 \qquad \text{for} \qquad -\infty < t < \infty$$

Therefore by Theorem 2.4, $\Phi(t)$ is a fundamental matrix of (2.20). This completes the proof of Theorem 2.7. ▌

• **EXERCISE**

5. Prove

$$(\exp At)' = A + \frac{A^2 t}{1!} + \frac{A^3 t^2}{2!} + \cdots + \frac{A^k t^{k-1}}{(k-1)!} + \cdots \qquad -\infty < t < \infty$$

It follows from Theorem 2.7 and Equation (2.10), that every solution $\boldsymbol{\phi}$ of the system (2.20) has the form

$$\boldsymbol{\phi}(t) = (\exp At)\mathbf{c} \qquad (-\infty < t < \infty) \tag{2.26}$$

for a suitably chosen constant vector $\mathbf{c}$.

• **EXERCISES**

6. Show that if ϕ is that solution of (2.20) satisfying $\phi(t_0) = \eta$, then

$$\phi(t) = [\exp A(t - t_0)]\eta \qquad (-\infty < t < \infty) \tag{2.27}$$

7. Show that if $\Phi(t) = e^{tA}$, then $\Phi^{-1}(t) = e^{-tA}$.

We now proceed to find some fundamental matrices in certain special cases, that is, we evaluate $\exp(tA)$ for certain matrices A.

Example 1. Find a fundamental matrix of the system $\mathbf{y}' = A\mathbf{y}$ if A is a diagonal matrix,

$$A = \begin{pmatrix} d_1 & & & 0 \\ & d_2 & & \\ & & \ddots & \\ 0 & & & d_n \end{pmatrix}$$

From (2.21)

$$\exp(At) = E + \begin{pmatrix} d_1 & & 0 \\ & \ddots & \\ 0 & & d_n \end{pmatrix} \frac{t}{1!} + \begin{pmatrix} d_1^{\,2} & & 0 \\ & \ddots & \\ 0 & & d_n^{\,2} \end{pmatrix} \frac{t^2}{2!} + \cdots$$

$$+ \begin{pmatrix} d_1^{\,k} & & 0 \\ & \ddots & \\ 0 & & d_n^{\,k} \end{pmatrix} \frac{t^k}{k} + \cdots$$

$$= \begin{pmatrix} \exp(d_1 t) & & & 0 \\ & \exp(d_2 t) & & \\ & & \ddots & \\ 0 & & & \exp(d_n t) \end{pmatrix}$$

and by Theorem 2.7 this is a fundamental matrix. This result is, of course, obvious since in the present case each equation of the system is $y_k' = d_k y_k$ $(k = 1, \ldots, n)$ and can be integrated.

Example 2. Find a fundamental matrix of $\mathbf{y}' = A\mathbf{y}$ if

$$A = \begin{pmatrix} 3 & 1 \\ 0 & 3 \end{pmatrix}$$

Solution.

Since

$$A = \begin{pmatrix} 3 & 0 \\ 0 & 3 \end{pmatrix} + \begin{pmatrix} 0 & 1 \\ 0 & 0 \end{pmatrix}$$

and since these two matrices commute, we have

$$\exp At = \exp \begin{pmatrix} 3 & 0 \\ 0 & 3 \end{pmatrix} t \cdot \exp \begin{pmatrix} 0 & 1 \\ 0 & 0 \end{pmatrix} t$$

$$= \begin{pmatrix} e^{3t} & 0 \\ 0 & e^{3t} \end{pmatrix} \left[E + \begin{pmatrix} 0 & 1 \\ 0 & 0 \end{pmatrix} t + \begin{pmatrix} 0 & 1 \\ 0 & 0 \end{pmatrix}^2 \frac{t^2}{2!} + \cdots \right]$$

But $\begin{pmatrix} 0 & 1 \\ 0 & 0 \end{pmatrix}^2 = \begin{pmatrix} 0 & 0 \\ 0 & 0 \end{pmatrix}$ and the infinite series terminates after two terms.

Therefore,

$$\exp At = e^{3t} \begin{pmatrix} 1 & t \\ 0 & 1 \end{pmatrix}$$

and by Theorem 2.7 this is a fundamental matrix.

• EXERCISES

8. Find a fundamental matrix of the system $\mathbf{y}' = A\mathbf{y}$ if

$$A = \begin{pmatrix} -2 & 1 & 0 \\ 0 & -2 & 1 \\ 0 & 0 & -2 \end{pmatrix}$$

and check your answer by direct integration of the given system.

9. Find a fundamental matrix of the system $\mathbf{y}' = A\mathbf{y}$ if

$$A = \begin{pmatrix} 0 & 1 & & 0 \\ & \ddots & \ddots & \\ & & \ddots & 1 \\ 0 & & & 0 \end{pmatrix}$$

where A is an n-by-n matrix.

10. Find a fundamental matrix of the system $\mathbf{y}' = A\mathbf{y}$ where A is the n-by-n matrix

$$A = \begin{pmatrix} 2 & 1 & & & 0 \\ & 2 & 1 & & \\ & & \ddots & \ddots & \\ & & & \ddots & 1 \\ 0 & & & & 2 \end{pmatrix}$$

The reader will have noticed that the examples and exercises presented so far, all of which involve the calculation of e^{tA}, are of a rather special form. In order to be able to handle more complicated problems and in order to obtain a general representation of solutions of (2.20) in a more explicit form than merely $\exp(tA)$ (that is, if we want to evaluate explicitly the entries of the matrix $\exp(tA)$), we need to introduce the notion of eigenvalue of a matrix.

Consider the system $\mathbf{y}' = A\mathbf{y}$, and look for a solution of the form

$$\boldsymbol{\phi}(t) = e^{\lambda t}\mathbf{c} \qquad (c \neq 0)$$

where the constant λ and the vector $\mathbf{c}$ are to be determined. Such a form is suggested by the above examples. Substitution shows that $e^{\lambda t}\mathbf{c}$ is a solution if and only if

$$\lambda e^{\lambda t}\mathbf{c} = Ae^{\lambda t}\mathbf{c}$$

Since $e^{\lambda t} \neq 0$, this condition becomes

$$(\lambda E - A)\mathbf{c} = \mathbf{0}$$

which can be regarded as a linear homogeneous algebraic system for the vector $\mathbf{c}$. This system has a nontrivial solution if and only if λ is chosen in such a way that

$$\det(\lambda E - A) = 0$$

This suggests the following definition.

If A is any n-by-n matrix, the polynomial in λ of degree n, $p(\lambda) = \det(\lambda E - A)$, is called **the characteristic polynomial of** A; its n roots (not necessarily distinct) are called the **eigenvalues** (also characteristic values) of A. We remark that if $\lambda = 0$ is not an eigenvalue of A, the constant term of $p(\lambda)$ is $p(0) = -\det A \neq 0$, and thus A is nonsingular. For a given eigenvalue λ_0, we say that a nonzero vector $\mathbf{c}_0$ is an **eigenvector** corresponding to the eigenvalue λ_0 if and only if $(\lambda_0 E - A)\mathbf{c}_0 = \mathbf{0}$.

Example 3. Find the eigenvalues of the matrix

$$A = \begin{pmatrix} 2 & 1 \\ -1 & 4 \end{pmatrix}$$

Consider the equation $\det(\lambda E - A) = 0$.

$$\begin{pmatrix} \lambda - 2 & -1 \\ 1 & \lambda - 4 \end{pmatrix} = (\lambda - 2)(\lambda - 4) + 1 = \lambda^2 - 6\lambda + 9 = 0$$

Thus $\lambda = 3$ is an eigenvalue of A of multiplicity two. To find a corresponding eigenvector we consider the system

$$(3E - A)\mathbf{c} = \mathbf{0}$$

or

$$\begin{pmatrix} 1 & -1 \\ 1 & -1 \end{pmatrix}\begin{pmatrix} c_1 \\ c_2 \end{pmatrix} = \begin{pmatrix} 0 \\ 0 \end{pmatrix} \qquad \text{or} \qquad \begin{matrix} c_1 - c_2 = 0 \\ c_1 - c_2 = 0 \end{matrix}$$

Any vector $\mathbf{c}$ with components $c_1 = c_2$, where c_1, c_2, is a solution Thus any vector $\mathbf{c} = \alpha\begin{pmatrix} 1 \\ 1 \end{pmatrix}$, where α is any scalar, is an eigenvector corresponding to the eigenvalue $\lambda = 3$. The reader will note that even though the eigenvalue has multiplicity two, the corresponding eigenvectors form a subspace of E_2 whose dimension is only one.

• EXERCISES

11. Compute the eigenvalues and corresponding eigenvectors of each of the following matrices. In each case, determine the subspace spanned by the eigenvectors.

(a) $\begin{pmatrix} 3 & 5 \\ -5 & 3 \end{pmatrix}$ (b) $\begin{pmatrix} -3 & 1 & 7 \\ 0 & 4 & -1 \\ 0 & 0 & 2 \end{pmatrix}$

(c) $\begin{pmatrix} 1 & 0 & 3 \\ 8 & 1 & -1 \\ 5 & 1 & -1 \end{pmatrix}$ (d) $\begin{pmatrix} 3 & -1 & -4 & 2 \\ 2 & 3 & -2 & -4 \\ 2 & -1 & -3 & 2 \\ 1 & 2 & -1 & -3 \end{pmatrix}$

12. Show that if A is a triangular matrix of the form

$$A = \begin{pmatrix} a_{11} & a_{12} & \cdots & a_{1n} \\ & a_{22} & & \\ 0 & & & \cdot \\ \cdot & 0 & \cdot & \cdot \\ \cdot & \cdot & \cdot & \cdot \\ \cdot & \cdot & & \cdot \\ & & \cdot & \\ 0 & \cdots & 0 & a_{nn} \end{pmatrix}$$

the eigenvalues of A are $\lambda = a_{ii}$ $(i = 1, \ldots, n)$, and the corresponding eigenvectors $\mathbf{v}_i$ $(i = 1, \ldots, n)$ form a basis for E_n.

Suppose the constant n-by-n matrix A has n linearly independent eigenvectors $\mathbf{v}_1, \ldots, \mathbf{v}_n$ corresponding to the eigenvalues $\lambda_1, \ldots, \lambda_n$. These vectors form a basis of E_n. In particular, if $\lambda_1, \ldots, \lambda_n$ are distinct, this will be the case (see Appendix 1, Theorem 1). It follows from our earlier discussion that $\exp(\lambda_j t)\mathbf{v}_j$ $(j = 1, \ldots, n)$ are solutions of $\mathbf{y}' = A\mathbf{y}$ (the reader may easily verify this directly by substitution). Since these solutions are obviously linearly independent on $-\infty < t < \infty$ it follows that the matrix $\Phi(t)$ whose columns are $\exp(\lambda_1 t)\mathbf{v}_1, \ldots, \exp(\lambda_n t)\mathbf{v}_n$ is a fundamental matrix on $-\infty < t < \infty$. We shall denote this matrix by

$$\Phi(t) = (\exp(\lambda_1 t)\mathbf{v}_1, \ldots, \exp(\lambda_n t)\mathbf{v}_n)$$

Incidentally, since e^{tA} is also a fundamental matrix (even though we have not calculated it explicitly), it follows from Theorem 2.5 that

$$e^{tA} = \Phi(t)C$$

for some nonsingular constant matrix C. **This technique can always be used to construct a fundamental matrix if the eigenvalues of A are distinct.**

If the eigenvalues are not all distinct, it may still be possible to find n linearly independent eigenvectors, for example, if A is a diagonal constant matrix whose diagonal entries are not all distinct or if A is the triangular matrix of Exercise 12. In this case we may again apply the above technique to find a fundamental matrix of $\mathbf{y}' = A\mathbf{y}$. However, the technique will fail if the eigenvectors do not form a basis for E_n. As shown in Example 2, there may then be solutions that cannot be expressed using only exponential functions and **constant vectors.**

Example 4. Find a fundamental matrix of the system $\mathbf{y}' = A\mathbf{y}$, where $A = \begin{pmatrix} 3 & 5 \\ -5 & 3 \end{pmatrix}$.

The eigenvalues of A are roots of the equation

$$\det(A - \lambda E) = \det\begin{pmatrix} 3-\lambda & 5 \\ -5 & 3-\lambda \end{pmatrix} = \lambda^2 - 6\lambda + 34 = 0$$

Thus $\lambda_{1,2} = 3 \pm 5i$. The eigenvector $\mathbf{u} = \begin{pmatrix} u_1 \\ u_2 \end{pmatrix}$ corresponding to the eigenvalue $\lambda_1 = 3 + 5i$ must satisfy the linear homogeneous algebraic system

$$(A - \lambda_1 E)\mathbf{u} = \begin{pmatrix} -5i & 5 \\ -5 & -5i \end{pmatrix}\begin{pmatrix} u_1 \\ u_2 \end{pmatrix} = 0$$

Thus u_1, u_2 satisfy the system of equations

$$-iu_1 + u_2 = 0$$

$$-u_1 - iu_2 = 0$$

and therefore

$$\mathbf{u} = \alpha\begin{pmatrix} 1 \\ i \end{pmatrix}$$

is an eigenvector for any constant α. Clearly it spans a one-dimensional subspace. Similarly, the eigenvector $\mathbf{v} = \begin{pmatrix} v_1 \\ v_2 \end{pmatrix}$ corresponding to the eigenvalue $\lambda_2 = 3 - 5i$ is found to be

$$\mathbf{v} = \beta\begin{pmatrix} i \\ 1 \end{pmatrix}$$

for any constant β. Clearly it also spans a one-dimensional subspace and since

$$\det\begin{vmatrix} 1 & i \\ i & 1 \end{vmatrix} = 2 \neq 0$$

the eigenvectors $\mathbf{u}$ and $\mathbf{v}$ are linearly independent; in fact, they form a basis for E_2.

Since $\lambda_1 = 3 + 5i$ and $\lambda_2 = 3 - 5i$ are eigenvalues of A having $\mathbf{u}_1 = \begin{pmatrix} 1 \\ i \end{pmatrix}$ and $\mathbf{u}_2 = \begin{pmatrix} i \\ 1 \end{pmatrix}$ as corresponding respective eigenvectors, it follows from the discussion after Exercise 12, p. 62, that

$$\boldsymbol{\phi}_1(t) = \exp t(3+5i)\begin{pmatrix}1\\ i\end{pmatrix} \quad \text{and} \quad \boldsymbol{\phi}_2(t) = \exp t(3-5i)\begin{pmatrix}i\\ 1\end{pmatrix}$$

are solutions. Moreover, $\boldsymbol{\phi}_1$ and $\boldsymbol{\phi}_2$ are linearly independent and therefore

$$\Phi(t) = \begin{pmatrix}\exp t(3+5i) & i\exp t(3-5i)\\ i\exp t(3+5i) & \exp t(3-5i)\end{pmatrix}$$

is a fundamental matrix. Observe that $\det \Phi(0) = 2$, which gives another proof of the linear independence of the solutions $\boldsymbol{\phi}_1$, $\boldsymbol{\phi}_2$ (by Theorem 2.4, p. 48).

To determine the form of e^{tA} when A is an arbitrary matrix, we require the following result from linear algebra (see Appendix 1). We note that the following method, which leads to the formula (2.30) below, is always applicable, and includes the technique described above as a special case.

Let A be a (complex) n-by-n matrix. Compute $\lambda_1, \lambda_2, \ldots, \lambda_k$, the distinct eigenvalues of A with respective multiplicities $n_1, n_2, \ldots, n_k$, where $n_1 + n_2 + \cdots + n_k = n$. Corresponding to each eigenvalue λ_j of multiplicity n_j consider the system of linear equations

$$(A - \lambda_j E)^{n_j}\mathbf{x} = 0 \qquad (j = 1, 2, \ldots, k) \tag{2.28}$$

The solutions of each such linear system obviously span a subspace of E_n which we call X_j $(j = 1, 2, \ldots, k)$. The result from linear algebra needed (see Appendix 1) tells us that **for every** $\mathbf{x} \in E_n$ **there exist unique vectors** $\mathbf{x}_1, \mathbf{x}_2, \ldots, \mathbf{x}_k$, where $\mathbf{x}_j \in X_j$, **such that**

$$\mathbf{x} = \mathbf{x}_1 + \mathbf{x}_2 + \cdots + \mathbf{x}_k \tag{2.29}$$

and $\mathbf{x}_j \in X_j$ $(j = 1, \ldots, k)$. It is important to know that the linear algebraic system (2.28) has n_j linearly independent solutions so that the dimension of X_j is n_j. We note that if all the eigenvalues of A are distinct, that is, if each $n_j = 1$ $(j = 1, \ldots, k)$ and $k = n$, then the vectors $\mathbf{x}_1, \mathbf{x}_2, \ldots, \mathbf{x}_n$ are suitable multiples of fixed eigenvectors that are linearly independent and span E_n. Thus, if $\mathbf{v}_1, \ldots, \mathbf{v}_n$ is a fixed set of linearly independent eigenvectors of A and if $\mathbf{x}$ is an arbitrary vector, the vectors $\mathbf{x}_j$ are given by $\mathbf{x}_j = c_j\mathbf{v}_j$ for some scalars c_j $(j = 1, \ldots, n)$.

In the language of linear algebra, E_n is the **direct sum** of subspaces $X_1, X_2, \ldots, X_k$,

$$E_n = X_1 \oplus X_2 \oplus \cdots \oplus X_k$$

such that the dimension of X_j is n_j, X_j is invariant under A, and the transformation defined by the matrix $A - \lambda_j E$ is **nilpotent** on X_j of index at most n_j (this is just another way of stating (2.28)).

It may, in fact, happen that in (2.28) $(A - \lambda_j E)^q \mathbf{x} = 0$ for all $\mathbf{x}$ in X_j and $q < n_j$, as the following example demonstrates.

Example 5. Consider the matrix

$$A = \begin{pmatrix} 4 & 1 & 0 & 0 & 0 \\ 0 & 4 & 1 & 0 & 0 \\ 0 & 0 & 4 & 0 & 0 \\ 0 & 0 & 0 & 4 & 0 \\ 0 & 0 & 0 & 0 & 4 \end{pmatrix}$$

then E_n is E_5, and $\lambda = 4$ is the only eigenvalue, with multiplicity 5; that is, $k = 1$. Since there is only one eigenvalue, no decomposition into subspaces is necessary, and $E_5 = X_1$. According to the theorem, $n_1 = 5$, so that certainly $(A - 4E)^5 = 0$. However, as the reader can easily verify $(A - 4E)^3 = 0$ but $(A - 4E)^2 \neq 0$. Therefore $q = 3 < n_1 = 5$.

To apply this theory to the linear system $\mathbf{y}' = A\mathbf{y}$, we look for that solution $\boldsymbol{\phi}(t)$ satisfying the initial condition $\boldsymbol{\phi}(0) = \boldsymbol{\eta}$. By Theorem 2.7., we know that $\boldsymbol{\phi}(t) = e^{tA}\boldsymbol{\eta}$ and our object is to evaluate $e^{tA}\boldsymbol{\eta}$ explicitly, that is, to see exactly what the components of $\boldsymbol{\phi}$ are. We compute $\lambda_1, \lambda_2, \ldots, \lambda_k$, the eigenvalues of A of multiplicities $n_1, n_2, \ldots, n_k$, respectively. We apply the theorem to the initial vector $\boldsymbol{\eta}$ and in accordance with (2.29) we have

$$\boldsymbol{\eta} = \mathbf{v}_1 + \mathbf{v}_2 + \cdots + \mathbf{v}_k$$

where $\mathbf{v}_j$ is some suitable vector in the subspace X_j ($j = 1, \ldots, k$). Since the subspace X_j is generated by the system (2.28), $\mathbf{v}_j$ must be some solution of (2.28). Now $e^{tA}\boldsymbol{\eta} = \sum_{j=1}^{k} e^{tA}\mathbf{v}_j$, and we may write

$$\begin{aligned} e^{tA}\mathbf{v}_j &= \exp(\lambda_j t) \exp[(A - \lambda_j E)t]\mathbf{v}_j \\ &= \exp(\lambda_j t)\left[E + t(A - \lambda_j E) + \frac{t^2}{2!}(A - \lambda_j E)^2 + \cdots \right. \\ &\quad \left. + \frac{t^{n_j - 1}}{(n_j - 1)!}(A - \lambda_j E)^{n_j - 1}\right]\mathbf{v}_j \end{aligned}$$

for $-\infty < t < \infty$, where the series in parentheses terminates because $\mathbf{v}_j$ is a solution of (2.28); thus the term $(A - \lambda_j E)^{n_j}\mathbf{v}_j = 0$ and all subsequent terms are zero. Observe that the vectors $\mathbf{w}_j = (A - \lambda_j E)^p \mathbf{v}_j$, for $p = 0, 1, \ldots, n_j - 1$, belong to the subspace X_j because

$$(A - \lambda_j E)^{n_j} \mathbf{w}_j = (A - \lambda_j E)^{n_j}[(A - \lambda_j E)^p \mathbf{v}_j] = (A - \lambda_j E)^{n_j + p} \mathbf{v}_j = \mathbf{0}$$

Thus the vector $e^{tA}\mathbf{v}_j$ remains in X_j for $-\infty < t < \infty$. Applying the above calculations to the solution $\boldsymbol{\phi}(t) = e^{tA}\boldsymbol{\eta}$ of $\mathbf{y}' = A\mathbf{y}$, we have

$$\begin{aligned}\boldsymbol{\phi}(t) &= e^{tA}\boldsymbol{\eta} = e^{tA}\sum_{j=1}^{k} \mathbf{v}_j = \sum_{j=1}^{k} e^{tA}\mathbf{v}_j \\ &= \sum_{j=1}^{k} \exp(\lambda_j t)\left[E + t(A - \lambda_j E) + \cdots + \frac{t^{n_j - 1}}{(n_j - 1)!}(A - \lambda_j E)^{n_j - 1}\right]\mathbf{v}_j\end{aligned}$$

or finally, the solution $\boldsymbol{\phi}$ satisfying $\boldsymbol{\phi}(0) = \boldsymbol{\eta}$ is

$$\boldsymbol{\phi}(t) = \sum_{j=1}^{k} \exp(\lambda_j t)\left[\sum_{i=0}^{n_j - 1} \frac{t^i}{i!}(A - \lambda_j E)^i\right]\mathbf{v}_j, \qquad (-\infty < t < \infty) \qquad (2.30)$$

We point out again that if $(A - \lambda_j E)^{q_j} = 0$ where $q_j < n_j$, then the sum on i in Equation (2.30) will contain only q_j, rather than n_j, terms. This formula also tells us precisely how the components of the solution behave as functions of t for any given coefficient matrix A.

Example 6. Solve the linear system $\mathbf{y}' = A\mathbf{y}$ if A is the matrix in Example 3 above. Also obtain a fundamental matrix.

From Example 3 we know that $\lambda_1 = 3$ is an eigenvalue of multiplicity 2. In the above notation, $n_1 = 2$. Therefore only the subspace X_1 in this case ($X_1 = E_2$) is relevant. We readily calculate

$$A - 3E = \begin{pmatrix} -1 & 1 \\ -1 & 1 \end{pmatrix}$$

and we also see that

$$(A - 3E)^2 = \begin{pmatrix} 0 & 0 \\ 0 & 0 \end{pmatrix}$$

so that (2.28) is satisfied for every vector in E_2. Substituting in (2.30) with $n_1 = 2$, $\boldsymbol{\eta} = \begin{pmatrix} \eta_1 \\ \eta_2 \end{pmatrix}$, we find

$$\boldsymbol{\phi}(t) = e^{3t}[E + t(A - 3E)]\boldsymbol{\eta}$$

and therefore

$$\begin{aligned}\boldsymbol{\phi}(t) &= e^{3t}\left[E + t\begin{pmatrix}-1 & 1\\ -1 & 1\end{pmatrix}\right]\begin{pmatrix}\eta_1\\ \eta_2\end{pmatrix}\\ &= e^{3t}\begin{bmatrix}\eta_1 + t(-\eta_1 + \eta_2)\\ \eta_2 + t(-\eta_1 + \eta_2)\end{bmatrix}\end{aligned} \tag{2.31}$$

is the solution with $\boldsymbol{\phi}(0) = \boldsymbol{\eta}$. To construct a fundamental matrix we may appeal to Theorem 2.7, which says that exp tA is a fundamental matrix. So far, we have computed, in Equation (2.31), exp $(tA)\boldsymbol{\eta}$ for an arbitrary constant vector $\boldsymbol{\eta}$. But

$$\exp tA = \exp tA\begin{pmatrix}1 & 0\\ 0 & 1\end{pmatrix} = \left(\exp tA\begin{pmatrix}1\\ 0\end{pmatrix}, \exp tA\begin{pmatrix}0\\ 1\end{pmatrix}\right),$$

where the solution vectors $\exp tA\begin{pmatrix}1\\ 0\end{pmatrix}$ and $\exp tA\begin{pmatrix}0\\ 1\end{pmatrix}$ are found from (2.31) by substituting first $\boldsymbol{\eta} = \begin{pmatrix}1\\ 0\end{pmatrix}$ and then $\boldsymbol{\eta} = \begin{pmatrix}0\\ 1\end{pmatrix}$, respectively. Therefore, a fundamental matrix is

$$\Phi(t) = \exp tA = e^{3t}\begin{pmatrix}1 - t & t\\ -t & 1 + t\end{pmatrix}$$

Example 7. Consider the system

$$\begin{aligned}x_1' &= 3x_1 - x_2 + x_3\\ x_2' &= 2x_1 \qquad\; + x_3\\ x_3' &= x_1 - x_2 + 2x_3\end{aligned}$$

which has coefficient matrix

$$A = \begin{pmatrix}3 & -1 & 1\\ 2 & 0 & 1\\ 1 & -1 & 2\end{pmatrix}$$

Find that solution $\boldsymbol{\phi}$ satisfying the initial condition

$$\boldsymbol{\phi}(0) = \begin{pmatrix}\eta_1\\ \eta_2\\ \eta_3\end{pmatrix} = \boldsymbol{\eta},$$

and also find a fundamental matrix.

The characteristic polynomial of A is $\det(\lambda E - A) = (\lambda - 1)(\lambda - 2)^2$, and therefore the eigenvalues are $\lambda_1 = 1$, $\lambda_2 = 2$ with multiplicities $n_1 = 1$, $n_2 = 2$, respectively.

In the notation of (2.28), we consider the systems of algebraic equations

$$(A - E)\mathbf{x} = 0 \quad \text{and} \quad (A - 2E)^2\mathbf{x} = 0$$

in order to determine the subspaces X_1 and X_2 of E. Taking these in succession, we have first

$$(A - E)\mathbf{x} = \begin{pmatrix} 2 & -1 & 1 \\ 2 & -1 & 1 \\ 1 & -1 & 1 \end{pmatrix}\mathbf{x} = \mathbf{0} \quad \text{or} \quad \begin{aligned} 2x_1 - x_2 + x_3 &= 0 \\ 2x_1 - x_2 + x_3 &= 0 \\ x_1 - x_2 + x_3 &= 0 \end{aligned}$$

Thus X_1 is the subspace spanned by the vectors $\begin{pmatrix} x_1 \\ x_2 \\ x_3 \end{pmatrix}$ with $x_1 = 0$, $x_2 = x_3$, and clearly $\dim X_1 = 1$. Next

$$(A - 2E)^2\mathbf{x} = \begin{pmatrix} 0 & 0 & 0 \\ -1 & 1 & 0 \\ -1 & 1 & 0 \end{pmatrix}\mathbf{x} = \mathbf{0} \quad \text{or} \quad \begin{aligned} -x_1 + x_2 &= 0 \\ -x_1 + x_2 &= 0 \end{aligned}$$

Thus X_2 is the subspace spanned by vectors $\begin{pmatrix} x_1 \\ x_2 \\ x_3 \end{pmatrix}$ with $x_1 = x_2$ and x_3 arbitrary; clearly, $\dim X_2 = 2$. The reader is advised to picture these subspaces in E_3. Observe that the rank of the matrix $A - E$ is 2. Thus by a well-known result of linear algebra, $\dim X_1 = 3 - 2 = 1$. Similarly, the rank of the matrix $(A - 2E)^2$ is clearly 1 and $\dim X_2 = 3 - 1 = 2$.

We now wish to find vectors $\mathbf{v}_1 \in X_1$, $\mathbf{v}_2 \in X_2$ such that we can write the initial vector $\boldsymbol{\eta}$ as

$$\boldsymbol{\eta} = \mathbf{v}_1 + \mathbf{v}_2$$

Since $\mathbf{v}_1 \in X_1$, $\mathbf{v}_1 = \begin{pmatrix} 0 \\ \alpha \\ \alpha \end{pmatrix}$ for some scalar α, and since $\mathbf{v}_2 \in X_2$, $\mathbf{v}_2 = \begin{pmatrix} \beta \\ \beta \\ \gamma \end{pmatrix}$ for some scalars β, γ. Therefore

$$\begin{pmatrix} \eta_1 \\ \eta_2 \\ \eta_3 \end{pmatrix} = \begin{pmatrix} 0 \\ \alpha \\ \alpha \end{pmatrix} + \begin{pmatrix} \beta \\ \beta \\ \gamma \end{pmatrix}$$

so that $\beta = \eta_1$, $\alpha + \beta = \eta_2$, $\alpha + \gamma = \eta_3$. Solving these equations for α, β, γ, we find that $\alpha = \eta_2 - \eta_1$, $\beta = \eta_1$, $\gamma = \eta_3 - \eta_2 + \eta_1$ and

$$\mathbf{v}_1 = \begin{pmatrix} 0 \\ \eta_2 - \eta_1 \\ \eta_2 - \eta_1 \end{pmatrix} \qquad \mathbf{v}_2 = \begin{pmatrix} \eta_1 \\ \eta_1 \\ \eta_3 - \eta_2 + \eta_1 \end{pmatrix}$$

Thus by the formula (2.30), we find that the solution $\boldsymbol{\phi}$ such that $\boldsymbol{\phi}(0) = \boldsymbol{\eta}$ is given by

$$\begin{aligned} \boldsymbol{\phi}(t) &= e^t\mathbf{v}_1 + e^{2t}(E + t(A - 2E))\mathbf{v}_2 \\ &= e^t\begin{pmatrix} 0 \\ \eta_2 - \eta_1 \\ \eta_2 - \eta_1 \end{pmatrix} + e^{2t}\left(E + t\begin{pmatrix} 1 & -1 & 1 \\ 2 & -2 & 1 \\ 1 & -1 & 0 \end{pmatrix}\right)\begin{pmatrix} \eta_1 \\ \eta_1 \\ \eta - \eta_2 + \eta_1 \end{pmatrix} \\ &= e^t\begin{pmatrix} 0 \\ \eta_2 - \eta_1 \\ \eta_2 - \eta_1 \end{pmatrix} + e^{2t}\begin{pmatrix} 1 + t & -t & t \\ 2t & 1 - 2t & t \\ t & -t & 1 \end{pmatrix}\begin{pmatrix} \eta_1 \\ \eta_1 \\ \eta_3 - \eta_2 + \eta_1 \end{pmatrix} \end{aligned}$$

To find a fundamental matrix, we proceed as in Example 6. Putting $\boldsymbol{\eta}$ successively equal to $\begin{pmatrix} 1 \\ 0 \\ 0 \end{pmatrix}$, $\begin{pmatrix} 0 \\ 1 \\ 0 \end{pmatrix}$, $\begin{pmatrix} 0 \\ 0 \\ 1 \end{pmatrix}$ in this formula, we obtain the three linearly independent solutions that we use as columns of the matrix

$$\Phi(t) = e^{tA} = \begin{pmatrix} (1 + t)e^{2t} & -te^{2t} & te^{2t} \\ -e^t + (1 + t)e^{2t} & e^t - te^{2t} & te^{2t} \\ -e^t + e^{2t} & e^t - e^{2t} & e^{2t} \end{pmatrix}$$

Hence $\Phi(t)$ is the fundamental matrix that reduces to the identity matrix when $t = 0$.

Example 8. Find a fundamental matrix for the system $\mathbf{y}' = A\mathbf{y}$ with

$$A = \begin{pmatrix} 4 & 1 & 0 & 0 & 0 \\ 0 & 4 & 1 & 0 & 0 \\ 0 & 0 & 4 & 0 & 0 \\ 0 & 0 & 0 & 4 & 0 \\ 0 & 0 & 0 & 0 & 4 \end{pmatrix}$$

Using the results of Example 5, we have $(A - 4E)^3 = 0$, so that $(A - 4E)^3\mathbf{x} = 0$ for any vector $\mathbf{x}$ in E_5 and the initial vector $\boldsymbol{\eta}$ remains arbitrary. Since there is only one eigenvalue ($\lambda = 4$), only the subspace $X_1 = E_5$ is relevant and we have, from (2.30),

$$\phi(t) = e^{4t}[E + t(A - 4E) + \frac{t^2}{2!}(A - 4E)^2]\eta.$$

Therefore

$$\phi(t) = e^{4t}\left[E + t\begin{pmatrix} 0 & 1 & 0 & 0 & 0 \\ 0 & 0 & 1 & 0 & 0 \\ 0 & 0 & 0 & 0 & 0 \\ 0 & 0 & 0 & 0 & 0 \\ 0 & 0 & 0 & 0 & 0 \end{pmatrix} + \frac{t^2}{2!}\begin{pmatrix} 0 & 0 & 1 & 0 & 0 \\ 0 & 0 & 0 & 0 & 0 \\ 0 & 0 & 0 & 0 & 0 \\ 0 & 0 & 0 & 0 & 0 \\ 0 & 0 & 0 & 0 & 0 \end{pmatrix}\right]\eta$$

Again letting η successively assume the values

$$\eta_1 = \begin{pmatrix} 1 \\ 0 \\ 0 \\ 0 \\ 0 \end{pmatrix}, \quad \eta_2 = \begin{pmatrix} 0 \\ 1 \\ 0 \\ 0 \\ 0 \end{pmatrix}, \quad \eta_3 = \begin{pmatrix} 0 \\ 0 \\ 1 \\ 0 \\ 0 \end{pmatrix}, \quad \eta_4 = \begin{pmatrix} 0 \\ 0 \\ 0 \\ 1 \\ 0 \end{pmatrix}, \quad \eta_5 = \begin{pmatrix} 0 \\ 0 \\ 0 \\ 0 \\ 1 \end{pmatrix}$$

in the above formula, the resulting solutions will be linearly independent and can be used as columns of a fundamental matrix. Thus

$$\Phi(t) = e^{4t}\begin{pmatrix} 1 & t & \frac{t^2}{2!} & 0 & 0 \\ 0 & 1 & t & 0 & 0 \\ 0 & 0 & 1 & 0 & 0 \\ 0 & 0 & 0 & 1 & 0 \\ 0 & 0 & 0 & 0 & 1 \end{pmatrix}$$

is a fundamental matrix.

• EXERCISES

Find a fundamental matrix for each of the following systems $\mathbf{y}' = A\mathbf{y}$ having the coefficient matrix given. Also find a particular solution satisfying the given initial condition.

13. $A = \begin{pmatrix} 1 & 2 \\ 4 & 3 \end{pmatrix}; \quad \eta = \begin{pmatrix} 3 \\ 3 \end{pmatrix}.$

14. $A = \begin{pmatrix} 2 & -3 \\ 3 & -4 \end{pmatrix}$; $\eta = \begin{pmatrix} 1 \\ 2 \end{pmatrix}$.

15. $A = \begin{pmatrix} 1 & 0 & 3 \\ 8 & 1 & -1 \\ 5 & 1 & -1 \end{pmatrix}$; $\eta = \begin{pmatrix} 0 \\ -2 \\ -7 \end{pmatrix}$.

16. $A = \begin{pmatrix} 3 & -1 & -4 & 2 \\ 2 & 3 & -2 & -4 \\ 2 & -1 & -3 & 2 \\ 1 & 2 & -1 & -3 \end{pmatrix}$; $\eta = \begin{pmatrix} 1 \\ 0 \\ -1 \\ 0 \end{pmatrix}$.

[*Note:* The characteristic polynomial is $(\lambda - 1)^2 \cdot (\lambda + 1)^2$.]

17. Find that solution of the system

$$y_1' = y_1 + y_2 + \sin t$$
$$y_2' = 2y_1 + \cos t$$

such that $y_1(0) = 1$, $y_2(0) = 1$. [*Hint:* Find a fundamental matrix of the homogeneous system; then use the variation of constants formula in Section 2.4.]

Consider the scalar linear differential equation of second order

$$y'' + py' + qy = 0 \tag{2.32}$$

where p and q are constants. We can solve this equation as a special case of the theory developed here as outlined in the following exercises.

• EXERCISES

18. Show that Equation (2.32) is equivalent to the system $\mathbf{y}' = A\mathbf{y}$ with

$$A = \begin{pmatrix} 0 & 1 \\ -q & -p \end{pmatrix}$$

and compute the eigenvalues λ_1, λ_2 of A.

19. Compute a fundamental matrix for the system in Exercise 18 if $\lambda_1 \neq \lambda_2$, that is, if $p^2 \neq 4q$, and construct the general solution of Equation (2.32) in this case.

20. Compute a fundamental matrix for the system in Exercise 18 in the case $\lambda_1 = \lambda_2 = \lambda$, that is, $p^2 = 4q$, and construct the general solution of (2.32) in the case $p^2 = 4q$. Note that $A - \lambda E$ is never zero in this case, so that the fundamental matrix, as well as the general solution of (2.32), must necessarily contain a term in $te^{\lambda t}$.

21. Generalize the results of Exercises 18, 19, and 20 to the scalar equation

$$y''' + p_1 y'' + p_2 y' + p_3 y = 0$$

where p_1, p_2, p_3 are constants. (Needless to say, you are not expected actually to solve a cubic equation.)

Using the variation of constants formula (Theorem 2.6), we may find a particular solution of the nonhomogeneous system

$$\mathbf{y}' = A\mathbf{y} + \mathbf{g}(t) \tag{2.33}$$

where A is a constant matrix and $\mathbf{g}$ is a given continuous function. The variation of constants formula with $\Phi(t) = \exp(tA)$ as a fundamental matrix of the homogeneous system now becomes particularly simple in appearance. We have $\Phi^{-1}(s) = \exp(-sA)$, $\Phi(t)\Phi^{-1}(s) = \exp[(t-s)A]$; if the initial condition is $\boldsymbol{\phi}(t_0) = \boldsymbol{\eta}$, $\boldsymbol{\phi}_h(t) = \exp[(t-t_0)A]\boldsymbol{\eta}$ and the solution of (2.33) is

$$\boldsymbol{\phi}(t) = \exp[(t-t_0)A]\boldsymbol{\eta} + \int_{t_0}^{t} \exp[(t-s)A]\mathbf{g}(s)\,ds \qquad (-\infty < t < \infty) \tag{2.34}$$

where e^{tA} is the fundamental matrix of the homogeneous system that we can construct by the method of this section. Note how easy it is to compute the inverse of Φ and also $\Phi(t)\Phi^{-1}(s)$ in this case.

Example 9. Solve the initial value problem

$$\mathbf{y}' = A\mathbf{y} + \mathbf{g}(t)$$

where A is the constant matrix in Example 6, and where $\mathbf{g}(t) = \begin{pmatrix} e^{3t} \\ 1 \end{pmatrix}$, with the initial condition $\boldsymbol{\phi}(0) = \boldsymbol{\eta}$. From Example 6 we have

$$\Phi(t) = e^{tA} = e^{3t}\begin{pmatrix} 1-t & t \\ -t & 1+t \end{pmatrix}$$

$$\Phi(t)\Phi^{-1}(s) = \exp[(t-s)A] = \exp[3(t-s)]\begin{pmatrix} 1-(t-s) & t-s \\ -(t-s) & 1+(t-s) \end{pmatrix}$$

$$\exp[(t-s)A]\mathbf{g}(s) = e^{3t}\begin{pmatrix} 1-(t-s)+e^{-3s}(t-s) \\ -(t-s)+e^{-3s}(1+t-s) \end{pmatrix}$$

Therefore

$$\boldsymbol{\phi}(t) = e^{3t}\begin{pmatrix} 1-t & t \\ -t & 1+t \end{pmatrix}\boldsymbol{\eta} + e^{3t}\int_0^t \begin{pmatrix} 1-(t-s)+e^{-3s}(t-s) \\ -(t-s)+e^{-3s}(1+t-s) \end{pmatrix} ds$$

and the integrals are easily evaluated.

• **EXERCISES**

22. Write the solution $\boldsymbol{\phi}(t)$ of the system (2.33) if A is the matrix in Example 7 and g is any vector with three continuous components, subject to the initial condition $\boldsymbol{\phi}(0) = \boldsymbol{\eta}$.

23. By converting to an equivalent system, find the general solution of the scalar equation

$$y'' - y = f(t)$$

where f is continuous, by using the theory of this section.

24. Given the matrix

$$A = \begin{pmatrix} 0 & 1 \\ -1 & 0 \end{pmatrix}$$

Show that $A^2 = -E$, $A^3 = -A$, $A^4 = E$, and compute A^m where m is an arbitrary positive integer.

25. Use the result of Exercise 24 and the definition (2.21) to show that

$$e^{tA} = \begin{pmatrix} \cos t & \sin t \\ -\sin t & \cos t \end{pmatrix}$$

[*Note:* In this case the above approach is easier than the one in which Equation (2.30) is used.]

26. Compute e^{tA} if $A = \begin{pmatrix} 2 & 1 \\ -1 & 2 \end{pmatrix}$. [*Hint:* Use Exercise 24 and 25.]

27. Use the results of this section and Exercise 25 to find the general solution of the scalar equation

$$y'' + y = f(t)$$

where f is continuous.

2.6 Similarity of Matrices and the Jordan Canonical Form

Consider the linear system

$$\mathbf{y}' = A\mathbf{y} \tag{2.20}$$

where A is an n-by-n constant matrix. The change of variable $\mathbf{y} = T\mathbf{z}$ where T is some nonsingular constant matrix, transforms the system (2.20) to the system

$$\mathbf{z}' = T^{-1}AT\mathbf{z} \tag{2.35}$$

Our object is to choose the matrix T in such a way as to make the coefficient matrix $B = T^{-1}AT$ of (2.35) as simple as possible. This should then facilitate the calculation of exp Bt, which is a fundamental matrix of (2.35). Since $A = TBT^{-1}$, Exercise 4, Section 2.5, shows that $\exp tA = \exp(tTBT^{-1}) = T(\exp tB)T^{-1}$. Thus, exp tA and (by Theorem 2.5) $T(\exp tB)$ are both fundamental matrices of (2.20). We will see that a great simplification is always possible by a proper choice of the matrix T. This will provide an alternative treatment, primarily of theoretical value, of linear systems with constant coefficients.

• EXERCISE

1. Show that transformation $\mathbf{y} = T(t)\mathbf{z}$, where T is a nonsingular differentiable matrix function on some interval I, reduces the linear system $\mathbf{y}' = A(t)\mathbf{y}$, where $A(t)$ is continuous on I, to the system

$$\mathbf{z}' = (T^{-1}(t)A(t)T(t) - T^{-1}(t)T'(t))\mathbf{z}$$

This is much more complicated than the system (2.35), which suggests that the method we are going to present is valuable primarily for systems with constant coefficients.

In order to present this topic we need a new concept. Let A and B be two n-by-n matrices of complex numbers. **We say that A and B are similar, notation $A \sim B$, if and only if there exists a nonsingular matrix T such that**

$$T^{-1}AT = B$$

An important fact is that **similar matrices have the same characteristic polynomial** (for this reason, the coefficients of the characteristic polynomial of a matrix A are the same as those of any matrix similar to A and are called **similarity invariants**) and hence the same eigenvalues. To prove this, we let A and B be similar; then

$$\begin{aligned}\det(\lambda E - B) &= \det(\lambda E - T^{-1}AT) = \det\{T^{-1}(\lambda E - A)T\}\\ &= (\det T^{-1})(\det \lambda E - A)\det T = \det(\lambda E - A)\end{aligned}$$

• EXERCISE

2. Justify each step of the above calculation.

We note, however, that matrices with the same eigenvalues need not be similar; for example,

$$A = \begin{pmatrix} 0 & 0 \\ 0 & 0 \end{pmatrix} \qquad B = \begin{pmatrix} 0 & 1 \\ 0 & 0 \end{pmatrix}$$

• **EXERCISE**

3. Show that the matrices A and B are not similar.

We remark also that the coefficient matrices of the systems (2.20) and (2.35) above, obtained by the change of variable $\mathbf{y} = T\mathbf{z}$ are similar.

Given a matrix A, our task is to see

(i) how simple we can make $T^{-1}AT$ by a suitable choice of the matrix T;
(ii) how we choose T to achieve this.

Both of these questions are intimately connected with the nature of the eigenvalues and eigenvectors of A. Let us begin with the simplest case in which the matrix A has n linearly independent eigenvectors $\mathbf{v}_1, \ldots, \mathbf{v}_n$ corresponding to the eigenvalues $\lambda_1, \ldots, \lambda_n$. (As we have seen, this case always arises if $\lambda_1, \lambda_2, \ldots, \lambda_n$ are distinct; see Appendix 1.)

Define the matrix

$$T = (\mathbf{v}_1, \mathbf{v}_2, \ldots, \mathbf{v}_n) \tag{2.36}$$

having these n linearly independent eigenvectors as columns. Thus T is a nonsingular constant matrix. We assert that

$$T^{-1}AT = D$$

where D is the diagonal matrix

$$D = \begin{pmatrix} \lambda_1 & & & 0 \\ & \lambda_2 & & \\ & & \ddots & \\ 0 & & & \lambda_n \end{pmatrix} \tag{2.37}$$

To see this, we compute (using the definition of eigenvalue and matrix multiplication):

$$\begin{aligned} T^{-1}AT &= T^{-1}A(\mathbf{v}_1, \mathbf{v}_2, \ldots, \mathbf{v}_n) \\ &= T^{-1}(A\mathbf{v}_1, A\mathbf{v}_2, \ldots, A\mathbf{v}_n) \\ &= T^{-1}(\lambda_1\mathbf{v}_1, \lambda_2\mathbf{v}_2, \ldots, \lambda_n\mathbf{v}_n) \\ &= (\lambda_1 T^{-1}\mathbf{v}_1, \lambda_2 T^{-1}\mathbf{v}_2, \ldots, \lambda_n T^{-1}\mathbf{v}_n) \end{aligned}$$

But for any j, since $\mathbf{v}_j$ is the jth column of T, we have

$$T^{-1}\mathbf{v}_j = j\text{th column of } T^{-1}T = j\text{th column of } E$$

and therefore $T^{-1}AT = D$. Summarizing, we have proved the following simple result.

Theorem 2.8. (Diagonal Canonical Form.) *If the n-by-n constant matrix A has n linearly independent eigenvectors $\mathbf{v}_1, \ldots, \mathbf{v}_n$, corresponding to the eigenvalues $\lambda_1, \lambda_2, \ldots, \lambda_n$, then A is similar to the diagonal matrix D given by* (2.37) *and the matrix T that accomplishes the similarity is given by* (2.36).

In the general case, the matrix A does not necessarily possess n linearly independent eigenvectors. However, the result from linear algebra, which we invoked in our treatment in Section 2.5, leads to the following simplification. Let $\lambda_1, \ldots, \lambda_k$ be the distinct eigenvalues of A with respective multiplicities $n_1, n_2, \ldots, n_k$. Let X_j be the subspace of E_n generated by the system (2.28) for $j = 1, 2, \ldots, k$. Let the matrix T be defined as follows:

$$T = (\mathbf{v}_{11}, \mathbf{v}_{12}, \ldots, \mathbf{v}_{1n_1}, \ldots, \mathbf{v}_{k1}, \mathbf{v}_{k2}, \ldots, \mathbf{v}_{k,n_k}) \tag{2.38}$$

where $\mathbf{v}_{j1}, \ldots, \mathbf{v}_{jn_j}$ is any basis for $X_j, j = 1, \ldots, k$. Clearly T is a nonsingular matrix. Then, because X_j is invariant under A, that is, $A\mathbf{v} \in X_j$ for every $\mathbf{v}$ in X_j (recall that X_j is the hyperplane generated by solutions of the system $(A - \lambda_j E)^{n_j}\mathbf{x} = 0$), we have

$$T^{-1}AT = B$$

where

$$B = \begin{pmatrix} B_1 & & & 0 \\ & B_2 & & \\ & & \ddots & \\ 0 & & & B_k \end{pmatrix} \tag{2.39}$$

in which each B_j is an n_j-by-n_j matrix.

• **EXERCISE**

4. Show that B has the form given by (2.39)

Thus we may say that A is similar to a **block diagonal matrix**, which, of course, reduces to Theorem 2.8 if the eigenvectors are linearly independent.

By a proper choice of the basis for each subspace $X_j, j = 1, \ldots, k$, we can make a significant further simplification, as shown in the following result (see, for example, [11] or [14] or Appendix 1 for the proof):

Theorem 2.9. (Jordan Canonical Form.) *For each complex constant n-by-n matrix A there exists a nonsingular matrix T such that the matrix* $J = T^{-1}AT$ *is in the canonical form*

$$J = \begin{pmatrix} J_0 & & & 0 \\ & J_1 & & \\ & & \ddots & \\ 0 & & & J_s \end{pmatrix} \tag{2.40}$$

where J_0 *is a diagonal matrix with diagonal elements* $\lambda_1, \lambda_2, \ldots, \lambda_k$ *(not necessarily distinct)*

$$J_0 = \begin{pmatrix} \lambda_1 & & & \\ & \lambda_2 & & \\ & & \ddots & \\ & & & \lambda_k \end{pmatrix} \tag{2.41}$$

and each J_p *is an* n_p-by-n_p *matrix of the form*

$$J_p = \begin{pmatrix} \lambda_{k+p} & 1 & 0 & \cdots & 0 \\ 0 & \lambda_{k+p} & 1 & \ddots & 0 \\ \vdots & & \ddots & \ddots & \vdots \\ \vdots & & & \ddots & 1 \\ 0 & & & 0 & \lambda_{k+p} \end{pmatrix} \qquad (p = 1, \ldots, s) \tag{2.42}$$

where λ_{k+p} *need not be different from* λ_{k+q} *if* $p \neq q$ *and* $k + n_1 + n_2 + \cdots + n_s = n$. *The numbers* λ_i $(i = 1, 2, \ldots, k + s)$ *are the eigenvalues of A. If* λ_i *is a simple eigenvalue of A, it appears in the block* J_0.

A matrix may be similar to a diagonal matrix without having simple eigenvalues; the identity matrix E is an example. While Theorem 2.9 is a very useful theoretical tool, it is not easy to apply to a specific matrix if there are multiple eigenvalues. Even in the simplest case $n = 2$ there are difficulties to be overcome. For example, suppose A is a 2-by-2 matrix having $\lambda = 3$ as a double eigenvalue. Then, according to Theorem 2.9, the canonical form J

could be either

$$J = \begin{pmatrix} 3 & 0 \\ 0 & 3 \end{pmatrix} \quad \text{or} \quad \tilde{J} = \begin{pmatrix} 3 & 1 \\ 0 & 3 \end{pmatrix}$$

and without further study, which we will not undertake, we simply do not know which of the two possibilities is correct. It should be noted that the matrices J and $\tilde{J}$ are **not** similar. Only in the case of distinct eigenvalues is there no doubt from what we have presented about the canonical form; it is the one given by (2.37). But notice that even in that case the work is not finished—we must still find the similarity transformation matrix T! Admittedly, if we know the proper canonical form, we can always find the matrix T by solving the system of n^2 equations

$$AT = TJ \tag{2.43}$$

for the elements of the matrix T. In case the matrix A has n linearly independent eigenvectors, Theorem 2.8 tells us how to find the similarity transformation matrix T.

• EXERCISES

5. Let $A = \begin{pmatrix} 1 & 1 \\ 2 & 0 \end{pmatrix}$; use Theorem 2.9 to determine the canonical form of A and use the system (2.43) to find the transformation matrix T. [*Hint:* $\lambda_1 = 2$, $\lambda_2 = -1$ are the eigenvalues; hence (2.43) is the system

$$\begin{pmatrix} 1 & 1 \\ 2 & 0 \end{pmatrix} T = T \begin{pmatrix} 2 & 0 \\ 0 & -1 \end{pmatrix}$$

Let $T = \begin{pmatrix} t_{11} & t_{12} \\ t_{21} & t_{22} \end{pmatrix}$; substitute and determine T. *Answer:* $T = \begin{pmatrix} 1 & 1 \\ 1 & -2 \end{pmatrix}$.] Also determine T by using the alternative procedure.

6. Show by direct computation that for the matrices A, J, T of Exercise 5, $T^{-1}AT = J$.

7. Use the result of Exercise 5 together with the change of variable $y = Tz$ to find a fundamental matrix of the system $\mathbf{y}' = A\mathbf{y}$, where $A = \begin{pmatrix} 1 & 1 \\ 2 & 0 \end{pmatrix}$.

In order to apply Theorem 2.9 to solve the system $\mathbf{y}' = A\mathbf{y}$ we make the change of variable $\mathbf{y} = T\mathbf{z}$ where we now choose T to be that matrix for which $T^{-1}AT = J$ is the Jordan canonical form of A (Theorem 2.9). Then the transformed system is

$$\mathbf{z}' = T^{-1}AT\mathbf{z} = J\mathbf{z} \tag{2.44}$$

Clearly, a fundamental matrix of (2.44) is $\exp(tJ)$, which can be evaluated explicitly and very concisely as follows. First, from (2.40) and the definition of the exponential of a matrix, we have

$$\exp(tJ) = \begin{pmatrix} \exp(tJ_0) & & & 0 \\ & \exp(tJ_1) & & \\ & & \ddots & \\ 0 & & & \exp(tJ_s) \end{pmatrix} \qquad (-\infty < t < \infty) \tag{2.45}$$

where, from the same definition,

$$\exp tJ_0 = \begin{pmatrix} \exp \lambda_1 t & & & \\ & \exp \lambda_2 t & & \\ & & \ddots & \\ & & & \exp \lambda_k t \end{pmatrix} \qquad (-\infty < t < \infty) \tag{2.46}$$

To evaluate $\exp tJ_p$ we proceed as follows (see Examples 1 and 2, and Exercises 8, 9, and 10, Section 2.5). From (2.42)

$$J_p = \lambda_{k+p} E_p + N_p \tag{2.47}$$

where E_p is the $n_p \times n_p$ identity matrix and N_p is the $n_p \times n_p$ matrix

$$N_p = \begin{pmatrix} 0 & 1 & & 0 \\ & \ddots & \ddots & \\ & & \ddots & 1 \\ 0 & & & 0 \end{pmatrix} \tag{2.48}$$

Clearly, $\lambda_{k+p} E$ and N_p commute, and therefore

$$\exp tJ_p = \exp(\lambda_{k+p} t) \exp N_p t \tag{2.49}$$

It is readily verified that N_p is nilpotent ($N_p^{n_p} = 0$), and thus the series defining $\exp(tN_p)$ terminates and (see also Exercise 9, Section 2.5).

$$\exp tJ_p = \exp(\lambda_{k+p} t) \begin{pmatrix} 1 & t & \cdots & \dfrac{t^{n_p - 1}}{(n_{p-1})!} \\ 0 & \ddots & \ddots & \vdots \\ \vdots & & \ddots & t \\ 0 & \cdots & 0 & 1 \end{pmatrix} \tag{2.50}$$

$(p = 1, 2, \ldots, s;\ -\infty < t < \infty)$. Substitution of (2.46) and (2.50) into (2.45) gives the fundamental matrix of (2.44), which reduces to the identity at $t = 0$, as well as the evaluation of $\exp(tJ)$. To evaluate the fundamental matrix $\exp tA$ of $\mathbf{y}' = A\mathbf{y}$ we use the formula (see Exercise 4, Section 2.5).

$$\exp tA = \exp[tTJT^{-1}] = T\{\exp(tJ)\}T^{-1}$$

and the formula for $\exp tJ$ derived above.

• **EXERCISE**

8. Show that if N is any nilpotent matrix (not merely the special matrix N_p defined by (2.48) above), then $\exp(Nt)$ is a matrix polynomial (that is, a matrix whose elements are polynomials in t).

2.7 Asymptotic Behavior of Solutions of Linear Systems with Constant Coefficients

In many problems, in order to apply formula (2.34), and others derivable from it, we need to obtain a useful estimate for the norm of $\exp(tA)$ for $t \geq 0$. For example, in order to measure the growth of solutions of (2.33) as $t \to \infty$ we need to estimate $|\boldsymbol{\phi}(t)|$ as $t \to \infty$ where $\boldsymbol{\phi}(t)$ is given by (2.34). This, however, cannot be done without some useful estimate for $|\exp(tA)|$.

Theorem 2.10. *If $\lambda_1, \lambda_2, \ldots, \lambda_k$ are the distinct eigenvalues of A, where λ_j has multiplicity n_j and $n_1 + n_2 + \cdots + n_k = n$ and if ρ is any number larger than the real part of $\lambda_1, \ldots, \lambda_k$, that is*

$$\rho > \max_{j=1,\ldots,k} (\mathscr{R}\lambda_j) \tag{2.51}$$

then there exists a constant $K > 0$ such that

$$|\exp(tA)| \leq K \exp(\rho t) \qquad (0 \leq t < \infty) \tag{2.52}$$

Proof. We have seen that e^{tA} is a fundamental matrix of the linear system $\mathbf{y}' = A\mathbf{y}$. Combining this fact with the formula (2.30) for any solution $\boldsymbol{\phi}(t)$ of this system, we see that every element of the matrix e^{tA} is of the form $\sum_{j=1}^{k} p_j(t) \exp(\lambda_j t)$, where $p_j(t)$ is a polynomial of degree not more than $(n_j - 1)$. If ρ is chosen to satisfy the inequality (2.51), then $|t^k \exp(\lambda_j t)| = t^k \exp[(\mathscr{R}\lambda_j)t] < e^{\rho t}$ for t large enough, and every term in the sum $\sum_{j=1}^{k} p_j(t)$

$\exp(\lambda_j t)$ is at most $Me^{\rho t}$ $(0 \le t < \infty)$ for some constant M. As there are at most n^2 such terms in the matrix e^{tA}, (2.52) holds with $K = \tilde{M}n^2$, where $\tilde{M}$ is the largest of the n^2 values of M.

We also remark that the constant ρ in (2.52) may be chosen as any number greater than or equal to the largest of $\mathscr{R}\lambda_1, \mathscr{R}\lambda_2, \ldots, \mathscr{R}\lambda_n$, whenever every eigenvalue whose real part is equal to this maximum is itself simple. In particular, this is always true if A has no multiple eigenvalues.

• **EXERCISE**

1. Prove Theorem 2.10 by transforming A to the Jordan canonical form. [*Hint:* Use formulas (2.45)–(2.50).]

As far as applications are concerned, the following consequence of Theorem 2.10 is of great importance.

Corollary to Theorem 2.10. If all eigenvalues of A have real parts negative, then every solution $\boldsymbol{\phi}(t)$ *of the system*

$$\mathbf{y}' = A\mathbf{y} \tag{2.20}$$

approaches zero as $t \to +\infty$. *More precisely, there exist constants* $\tilde{K} > 0$, $\sigma > 0$ *such that*

$$|\boldsymbol{\phi}(t)| \le \tilde{K}e^{-\sigma t} \qquad (0 \le t < \infty) \tag{2.53}$$

It is, of course, also true that under the hypothesis of the corollary, there exist constants $K > 0$, $\sigma > 0$ such that

$$|\exp tA| \le Ke^{-\sigma t} \qquad (0 \le t < \infty) \tag{2.54}$$

To prove the corollary we choose $-\sigma$ $(\sigma > 0)$ as any number larger than the real part of every eigenvalue ($-\sigma$ plays the role of ρ in Theorem 2.10). By Theorem 2.7 every solution has the form $\exp(tA)\mathbf{c}$ for some constant vector $\mathbf{c}$. By Theorem 2.10

$$|\boldsymbol{\phi}(t)| \le |\exp(tA)|\,|\mathbf{c}| \le K|\mathbf{c}|\,e^{-\sigma t} = \tilde{K}e^{-\sigma t} \qquad (0 \le t < \infty)$$

• **EXERCISE**

2. Show that if all eigenvalues have real part negative or zero and if those eigenvalues with zero real part are simple, there exists a constant $K > 0$ such that $|\exp tA| \le K, (0 \le t < \infty)$, and hence every solution of $\mathbf{y}' = A\mathbf{y}$ is bounded on $0 \le t < \infty$.

We also remark, that using Theorem 2.10 in (2.34), for example, we obtain the following estimate for solutions of $\mathbf{y}' = A\mathbf{y} + \mathbf{g}(t)$.

$$|\boldsymbol{\phi}(t)| \leq K\,|\mathbf{y}| \exp\,[\rho(t - t_0)] + K \int_{t_0}^{t} \exp\,[\rho(t - s)]\,|\mathbf{g}(s)|\, ds \qquad (0 \leq t_0 \leq t < \infty)$$

where K, ρ are defined by the theorem. If we have more information on $\mathbf{g}(t)$ and on the characteristic roots of A, we can deduce more from this estimate. An example is given by the following exercise.

• **EXERCISE**

3. Show that if the hypothesis of Exercise 2 is satisfied and if $\int_{t_0}^{\infty} |\mathbf{g}(s)|\, ds < \infty$, then every solution $\boldsymbol{\phi}(t)$ of (2.33) on $0 \leq t_0 \leq t < \infty$ is bounded.

Another consequence of Theorem (2.10) is that every solution of a linear system of differential equations grows no faster than an exponential function. This fact is needed to solve linear systems with constant coefficients by the method of Laplace transforms (see, for example [2, Ch. 9]).

Theorem 2.11. *Suppose that in the linear nonhomogeneous system*

$$\mathbf{y}' = A\mathbf{y} + \mathbf{g}(t) \tag{2.33}$$

the function $\mathbf{g}(t)$ *grows no faster than an exponential function, that is, there exist real constants* $M > 0$, $T \geq 0$, a *such that*

$$|\mathbf{g}(t)| \leq Me^{at} \qquad (t \geq T)$$

Then every solution $\boldsymbol{\phi}$ *of* (2.33) *grows no faster than an exponential function, that is, there exist real constants* $K > 0$, b *such that*

$$|\boldsymbol{\phi}(t)| \leq Ke^{bt} \qquad (t \geq T)$$

The derivative $\boldsymbol{\phi}'(t)$ *also grows no faster than an exponential function.*

Proof. It follows from Theorem 2.1 that every solution ϕ of (2.33) exists on $0 \leq t < \infty$. By the variation of constants formula (2.34), every solution of (2.33) has the form

$$\boldsymbol{\phi}(t) = e^{tA}\mathbf{c} + \int_0^t e^{(t-s)A}\mathbf{g}(s)\, ds$$

for a suitably chosen constant vector $\mathbf{c}$. By Theorem 2.10, there exists a real number ρ and a constant $K_1 \geq 0$ such that

$$|e^{tA}| \leq K_1 e^{\rho t} \qquad (0 \leq t < \infty)$$

Here ρ may be any number larger than the real part of every eigenvalue of A as defined in the inequality (2.51). We now write the variation of constants formula in the form

$$\boldsymbol{\phi}(t) = e^{tA}\mathbf{c} + \int_0^T e^{(t-s)A}\mathbf{g}(s)\, ds + \int_T^t e^{(t-s)A}\mathbf{g}(s)\, ds$$

Letting $K_2 = \sup_{0 \leq t \leq T} |\mathbf{g}(s)|$ and using the known bounds on $|\mathbf{g}(s)|$ for $t \geq T$ and on $|e^{tA}|$, we obtain the estimate

$$|\boldsymbol{\phi}(t)| \leq K_1 |\mathbf{c}| e^{\rho t} + K_1 \int_0^T e^{\rho(t-s)} K_2\, ds + K_1 \int_T^t e^{\rho(t-s)} M e^{as}\, ds$$

From this, we obtain

$$|\boldsymbol{\phi}(t)| \leq K e^{bt} \qquad (t \geq T)$$

where b is the larger of a and ρ. Using this estimate in the system (2.33), we see that $\boldsymbol{\phi}'(t)$ also satisfies an inequality of the same type. Of course, the constant K does not have the same value as (2.52).

• **EXERCISES**

4. Evaluate the constant K in Theorem 2.11.
5. Obtain an explicit estimate for $|\boldsymbol{\phi}'(t)|$ in Theorem 2.11.
6. Let ϕ be a solution of the nth-order linear equation

$$u^{(n)} + a_1 u^{(n-1)} + \cdots + a_n u = f(t)$$

where $a_1, a_2, \ldots, a_n$ are constants, and where f is continuous on $0 \leq t < \infty$ and grows no faster than an exponential function. Show that $\phi(t), \phi'(t), \ldots, \phi^{(n)}(t)$ all grow no faster than an exponential function.

2.8 Autonomous Systems—Phase Space—Two-Dimensional Systems

In the system $\mathbf{y}' = \mathbf{f}(t, \mathbf{y})$ a case of considerable importance in applications occurs if $\mathbf{f}$ does not depend on t explicitly. Such systems are called **autonomous.** We will therefore study the system

$$\mathbf{y}' = \mathbf{g}(\mathbf{y}) \qquad (2.55)$$

where $\mathbf{g}$ is a real function defined in some real n-dimensional domain D in y-space. The reader will note that D is now n-dimensional, not $(n+1)$-dimensional. The implication of this will be clear in a moment. We shall assume throughout that $\mathbf{g}$ and $\partial\mathbf{g}/\partial y_j$ $(j = 1, \ldots, n)$ are continuous in D, so that by Theorem 1.1 (p. 26), given any $(t_0, \boldsymbol{\eta})$, $\boldsymbol{\eta}$ in D, there exists a unique real solution $\boldsymbol{\phi}$ of (2.55) satisfying the initial condition $\boldsymbol{\phi}(t_0) = \boldsymbol{\eta}$. This solution exists on some interval I and is a continuous function of $(t, t_0, \boldsymbol{\eta})$ with t, t_0 on I, $\boldsymbol{\eta}$ in D. We observe that a linear system with constant coefficients $\mathbf{y}' = A\mathbf{y}$ is autonomous, as is any scalar differential equation in which the independent variable does not enter explicitly, for example, $\theta'' + (g/L)\sin\theta = 0$, the pendulum equation derived in Section 1.1 (p. 6).

Autonomous systems have several important properties. If $\boldsymbol{\phi}(t)$ is a solution of (2.55) existing for $-\infty < t < \infty$, then it is easily verified by direct substitution that for any constant a, $\boldsymbol{\phi}(t + a)$ is also a solution of (2.55). We therefore say that **an autonomous system is invariant under translations of the independent variable.** In particular, if $\boldsymbol{\phi}(t)$ is that solution for which $\boldsymbol{\phi}(0) = \boldsymbol{\eta}$, then $\boldsymbol{\phi}(t - t_0)$ is that solution of (2.55) satisfying the initial condition $\boldsymbol{\phi}(t_0) = \boldsymbol{\eta}$. (For linear systems with constant coefficients this was established in Equation (2.27), p. 58.)

Solutions of autonomous systems are conveniently represented by curves in n-space rather than $(n+1)$-dimensional space. The reason for this is that to every autonomous system such as (2.55) there corresponds a unique vector field $\mathbf{g}(\mathbf{y})$ at points of the domain D in Euclidean n-space and this vector field is independent of t. Thus, if we think of (2.55) as representing the equations of motion of a moving particle, then to each point $\mathbf{y}$ in D there corresponds the vector $\mathbf{g}(\mathbf{y})$ which is the velocity vector of the particle at $\mathbf{y}$, and this velocity vector does not depend on t. Let $\boldsymbol{\phi}(t, \boldsymbol{\eta})$ be that solution of (2.55), satisfying the initial condition $\boldsymbol{\phi}(t_0, \boldsymbol{\eta}) = \boldsymbol{\eta}$, where $\boldsymbol{\eta}$ is any point of D. Then $\boldsymbol{\phi}$ represents the motion of the particle obeying the law (2.55), passing through the point $\boldsymbol{\eta}$ in D at time $t = t_0$, and we can completely characterize this motion by the curve C (see Figure 2.1) in the n-dimensional region D prescribed

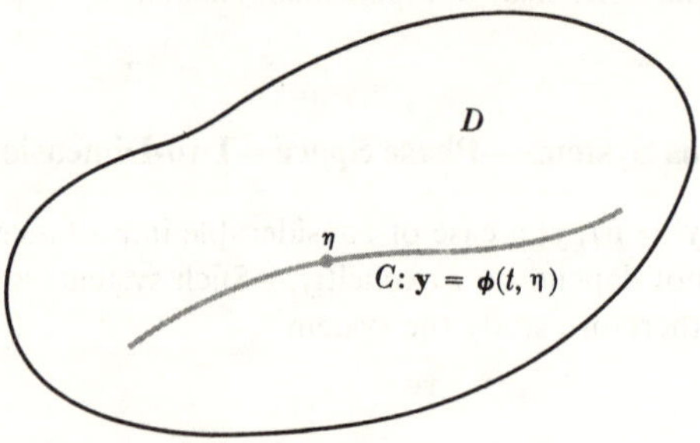

Figure 2.1

by the parametric equation $\mathbf{y} = \boldsymbol{\phi}(t, \boldsymbol{\eta})$, where the time t plays the role of a parameter. We can assign a direction to the curve C corresponding to the direction of increasing t, indicated by an arrow in Figure 2.1.

Example 1. In the special case of a dynamical system with one degree of freedom (mass-spring, pendulum, and so on) in which x is the displacement and the acceleration x'' is determined by a scalar equation of the form

$$x'' = F(x, x')$$

we write $y_1 = x$, $y_2 = x'$ and obtain the autonomous system

$$\begin{aligned} y_1' &= y_2 \\ y_2' &= F(y_1, y_2) \end{aligned}$$

Now the solutions (motions) can be represented by curves in the (y_1, y_2) plane. This displacement-velocity plane is often referred to as the **Poincaré*** **phase plane.** More generally, if (2.55) represents the motion of a dynamical system of n degrees of freedom. Its motion can be represented by curves in the **phase space** of $2n$ dimensions.

With reference to the above discussion and because the system (2.55) is invariant under translation of time, $\boldsymbol{\phi}(t + t_0, \boldsymbol{\eta})$ is, for any constant t_0, also a solution (2.55). It may be interpreted as that solution (motion) which passes through the point $\boldsymbol{\eta}$ at time $t = 0$ (not at time t_0). This is not the same motion as that represented by the solution $\boldsymbol{\phi}(t, \boldsymbol{\eta})$. The two solutions differ by the **phase** t_0. But the motions $\boldsymbol{\phi}(t, \boldsymbol{\eta})$ and $\boldsymbol{\phi}(t + t_0, \boldsymbol{\eta})$ are represented by the same curve C in D. In fact, for any constant a, the motion $\boldsymbol{\phi}(t + a, \boldsymbol{\eta})$ is also represented by the curve C. The curve C in D is called an **orbit, trajectory**, or **path** of (2.55). From the above remarks we see that for a given orbit C there are infinitely many motions—in fact, a one-parameter family of solutions—differing from one another by a phase.

A point a in D is called a critical point of the system

$$\mathbf{y}' = \mathbf{g}(\mathbf{y}) \tag{2.55}$$

if and only if $\mathbf{g}(\mathbf{a}) = \mathbf{0}$.

Thus the critical points are those points of D at which the vector field $\mathbf{g}(\mathbf{y})$ vanishes. If $\mathbf{a}$ is a critical point of (2.55), then the constant vector $\boldsymbol{\phi}(t) \equiv \mathbf{a}$ is a solution of (2.55). The orbit of this solution is the single point

---

* The eminent French mathematician Henri Poincaré pioneered much of the research in the qualitative theory of differential equations.

a of D in phase space. From a physical viewpoint, critical points correspond to rest or equilibrium states: For the simple undamped pendulum equation

$$\theta'' + \frac{g}{L}\sin\theta = 0$$

(see Exercise 3, Section 1.1), the autonomous system corresponding to (2.55) is

$$y_1' = y_2$$

$$y_2' = -\frac{g}{L}\sin y_1$$

Its critical points are at $(n\pi, 0)$, where $n = 0, \pm 1, \pm 2, \ldots,$ and D is the whole (y_1, y_2) plane. We shall have much more to say about this system and the behavior of its solutions in Section 6.2.

Example 2. Consider the system $\mathbf{y}' = A\mathbf{y}$ where

$$A = \begin{pmatrix} -2 & 0 \\ 0 & -3 \end{pmatrix} \qquad \mathbf{y} = \begin{pmatrix} y_1 \\ y_2 \end{pmatrix}$$

Then in the above notation $\mathbf{g}(y) = A\mathbf{y}$; D is the whole (y_1, y_2) (phase) plane; and since $\det A \neq 0$, the origin is the only critical point. A fundamental matrix is

$$\begin{pmatrix} e^{-2t} & 0 \\ 0 & e^{-3t} \end{pmatrix}$$

and thus every solution is of the form

$$\boldsymbol{\phi}(t, \boldsymbol{\eta}) = \begin{pmatrix} e^{-2t} & \eta_1 \\ e^{-3t} & \eta_2 \end{pmatrix}$$

for some constant vector $\boldsymbol{\eta} = \begin{pmatrix} \eta_1 \\ \eta_2 \end{pmatrix}$. Here we have arbitrarily chosen $t_0 = 0$. Notice that $\boldsymbol{\phi}(t - t_0, \boldsymbol{\eta})$ is that solution passing through the point $\boldsymbol{\eta}$ at $t = t_0$. Let $P_{\boldsymbol{\eta}} = (\eta_1, \eta_2)$ be any point in the (y_1, y_2) plane. Then the solution $\boldsymbol{\phi}(t, \boldsymbol{\eta})$ for $t > 0$ is represented by the parametric equations $y_1 = \phi_1(t) = e^{-2t}\eta_1$, $y_2 = \phi_2(t) = e^{-3t}\eta_2$ for $t > 0$ and this represents the portion of the curve shown in Figure 2.2 between $P_{\boldsymbol{\eta}}$ and the origin, as is verified by elementary calculus; the arrow indicates the direction of increasing t. Notice that the slope of the tangent to this curve, dy_2/dy_1, also tends to zero as $t \to +\infty$.

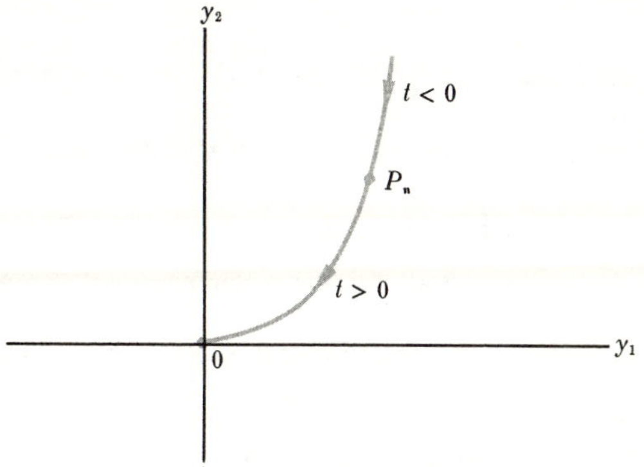

Figure 2.2

Similarly, $t < 0$ represents the portion of the curve in Figure 2.2 above the point P_η. It should be noted that $\lim_{t \to +\infty} \boldsymbol{\phi}(t, \boldsymbol{\eta}) = \mathbf{0}$, that is, both the solution and the orbit approach the origin as $t \to +\infty$. Proceeding in this way by choosing various points of the phase plane as initial points, we obtain the so-called **phase portrait** of the system, shown in Figure 2.3. Again, the arrow on every orbit approaches the origin (as $t \to +\infty$).

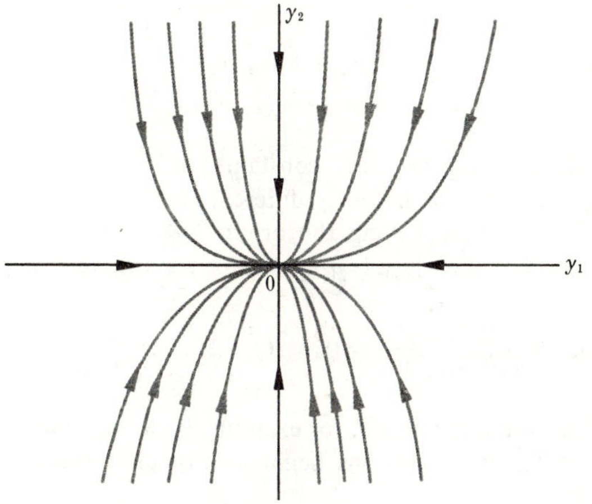

Figure 2.3

• EXERCISES

1. Obtain the phase portrait of the system $\mathbf{y}' = A\mathbf{y}$ where $A = \begin{pmatrix} 2 & 0 \\ 0 & 3 \end{pmatrix}$.
2. Obtain the phase portrait of the system $\mathbf{y}' = A\mathbf{y}$ where $A = \begin{pmatrix} -2 & 1 \\ 0 & -2 \end{pmatrix}$.
3. Obtain the phase portrait of the system $\mathbf{y}' = A\mathbf{y}$ where $A = \begin{pmatrix} 2 & 0 \\ 0 & -3 \end{pmatrix}$.
4. Obtain the phase portrait for the scalar equation $x'' + 4x = 0$. [*Hint:* Use the system $y_1' = y_2$, $y_2' = -4y_1$.]
5. Write the simple pendulum equation $\theta'' + g/L \sin\theta = 0$ as the system

$$y_1' = y_2$$
$$y_2' = \frac{-g}{L} \sin y_1 \qquad \text{where } y_1 = \theta,\ y_2 = \theta'.$$

Assuming that $|y_1|$ is small, replace $\sin y_1$ by y_1 and draw the resulting phase portrait.

We conclude this section with a complete discussion of the phase portraits of linear two-dimensional autonomous systems. The motivation for these considerations is the following: Suppose that in the case $n = 2$, $\mathbf{g}(\mathbf{y})$ in (2.55) has the form

$$\mathbf{g}(y) = \begin{pmatrix} g_1(y_1, y_2) \\ g_2(y_1, y_2) \end{pmatrix}$$

where

$$g_1(y_1, y_2) = a_{11}y_1 + a_{12}y_2 + h_1(y_1, y_2)$$
$$g_2(y_1, y_2) = a_{21}y_1 + a_{22}y_2 + h_2(y_1, y_2)$$

where $a_{11}, a_{12}, a_{21}, a_{22}$ are real constants with $a_{11}a_{22} - a_{12}a_{21} \neq 0$ and where h_1, h_2 are real, continuously differentiable functions defined in some domain D in the plane having the origin $y_1 = y_2 = 0$ in its interior and h_1, h_2 are "small" when $|y_1|, |y_2|$ are small; for example, in the sense that

$$\lim_{y_1^2 + y_2^2 \to 0} \frac{h_j(y_1, y_2)}{(y_1^2 + y_2^2)^{1/2}} = 0 \qquad (j = 1, 2).$$

This condition certainly holds if, for example, h_1, h_2 are **analytic** at $y_1 = y_2 = 0$ and if their Taylor expansions begin with quadratic terms. Then (2.55) becomes

$$\mathbf{y}' = A\mathbf{y} + \mathbf{h}(\mathbf{y}) \tag{2.56}$$

where

$$A = \begin{pmatrix} a_{11} & a_{12} \\ a_{21} & a_{22} \end{pmatrix} \quad \mathbf{y} = \begin{pmatrix} y_1 \\ y_2 \end{pmatrix} \quad \mathbf{h}(\mathbf{y}) = \begin{pmatrix} h_1(y_1, y_2) \\ h_2(y_1, y_2) \end{pmatrix}$$

and with $\det A \neq 0$ and $|\mathbf{h}(\mathbf{y})|$ small when $|\mathbf{y}|$ is small in some appropriate sense. We shall call $\mathbf{h}(\mathbf{y})$ the **perturbation term** and (2.56) the **perturbed system.** Many problems arising in applications have the form (2.56).

Example 3. Consider the simple pendulum equation $\theta'' + g/L \sin\theta = 0$. Then if $y_1 = \theta$, $y_2 = \theta'$,

$$y_1' = y_2$$

$$y_2' = -\frac{g}{L}\sin y_1 = -\frac{g}{L}y_1 + \frac{g}{L}\left(\frac{y_1^{\,3}}{3!} + \frac{y_1^{\,5}}{5!} + \cdots\right)$$

Thus here, in the notation of (2.56),

$$A = \begin{pmatrix} 0 & 1 \\ -\dfrac{g}{L} & 0 \end{pmatrix}, \quad h_1(y_1, y_2) \equiv 0 \quad h_2(y_1, y_2) = \frac{g}{L}\left(\frac{y_1^{\,3}}{3!} - \frac{y_1^{\,5}}{5!} + \cdots\right)$$

and the above hypotheses are satisfied.

Intuitively, we expect that if $|\mathbf{h}(\mathbf{y})|$ is small for small $|\mathbf{y}|$, then the behavior of solutions of (2.56) **near enough to the origin** would be similar to the behavior of the system in which the perturbations are zero. This latter system is linear with constant coefficients:

$$\mathbf{y}' = A\mathbf{y}$$

and thus easy to analyze. It can be shown that this intuition is essentially, but not completely, correct. (For the case of the nonlinear pendulum equation, see Sections 6.1 and 6.2.)

Here we shall be content to analyze the general two-dimensional linear system $\mathbf{y}' = A\mathbf{y}$ with constant coefficients. Let us make the change of variable $\mathbf{y} = T\mathbf{z}$ where T is a nonsingular constant matrix (to be determined), and substitute, obtaining the system

$$\mathbf{z}' = (T^{-1}AT)\mathbf{z} \tag{2.57}$$

whose coefficient matrix $T^{-1}AT$ is similar to A. For simplicity, and because this is the case that arises in applications most frequently, we consider only the case det $A \neq 0$. This means that zero is not an eigenvalue of A, and that the origin is the only critical point. For the case when det $A = 0$, the reader is referred to Exercises 19 and 20, p. 95.

As is done in Appendix 2, we may show that there is a **real** nonsingular matrix T such that $T^{-1}AT$ is equal to one of the following six matrices:

(i) $\begin{pmatrix} \lambda & 0 \\ 0 & \mu \end{pmatrix}$ where $(\mu < \lambda < 0)$ or $(0 < \mu < \lambda)$

(ii) $\begin{pmatrix} \lambda & 0 \\ 0 & \lambda \end{pmatrix}$ where $\lambda > 0$ or $\lambda < 0$

(iii) $\begin{pmatrix} \lambda & 0 \\ 0 & \mu \end{pmatrix}$ $\mu < 0 < \lambda$

(iv) $\begin{pmatrix} \lambda & 1 \\ 0 & \lambda \end{pmatrix}$ where $\lambda > 0$ or $\lambda < 0$

(v) $\begin{pmatrix} \sigma & \nu \\ -\nu & \sigma \end{pmatrix}$ $\sigma, \nu \neq 0$ $\sigma > 0$ or $\sigma < 0$

(vi) $\begin{pmatrix} 0 & \nu \\ -\nu & 0 \end{pmatrix}$ $\nu \neq 0$

The cases (v), (vi) correspond to complex conjugate eigenvalues of A, $\sigma \pm i\nu$ and $\pm i\nu$, respectively. In the four remaining cases the eigenvalues λ, μ are real. We obtain the possible phase portraits of (2.57) by assuming that $T^{-1}AT$ is one of the forms (i)–(vi). We emphasize that at this stage we in no way imply that these phase portraits of (2.57) also represent the phase portraits of the perturbed system (2.56), although such would be the hope, at least for $|\mathbf{y}|$ small. We also remark that the actual phase portraits of the linear system $\mathbf{y}' = A\mathbf{y}$ differ from those constructed below for the system (2.57) by the fact that the nonsingular transformation matrix T distorts but does not change the character of these portraits.

Case (*i*). (This is essentially Example 2 and Exercise 1 above.) The solution of (2.57) through the point $(\eta_1, \eta_2) \neq (0, 0)$ at $t = 0$ is

$$\boldsymbol{\phi}(t) = \begin{pmatrix} e^{\lambda t}\eta_1 \\ e^{\mu t}\eta_2 \end{pmatrix}$$

If $\mu < \lambda < 0$, we have $\boldsymbol{\phi}(t) \to \mathbf{0}$ as $t \to +\infty$ and we obtain the phase portrait in Figure 2.4 with every orbit tending to the origin as $t \to +\infty$. If $0 < \mu < \lambda$, we obtain the phase portrait in Figure 2.5 with every orbit tending away from the origin as $t \to \infty$. Arrows indicate the direction of increasing t. The origin in Figures 2.4 and 2.5 corresponding to Case (i) is called an **improper node**.

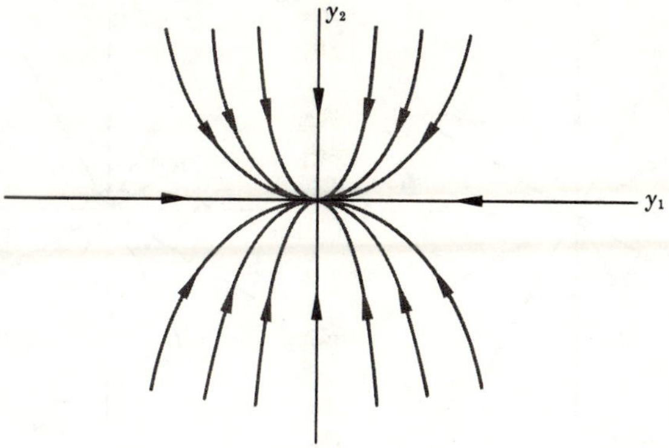

Figure 2.4

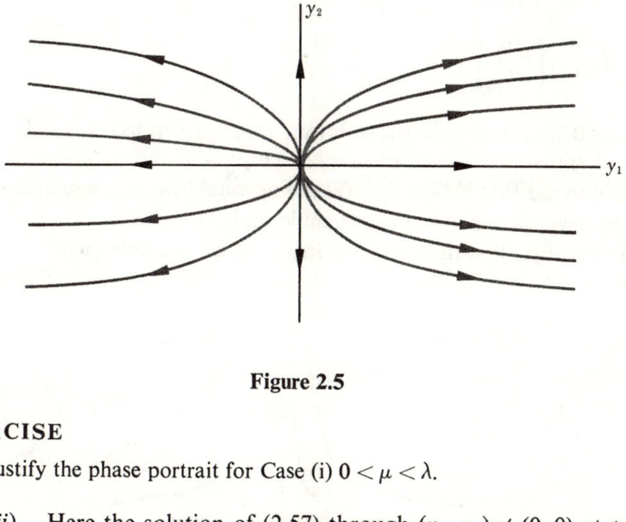

Figure 2.5

• **EXERCISE**

6. Justify the phase portrait for Case (i) $0 < \mu < \lambda$.

Case (*ii*). Here the solution of (2.57) through $(\eta_1, \eta_2) \neq (0, 0)$ at $t = 0$ is

$$\boldsymbol{\phi}(t) = \begin{pmatrix} e^{\lambda t}\eta_1 \\ e^{\lambda t}\eta_2 \end{pmatrix}$$

and if $\lambda > 0$, we obtain the phase portrait in Figure 2.6, whereas the case $\lambda < 0$ corresponds to Figure 2.7. Note that all orbits are straight lines tending away from the origin if $\lambda > 0$ and toward the origin if $\lambda < 0$.

The ratio $\phi_2(t)/\phi_1(t)$ if $\eta_1 \neq 0$ is constant, as is $\phi_1(t)/\phi_2(t)$ if $\eta_2 = 0$. The origin in Case (ii) is called a **proper node**.

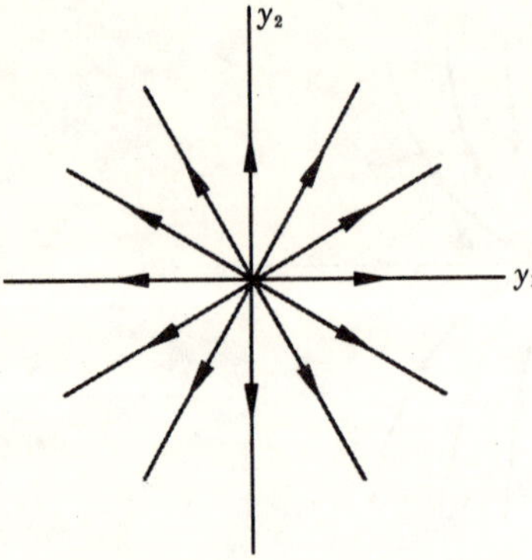

Figure 2.6

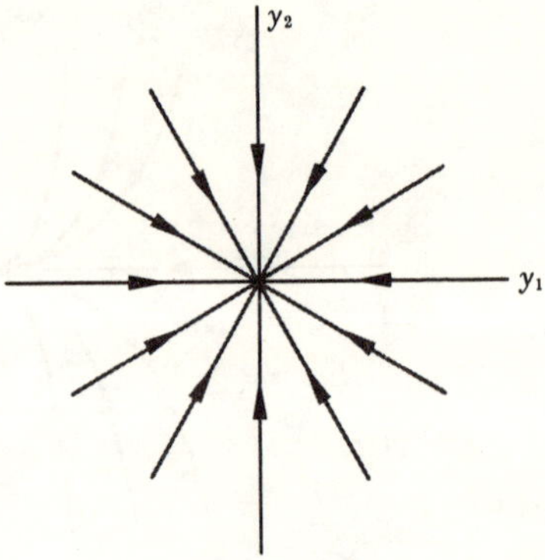

Figure 2.7

Case (iii). Here

$$\phi(t) = \begin{pmatrix} e^{\lambda t}\eta_1 \\ e^{\mu t}\eta_2 \end{pmatrix}$$

with $\mu < 0$ and $\lambda > 0$, is the solution through (n_1, η_2) at $t = 0$. Now as $t \to \infty$, $\phi_1(t) \to \pm\infty$ according as $\eta_1 > 0$ or $\eta_1 < 0$ and $\phi_2(t) \to 0$ as $t \to +\infty$. It is easy to see that if $|\lambda| = |\mu|$, the orbits would be rectangular hyperbolas; for arbitrary $\lambda > 0$, $\mu < 0$ they resemble these curves as shown in Figure 2.8. Quite naturally, the origin in Case (iii) is called a **saddle point**.

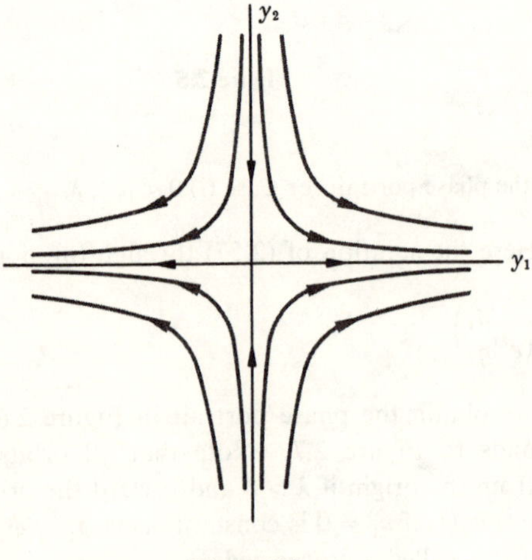

Figure 2.8

• **EXERCISE**

7. Construct the phase portrait in Case (iii) if $\lambda < 0$ and $\mu > 0$.

Case (*iv*). Here

$$\phi(t) = \begin{pmatrix} \eta_1 + \eta_2 t \\ \eta_2 \end{pmatrix} e^{\lambda t}$$

is that solution passing through (η_1, η_2) at $t = 0$ and if $\lambda < 0$ the phase portrait is easily characterized by the fact that every orbit tends to the origin as $t \to +\infty$ and has the same limiting direction at (0, 0). For, $dy_2/dy_1 = \phi_2'/\phi_1' = (\lambda\phi_2/(\lambda\phi_1 + \phi_2)) \to 0$ as $t \to +\infty$ (see Figure 2.9). The origin in Case (iv) is called (as in Case (i)) an **improper node.**

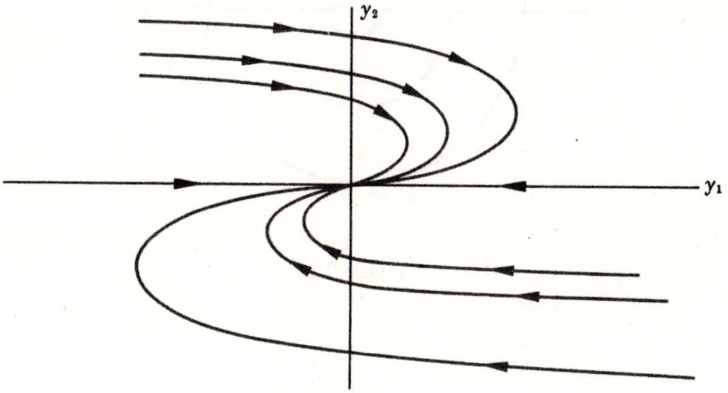

Figure 2.9

• **EXERCISE**

8. Construct the phase portrait in Case (iv) with $\lambda > 0$.

Case (*v*). Here the solution, for the case $\sigma > 0$, passing through the point (η_1, η_2) at $t = 0$ is

$$\phi(t) = e^{\sigma t}\begin{pmatrix} \eta_1 \cos \nu t + \eta_2 \sin \nu t \\ -\eta_1 \sin \nu t + \eta_2 \cos \nu t \end{pmatrix}$$

Let $\rho = (\eta_1{}^2 + \eta_2{}^2)^{1/2}$, $\cos \alpha = \eta_1/\rho$, $\sin \alpha = \eta_2/\rho$. Then

$$\phi(t) = e^{\sigma t}\begin{pmatrix} \rho \cos (\nu t - \alpha) \\ -\rho \sin (\nu t - \alpha) \end{pmatrix}$$

Letting r, θ be the polar coordinates, $y_1 = r\cos\theta$, $y_2 = r\sin\theta$, we may write the solution in polar form $r(t) = \rho e^{\sigma t}$, $\theta(t) = -(\nu t - \alpha)$. Eliminating the parameter t, we have $r = A\exp(-\sigma/\nu)\theta$ where $A = \rho\exp[(\sigma/\nu)\alpha]$. Thus the phase portrait is a family of spirals, as shown in Figure 2.10, for the case $\sigma > 0$, $\nu > 0$ and the origin is called a **spiral point**. In this case the orbits tend away from zero as $t \to +\infty$ (or, equivalently, approach zero as $t \to -\infty$).

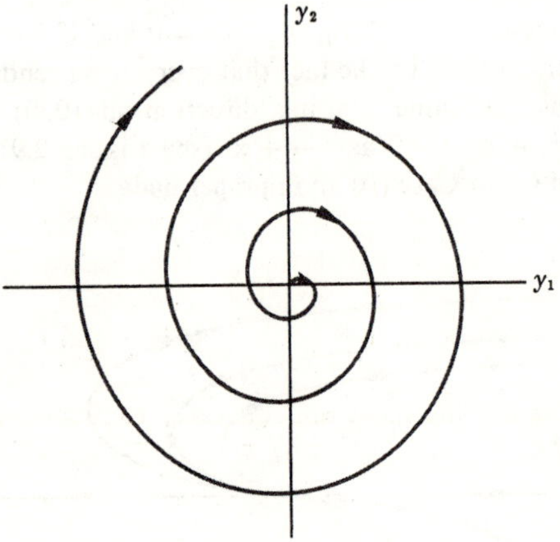

Figure 2.10

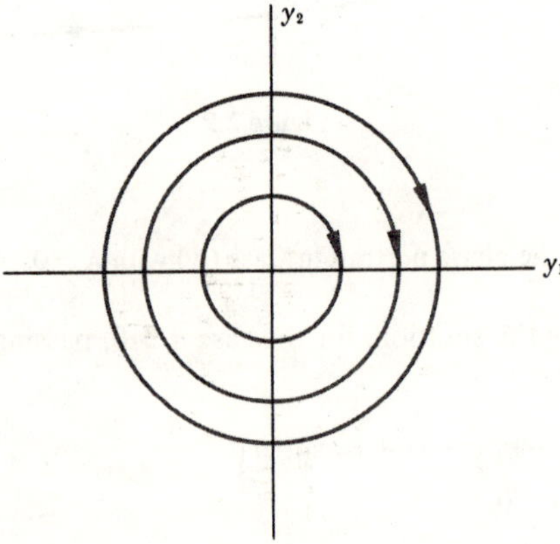

Figure 2.11

• **EXERCISE**

9. Sketch the phase portrait for the Case (v) in case $\sigma < 0$, $\nu < 0$.

Case (vi). This is a special case of Case (v) with $\sigma = 0$. From the above formulas we see that the orbits are concentric circles of radius ρ oriented as shown for $\nu > 0$ in Figure 2.11. The origin is called a **center**.

• **EXERCISE**

10. Sketch the phase portrait in Case (vi) when $\nu < 0$.

We observe from the possible cases considered above that all solutions of (2.57) and also their orbits tend to the origin as $t \to +\infty$ if and only if both eigenvalues of A have negative real parts; in this case we say that the origin is an **attractor of the linear system** (2.57). One of the results that can be established is that the origin remains an attractor when we add the perturbation terms (see Theorem 4.3, p. 161, and the application, p. 163). Notice that in case of a saddle point or center, the origin is not an attractor and, as might be expected, these are the most difficult cases to treat when perturbation terms are added.

• **EXERCISES**

Sketch the phase portrait of each of the following scalar equations by converting to an equivalent system. Identify the origin and decide whether it is an attractor.

11. $x'' + x = 0$.
12. $x'' - 3x' + x = 0$.
13. $x'' + 3x' + x = 0$.
14. $x'' + 3x' - x = 0$.
15. $x'' - 3x' + 2x = 0$.
16. $x'' + 3x' + 2x = 0$.
17. $x'' - 2x' + x = 0$.
18. $x'' - x' - 6x = 0$.
19. To illustrate the complexity of the case when the origin is not the only critical point of an linear system, consider the system

$$y_1' = y_1 - y_2$$
$$y_2' = 2y_1 - 2y_2$$

(a) Show that there is a line of critical points.
(b) Sketch the phase portrait.
[*Hint:* $y_2' = 2y_1'$, and the eigenvalues of the coefficient matrix are 0 and -1.]

20. Repeat as much as you can of Exercise 19 for the system

$$y_1' = a_{11}y_1 + a_{12}\, y_2$$
$$y_2' = a_{21}y_1 + a_{22}\, y_2$$

where $a_{11}a_2 - a_{12}a_{21} = 0$, but not all of $a_{11}, a_{12}, a_{21}, a_{22}$ are zero.

2.9 Linear Systems with Periodic Coefficients

We first consider the linear system

$$\mathbf{y}' = A(t)\mathbf{y} \tag{2.58}$$

where $A(t)$ is a continuous periodic n-by-n matrix of period ω [that is, $A(t + \omega) = A(t)$, $-\infty < t < \infty$]. We observe (see also Section 2.3, Exercise 26) that if $\Phi(t)$ is a fundamental matrix for (2.58) then

$$\Phi'(t + \omega) = A(t + \omega)\Phi(t + \omega) = A(t)\Phi(t + \omega)$$

Thus $\Phi(t + \omega)$ is a solution matrix of (2.58) and by Abel's formula (Theorem 2.3) $\det \Phi(t + \omega) \neq 0$. Thus (by Theorem 2.4) $\Phi(t + \omega)$ is also a fundamental matrix of (2.58) and therefore (by Theorem 2.5) there exists a nonsingular constant matrix C such that

$$\Phi(t + \omega) = \Phi(t)C \qquad (-\infty < t < \infty) \tag{2.59}$$

It can be shown (see Appendix 3) that corresponding to every nonsingular constant matrix C there exists a matrix R such that $C = \exp(\omega R)$. We note that if $\Phi(0) = E$, then from (2.59)

$$\Phi(\omega) = \exp \omega R \tag{2.60}$$

We may now establish the following result.

Theorem 2.12 (Floquet's Theorem.) *Let $A(t)$ be a continuous periodic matrix of period ω and let $\Phi(t)$ be any fundamental matrix of the system* (2.58). *Then there exists a periodic nonsingular matrix $P(t)$ of period ω and a constant matrix R such that*

$$\Phi(t) = P(t) \exp(tR) \tag{2.61}$$

The reader should observe that if $A(t)$ is a constant matrix, hence periodic of any period ω, the above result reduces to the well-known one for constant coefficients with $P(t) = E$, and $R = A$.

Proof of Theorem 2.12. Let Φ be an arbitrary given fundamental matrix of (2.58). Let R be the matrix determined by Φ as above from (2.59). Define

$$P(t) = \Phi(t) \exp(-tR) \qquad (-\infty < t < \infty) \tag{2.62}$$

Clearly, $P(t)$, being the product of two nonsingular matrices, is nonsingular. Moreover,

$$\begin{aligned} P(t+\omega) &= \Phi(t+\omega)\exp(-(t+\omega)R) \\ &= \Phi(t)\exp(\omega R)\exp(-(t+\omega)R) \\ &= \Phi(t)\exp(-tR) = P(t) \qquad (-\infty < t < \infty) \end{aligned}$$

Thus $P(t)$ has period ω and, solving (2.62) for $\Phi(t)$, we obtain (2.61) as asserted. ∎

REMARK If $A(t)$ is real (of course, its period ω is real), then even if $\Phi(t)$ is real, it is not necessarily true that the matrix R is real. (Note, for example, that if $n = 1$ and if C in (2.59) turns out to be any negative real number, then C does not have a real logarithm, so that R cannot be real.)

In this case we use

$$\Phi(t+2\omega) = \Phi(t+\omega)C = \Phi(t)C^2$$

and we define S by

$$C^2 = e^{2\omega S} \tag{2.63}$$

and it can be shown (see [4, p. 8]) that S is real (see also Appendix 3). It can then further be shown that there exists a real nonsingular matrix $Q(t)$ of period 2ω such that

$$\Phi(t) = Q(t)\exp(tS) \tag{2.64}$$

• **EXERCISE**

1. Assuming the validity of (2.63), prove this statement.

The Floquet theorem can be used to transform the system (2.58) to a linear system with constant coefficients as follows. Let

$$\mathbf{y} = P(t)\mathbf{u} \tag{2.65}$$

where P is the periodic matrix of Theorem 2.12. Since, from (2.58)

$$\begin{aligned} [P(t)\exp(tR)]' &= P'(t)\exp(tR) + P(t)R\exp(tR) \\ &= A(t)P(t)\exp(tR) \end{aligned}$$

it follows that

$$P'(t) = A(t)P(t) - P(t)R$$

Thus

$$\begin{aligned}\mathbf{y}' &= P(t)\mathbf{u}' + P'(t)\mathbf{u} = P(t)\mathbf{u}' + (A(t)P(t) - P(t)R)\mathbf{u}\\ &= A(t)P(t)\mathbf{u}\end{aligned}$$

and therefore

$$P(t)\mathbf{u}' - P(t)R\mathbf{u} = \mathbf{0}$$

or

$$\mathbf{u}' = R\mathbf{u} \tag{2.66}$$

which is a linear system with constant coefficients. This establishes the following result.

Corollary 1 to Theorem 2.12. The change of variable $\mathbf{y} = P(t)\mathbf{u}$ *transforms the periodic system* (2.58) *to the system* (2.66) *with constant coefficients.*

The reader should note, however, that this pleasant fact requires complete knowledge of the matrices $P(t)$ and R—by no means a trivial requirement.

It is customary to call the eigenvalues λ_i of the nonsingular matrix $\exp(\omega R)$ the **multipliers of the system** (2.58), and to call the eigenvalues ρ_i of the matrix R the **characteristic exponents of the system** (2.58). It follows from Appendix 3 that

$$\rho_i = \frac{1}{\omega} \log \lambda_i, \qquad (i = 1, \ldots, n) \tag{2.67}$$

The multipliers have the following interesting property, which justifies their name.

Corollary 2 to Theorem 2.12. A solution $\boldsymbol{\phi}(t)$ *of the system* (2.58) *has the property*

$$\boldsymbol{\phi}(t + \omega) = k\boldsymbol{\phi}(t) \qquad (-\infty < t < \infty)$$

where k is a constant, if and only if k is an eigenvalue of $\Phi(\omega) = \exp(\omega R)$.

• **EXERCISES**

2. Prove Corollary 2. [*Hint:* let $\boldsymbol{\phi}(t) = \Phi(t)\boldsymbol{\phi}(0)$ where $\Phi(t)$ is the fundamental matrix of (2.58) that is the identity at $t = 0$.]

3. Use Corollary 2 to deduce that there is a solution $\boldsymbol{\phi}(t)$ of the system (2.1) of period ω if and only if 1 is a multiplier of the system (2 58).

4. Show that if -1 is a multiplier of the system (2.58), then there is a solution of (2.58) of least period 2ω.

The relation (2.67) between the multipliers and the characteristic exponents combined with Theorem 2.12 leads to the following useful result.

Corollary 3 to Theorem 2.12. If the characteristic exponents of the system (2.58) *have negative real parts* (*or equivalently, if the multipliers of the system* (2.58) *have magnitude strictly less than* 1), *then all solutions of the system* (2.58) *approach zero as* $t \to +\infty$.

The reader should note that if $A(t)$ is a constant matrix, this result is the corollary to Theorem 2.10. Moreover, relation (2.53) holds, with practically the same proof for the system (2.58).

• **EXERCISE**

5. Prove Corollary 3. [*Hint:* Use (2.61), (2.67), and the fact that $P(t)$ periodic implies $|P(t)|$ bounded.]

The results of Corollaries 1, 2, 3 hinge on specific knowledge of the multipliers (or equivalently, of the characteristic exponents) of the system (2.58). That this is by no means a simple problem can be seen in the following special case.

Example 1. Consider the scalar equation

$$u'' + p(t)u' + q(t)u = 0 \tag{2.68}$$

where p, q are continuous functions of period ω in t. Determine an equation satisfied by the multipliers.

Letting $y_1 = u$, $y_2 = u'$, we obtain the equivalent system

$$\begin{aligned} y_1' &= y_2 \\ y_2' &= -q(t)y_1 - p(t)y_2 \end{aligned} \tag{2.69}$$

Let $\Phi(t)$ be the fundamental matrix that is the identity at $t = 0$. Then

$$\Phi(t) = \begin{pmatrix} \phi_1(t) & \phi_2(t) \\ \phi_1'(t) & \phi_2'(t) \end{pmatrix} \tag{2.70}$$

where ϕ_1, ϕ_2 are the solutions of (2.68) such that $\phi_1(0) = \phi_2'(0) = 1$, $\phi_2(0) = \phi_1'(0) = 0$. According to the definition of multipliers and Equation (2.60), the multipliers are the eigenvalues of $\Phi(\omega)$ and therefore, using (2.70), they satisfy the equation

$$\det(\Phi(\omega) - \lambda E) = \lambda^2 - (\phi_1(\omega) + \phi_2'(\omega))\lambda + \det \Phi(\omega) = 0$$

By Abel's formula (Theorem 2.3, p. 46),

$$\det \Phi(\omega) = \det \Phi(0) \exp\left(-\int_0^\omega p(s)\, ds\right) = \exp\left(-\int_0^\omega p(s)\, ds\right)$$

Thus the multipliers are the roots of the equation

$$\lambda^2 - (\phi_1(\omega) + \phi_2'(\omega))\lambda + \exp\left(-\int_0^\omega p(s)\, ds\right) = 0 \tag{2.71}$$

The reader should note that we cannot compute the coefficient $(\phi_1(\omega) + \phi_2'(\omega))$ without knowledge of the solutions ϕ_1, ϕ_2 of (2.68), and these are in general impossible to find. Nevertheless, the equation (2.71) for the multipliers gives some useful information, as is shown by the following exercises.

• EXERCISES

6. Given the scalar equation

$$u'' + (a + b\rho(t))u = 0 \tag{2.72}$$

where a, b are real constants and ρ is a real continuous function of period ω.

(a) Use the result of Example 1 to show that the multipliers are determined from the equation

$$\lambda^2 - A(a, b)\lambda + 1 = 0 \tag{2.73}$$

where in the notation of Example 1

$$A(a, b) = \phi_1(\omega) + \phi_2'(\omega)$$

A depends on the constants a, b (in fact continuously, as can be shown from Theorem 3.8, p. 137) because the solutions ϕ_1, ϕ_2 do.

(b) Show that if $-2 < A(a, b) < 2$, then the multipliers are complex conjugate and have magnitude 1. [*Hint:* Solve (2.73) by the quadratic formula.]

(c) Show that if $-2 < A(a, b) < 2$, then all solutions of (2.72) together with their first derivatives are bounded on $-\infty < t < \infty$. [*Hint:* From (b), compute the characteristic exponents in terms of $A(a, b)$, showing that they are pure imaginary; then apply Theorem 2.12, specifically (2.61), to the system (2.66).]

(d) If either $A(a, b) < -2$ or $A(a, b) > 2$, show by the technique of part (c) that the multipliers are both real and at least one of them has magnitude greater than 1. Deduce that at least one characteristic exponent has a positive real part so that there is, in either case, at least one unbounded solution of (2.72) on $-\infty < t < \infty$.

(e) If $A(a, b) = 2$, show that (2.72) has at least one solution of period ω. [*Hint:* See Exercise 3.]

(f) If $A(a, b) = -2$, show that (2.72) has at least one solution of period 2ω. [*Hint:* See Exercise 4.]

7. In Equation (2.72), take $\omega = 1$, and suppose that $a > 0$, $a \neq n^2\pi^2$ for any integer n.

(a) Consider first the case $b = 0$; show that

$$A(a, 0) = 2 \cos \sqrt{a}$$

and deduce that $-2 < A(a, 0) < 2$. [*Hint:* Find the solutions ϕ_1, ϕ_2 of Exercise 6 by solving the equation $u'' + au = 0$, then compute $A(a, 0)$.]

(b) Assuming the continuity of $A(a, b)$ as a function of the pair (a, b), show that the multipliers and characteristic exponents of (2.72) are continuous functions of (a, b).

(c) Show that all solutions of (2.72) are bounded if $a \neq n^2\pi^2$ and if b is small. [*Hint:* Show that $-2 < A(a, b) < 2$ for $a \neq n^2\pi^2$, and for b small. Then apply Exercise 6c.]

We turn briefly to the nonhomogeneous system

$$\mathbf{y}' = A(t)\mathbf{y} + \mathbf{g}(t) \tag{2.74}$$

where we assume throughout that $A(t)$ and $\mathbf{g}(t)$ are continuous and periodic in t of the same period ω. Note that the case $A(t)$ a constant matrix is not excluded. An important special case of (2.74) is the scalar equation

$$u'' + a_1 u' + a_2 u = A \cos kt$$

where a_1, a_2 are constants, which arises in several applications.

We wish to study the question of the existence of periodic solutions of (2.74). The following general result holds.

Theorem 2.13. *A solution $\boldsymbol{\phi}(t)$ of (2.74) is periodic of period ω in t if and only if $\boldsymbol{\phi}(\omega) = \boldsymbol{\phi}(0)$.*

Proof. If $\boldsymbol{\phi}$ is periodic of period ω, it is obvious that $\boldsymbol{\phi}(\omega) = \boldsymbol{\phi}(0)$. Conversely suppose $\boldsymbol{\phi}(t)$ is a solution of (2.74) such that $\boldsymbol{\phi}(\omega) = \boldsymbol{\phi}(0)$. Consider the functions $\boldsymbol{\phi}(t)$ and $\boldsymbol{\psi}(t) = \boldsymbol{\phi}(t + \omega)$. Then $\boldsymbol{\phi}$ and $\boldsymbol{\psi}$ are both solutions of (2.74) and $\psi(0) = \phi(\omega) = \phi(0)$. Thus the solutions ϕ and ψ have the same initial values and by the uniqueness theorem (Theorem 2.1) $\phi(t) \equiv \psi(t) = \phi(t + \omega)$, $-\infty < t < \infty$, which proves the periodicity. ▮

A more useful criterion for periodicity of solution is the following one.

Theorem 2.14. *The system* (2.74) *has a periodic solution of period* ω *for any periodic forcing vector* $\mathbf{g}$ *of period* ω *if and only if the homogeneous system* $\mathbf{y}' = A(t)\mathbf{y}$ *has no periodic solution of period* ω *except the trivial solution* $\mathbf{y} \equiv \mathbf{0}$.

Proof. Let $\Phi(t)$ be the fundamental matrix of the homogeneous system (2.58) that is the identity matrix at $t = 0$. In fact, Φ is given by (2.61), p. 96. By the variation of constants formula (Theorem 2.6) every solution $\psi(t)$ of (2.74) has the form

$$\psi(t) = \Phi(t)\psi(0) + \Phi(t) \int_0^t \Phi^{-1}(s)\mathbf{g}(s)\, ds$$

By Theorem 2.13, the solution ψ will be periodic if and only if $\psi(0) = \psi(\omega)$. But

$$\psi(\omega) = \Phi(\omega)\psi(0) + \Phi(\omega) \int_0^\omega \Phi^{-1}(s)\mathbf{g}(s)\, ds$$

and the periodicity condition $\psi(\omega) = \psi(0)$ becomes

$$[E - \Phi(\omega)]\psi(0) = \Phi(\omega) \int_0^\omega \Phi^{-1}(s)\mathbf{g}(s)\, ds$$

This is a linear nonhomogeneous system of algebraic equations for the components of the vector $\psi(0)$, which must be solvable for every periodic forcing vector $\mathbf{g}$. This is possible if and only if $\det(E - \Phi(\omega)) \neq 0$. Thus (2.60) and the definition of multipliers says that the solution ψ of (2.74) is periodic if and only if 1 is not a multiplier of the homogeneous system $\mathbf{y}' = A(t)\mathbf{y}$. This combined with Corollary 2 to Theorem 2.12, using $k = 1$, implies the result. ▮

The reader should note that Theorem 2.14 is a result about all periodic forcing terms of fixed period. It may happen that the homogeneous system has a periodic solution, and yet for **some** forcing terms $\mathbf{g}$, so does the nonhomogeneous system. This is exhibited by the following example.

• EXERCISE

8. Discuss this phenomenon for the scalar equation $u'' + u = \sin 2t$, which has $u = -\frac{1}{3}\sin 2t$ as a periodic solution of period 2π (note the least period is π, but this is not the issue). Can you suggest other examples?

• MISCELLANEOUS EXERCISES

1. What is wrong with the following calculation for an arbitrary continuous matrix $A(t)$?

$$\frac{d}{dt}\left[\exp \int_{t_0}^{t} A(s)\,ds\right] = A(t)\exp\left[\int_{t_0}^{t} A(s)\,ds\right]$$

so that $\exp\left(\int_{t_0}^{t} A(s)\,ds\right)$ is a fundamental matrix of $\mathbf{y}' = A(t)\mathbf{y}$ for an arbitrary continuous matrix $A(t)$.

2. Find a fundamental matrix for the system $\mathbf{y}' = A\mathbf{y}$, where A is the matrix.

(a) $A = \begin{pmatrix} 2 & 1 \\ 3 & 4 \end{pmatrix}$ (b) $A = \begin{pmatrix} -1 & 8 \\ 1 & 1 \end{pmatrix}$

(c) $A = \begin{pmatrix} 1 & 1 \\ 3 & -2 \end{pmatrix}$ (d) $A = \begin{pmatrix} 2 & -3 \\ 1 & -2 \end{pmatrix}$

(e) $A = \begin{pmatrix} 5 & 3 \\ -3 & -1 \end{pmatrix}$ (f) $A = \begin{pmatrix} 1 & -1 & 1 \\ 1 & 1 & -1 \\ 2 & -1 & 0 \end{pmatrix}$

(g) $A = \begin{pmatrix} -3 & 4 & -2 \\ 1 & 0 & 1 \\ 6 & -6 & 5 \end{pmatrix}$ (h) $A = \begin{pmatrix} 2 & 1 & 0 \\ 1 & 3 & -1 \\ -1 & 2 & 3 \end{pmatrix}$

(i) $A = \begin{pmatrix} 4 & -1 & -1 \\ 1 & 2 & -1 \\ 1 & -1 & 2 \end{pmatrix}$ (j) $A = \begin{pmatrix} -1 & 1 & -2 \\ 4 & 1 & 0 \\ 2 & 1 & -1 \end{pmatrix}$

(k) $A = \begin{pmatrix} 2 & 1 & 0 \\ 0 & 2 & 4 \\ 1 & 0 & -1 \end{pmatrix}$ (l) $A = \begin{pmatrix} 2 & -1 & -1 \\ 2 & -1 & -2 \\ -1 & 1 & 2 \end{pmatrix}$

(m) $A = \begin{pmatrix} 4 & -1 & 0 \\ 3 & 1 & -1 \\ 1 & 0 & 1 \end{pmatrix}$

3. Sketch the phase portrait for each of the systems in Exercises 2a–2e and determine in each case whether the origin is a node, saddle point, spiral point, or center. For which of these is the origin an attractor?

4. Find the general solution of the system $\mathbf{y}' = A\mathbf{y} + \mathbf{b}(t)$ in each of the following cases

(a) $A = \begin{pmatrix} 0 & 1 \\ 1 & 0 \end{pmatrix}$, $\mathbf{b}(t) = \begin{pmatrix} 2e^t \\ t^2 \end{pmatrix}$.

(b) $A = \begin{pmatrix} -1 & 2 \\ -2 & 3 \end{pmatrix}$, $\mathbf{b}(t) = \begin{pmatrix} 1 \\ 0 \end{pmatrix}$.

(c) $A = \begin{pmatrix} 4 & -3 \\ 2 & -1 \end{pmatrix}$, $\mathbf{b}(t) = \begin{pmatrix} \sin t \\ -2\cos t \end{pmatrix}$.

(d) $A = \begin{pmatrix} 2 & 1 \\ 1 & 2 \end{pmatrix}$, $\mathbf{b}(t) = \begin{pmatrix} 2e^t \\ -3e^{4t} \end{pmatrix}$.

(e) $A = \begin{pmatrix} 1 & -1 \\ 2 & -1 \end{pmatrix}$, $\mathbf{b}(t) = \begin{pmatrix} \dfrac{1}{\cos t} \\ 0 \end{pmatrix}$.

5. Suppose m is not an eigenvalue of the matrix A. Show that the nonhomogeneous system

$$\mathbf{y}' = A\mathbf{y} + \mathbf{c}e^{mt}$$

has a solution of the form

$$\boldsymbol{\phi}(t) = \mathbf{p}e^{mt}$$

and calculate the vector $\mathbf{p}$ in terms of A and $\mathbf{c}$.

6. Suppose m is not an eigenvalue of the matrix A. Show that the nonhomogeneous system

$$\mathbf{y}' = A\mathbf{y} + \sum_{j=0}^{k} \mathbf{c}_j t^j e^{mt}$$

has a solution of the form

$$\boldsymbol{\phi}(t) = \sum_{j=0}^{k} \mathbf{p}_j t^j e^{mt}$$

[*Hint:* Show that $\mathbf{p}_j$ satisfies the algebraic system

$$(A - mE)\mathbf{p}_k = -\mathbf{c}_k$$

$$(A - mE)\mathbf{p}_j = (j+1)\mathbf{p}_{j+1} - \mathbf{c}_j \qquad (j = 0, 1, \ldots, k-1)$$

and that these systems can be solved recursively.]

7. Consider the system

$$t\mathbf{y}' = A\mathbf{y}$$

where A is a constant matrix. Show that $t^A = e^{A \log t}$ is a fundamental matrix for $t \neq 0$ in two ways: (i) by direct substitution; (ii) by making the change of variable $|t| = e^s$.

8. Find the general solution of the system

$$ty' = Ay + \mathbf{b}(t)$$

9. Consider the system of differential equations

$$y_1'' - 3y_1' + 2y_1 + y_2' - y_2 = 0$$

$$y_1' - 2y_1 + y_2' + y_2 = 0$$

(a) Show that this system is equivalent to the system of first-order equations $\mathbf{u}' = A\mathbf{u}$, where

$$\mathbf{u} = \begin{pmatrix} u_1 \\ u_2 \\ u_3 \end{pmatrix} = \begin{pmatrix} y_1 \\ y_1' \\ y_2 \end{pmatrix}, \qquad A = \begin{pmatrix} 0 & 1 & 0 \\ -4 & 4 & 2 \\ 2 & -1 & -1 \end{pmatrix}$$

(b) Find a fundamental matrix for the system in part (a).
(c) Find the general solution of the original system.
(d) Find the solution of the original system satisfying the initial conditions

$$y_1(0) = 0,\ y_1'(0) = 1,\ y_2(0) = 0.$$

10. Repeat the procedure of Exercise 8 for the system

$$y_1'' + y_2'' - y_2' + y_2 = 0$$

$$y_1' + y_1 + y_2'' + y_2' = 0$$

In part (d) find that solution satisfying the initial conditions $y_1(0) = 0$, $y_1'(0) = 1$, $y_2(0) = 0$, $y_2'(0) = 2$.

11. Consider the matrix differential equation

$$Y' = AY + YB$$

where A, B, and Y are $n \times n$ matrices.

(a) Show that the solution satisfying the initial condition $Y(0) = C$, where C is a given $n \times n$ matrix, is given by

$$Y(t) = e^{At}Ce^{Bt}$$

(b) Show that

$$Z = -\int_0^\infty e^{At}Ce^{Bt}\,dt$$

is the unique solution of the matrix equation

$$AX + XB = C$$

whenever the integral exists.

(c) Show that the integral for Z in part (b) exists if all eigenvalues of both A and B have negative real parts.

12. Let $Y(t)$ be the solution of the matrix differential equation

$$Y' = A(t)Y, \qquad Y(0) = E$$

and let $Z(t)$ be the solution of the matrix differential equation

$$Z' = ZB(t) \qquad Z(0) = E$$

where $A(t)$ and $B(t)$ are continuous on an interval I containing the origin. Show that the solution of the matrix differential equation

$$X' = A(t)X + XB(t) \qquad X(0) = C$$

for any given constant matrix C is $Y(t)CZ(t)$.

13. (a) Consider the electrical circuit shown in Figure 2.12, with currents and voltages (with polarities) as shown. Use Kirchhoff's laws successively at the nodes A, B, C, and show that the circuit is governed by the system

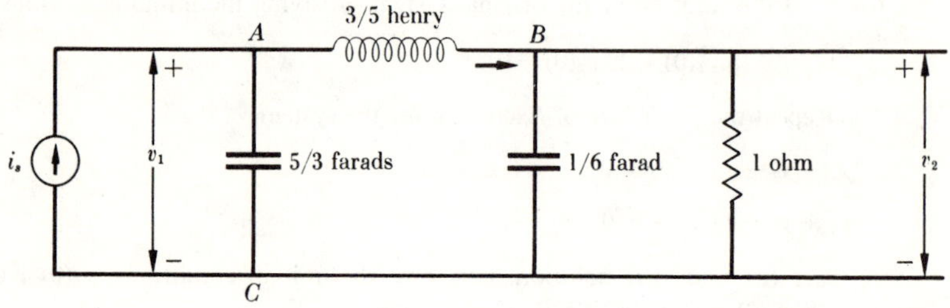

Figure 2.12

$$\frac{5}{3} v_1' = -i_1 + i_s$$

$$\frac{1}{6} v_2' = i_1 - v_2$$

$$\frac{3}{5} i_1' = v_1 - v_2$$

for the unknowns v_1, v_2, i_1; i_s is known.

(b) Find the general solution of the system derived in part (a).

REMARK. Consider the mechanical system shown in Figure 2.13, consisting of two masses m_1 and m_2 connected by a spring with a spring constant k, sliding on frictionless supports. A force $F(t)$ is applied to m_2 and m_1 is connected to a rigid wall by a dashpot (resistance) with damping constant p. Let y_1 and y_2 denote the positions of m_1 and m_2, respectively, and define $v_1 = y_1'$, $v_2 = y_2'$,

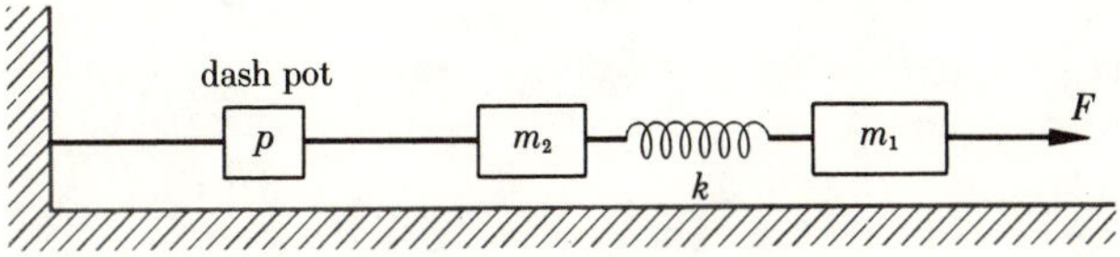

Figure 2.13

$z = y_2 - y_1$. Then the motion of the system is governed by the system of differential equations

$$m_1 v_1' = kz + F(t)$$

$$z' = v_2 - v_1$$

$$m_2 v_2' = pv_2 + kz = 0$$

which is equivalent to the circuit equations of part (a) if we make the identifications $F = i_S$, $-zk = i_1$, $m_1 = 5/3$, $m_2 = 1/6$, $p = 1$, $k = 5/3$.

14. A weight of mass m is connected to a rigid wall by a spring with spring constant k. A second weight of mass m is connected to this weight by a second spring with spring constant k. A force F is applied to this second weight. The whole system slides on a frictionless table (see Figure 2.14). Let y denote the displacement of the first weight from equilibrium and let z denote the displacement of the second weight from equilibrium.

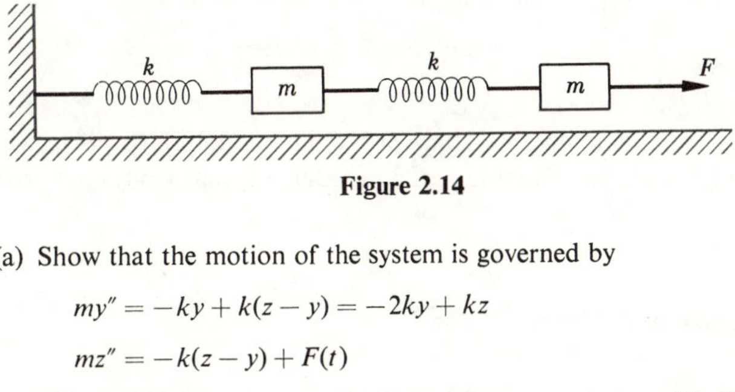

Figure 2.14

(a) Show that the motion of the system is governed by

$$my'' = -ky + k(z - y) = -2ky + kz$$

$$mz'' = -k(z - y) + F(t)$$

(b) Show that the solution of the homogeneous system, with $F(t) \equiv 0$, is a superposition of two simple harmonic motions with natural frequencies

$$\frac{1}{2\pi}\left(\frac{3+\sqrt{5}}{2}\right)^{1/2}\left(\frac{k}{m}\right)^{1/2} \quad \text{and} \quad \frac{1}{2\pi}\left(\frac{3-\sqrt{5}}{2}\right)^{1/2}\left(\frac{k}{m}\right)^{1/2}$$

(c) Obtain an expression for the general solution of the nonhomogeneous system.

Chapter 3 EXISTENCE THEORY

This chapter is devoted to the study of existence, uniqueness, and continuity properties of solutions of the system

$$\mathbf{y}' = \mathbf{f}(t, \mathbf{y})$$

In the course of this study, we shall prove Theorem 1.1 (p. 26), which is applicable to the study of mathematical models for many physical problems.

3.1 Existence in the Scalar Case

To simplify the exposition, we begin with the problem of proving the existence of a solution ϕ of the scalar differential equation

$$y' = f(t, y) \tag{3.1}$$

satisfying the initial conditions

$$\phi(t_0) = y_0 \tag{3.2}$$

on some interval containing t_0. This is called a **local problem**, since it is concerned only with existence of solutions near the initial point (t_0, y_0). To treat this problem we make certain hypotheses on f in some rectangle centered at (t_0, y_0). This will mean that we can apply the local result of this section at every point in a region D in which f satisfies these hypotheses.

Suppose f is continuous in D and that (t_0, y_0) is an arbitrary point of D. The first step in our development is the observation that the initial value problem (3.1), (3.2) is **equivalent** to the problem of finding a **continuous** function $y(t)$, defined in some interval I containing t_0, such that $y(t)$ satisfies **the integral equation***

$$y(t) = y_0 + \int_{t_0}^{t} f(s, y(s))\, ds \qquad (t \in I) \tag{3.3}$$

This equivalence is made precise as follows.

Lemma 3.1. *If ϕ is a solution of the initial value problem* (3.1), (3.2) *on an interval I, then ϕ satisfies* (3.3) *on I. Conversely, if $y(t)$ is a solution of* (3.3) *on some interval J containing t_0, then $y(t)$ satisfies* (3.1) *on J and also the initial condition* (3.2).

Proof. If ϕ is a solution of (3.1) on I satisfying (3.2), we have

$$\phi'(t) = f(t, \phi(t)) \qquad (t \in I)$$

and integrating from t_0 to any t on I, we obtain

$$\phi(t) - \phi(t_0) = \int_{t_0}^{t} f(s, \phi(s))\, ds$$

Imposing the initial condition (3.2) we see that ϕ satisfies (3.3).

Conversely, if $y(t)$ is a continuous solution of (3.3), then by the continuity of the function $f(s, y(s))$ under the integral in (3.3) $y(t)$ is differentiable. Thus by the fundamental theorem of calculus applied to (3.3) we have that $y(t)$ satisfies

$$y'(t) = f(t, y(t)) \qquad (t \in J)$$

and putting $t = t_0$ in (3.3), we have $y(t_0) = y_0$. This completes the proof. ∎

---

* Equation (3.3) is called an integral equation (of Volterra type) because the unknown function appears both under and outside the integral sign.

Lemma 3.1 permits us to establish existence of a solution of (3.1), (3.2) by proving existence of a solution of (3.3). This is important because integrals are in general easier to estimate than derivatives.

• EXERCISES

1. Determine the integral equation equivalent to the initial value problem

$$y' = t^2 + y^4 \qquad y(0) = 1$$

2. Prove that the initial value problem

$$y'' + g(t, y) = 0 \quad y(0) = y_0 \quad y'(0) = z_0 \tag{3.4}$$

where g is continuous in some region D containing $(0, y_0)$, is equivalent to the integral equation

$$y(t) = y_0 + z_0 t - \int_0^t (t-s)g(s, y(s))\, ds \tag{3.5}$$

[*Hint:* To show that if ϕ is a solution of (3.4) on I, then ϕ satisfies (3.5) on I, integrate (3.4) twice and use the fact that

$$\int_0^t \left\{ \int_0^s g(\tau, \phi(\tau))\, d\tau \right\} ds = \int_0^t \left\{ \int_\tau^t ds \right\} g(\tau, \phi(\tau))\, d\tau$$

$$= \int_0^t (t-\tau) g(\tau, \phi(\tau))\, d\tau$$

To prove that a solution of (3.5) is a solution of (3.4), proceed as in the proof of Lemma 3.1. But now you will need to use the formula

$$\frac{d}{dt} \int_0^t H(t, s)\, ds = H(t, t) + \int_0^t \frac{\partial H}{\partial t}(t, s)\, ds$$

which is easily proved by the chain rule, assuming only that H, $\partial H/\partial t$ are continuous on some rectangle containing $s = t = 0$.]

3. Construct an equivalent integral equation to the initial value problem

$$y'' + \mu^2 y = g(t, y), \quad y(0) = y_0, \quad y'(0) = z_0$$

assuming that g is continuous in a region D containing $(0, y_0)$ and where $\mu > 0$ is a constant. [*Hint*: Assuming a solution ϕ of the differential equation on an interval I that satisfies the initial conditions, apply the variation of constants formula (Exercise 6, Section 2.4, p. 54). To prove the converse, proceed as in Exercise 2.

Answer:

$$y(t) = y_0 \cos \mu t + \frac{z_0}{\mu} \sin \mu t + \int_0^t \frac{\sin \mu(t-s)}{\mu} g(s, y(s))\, ds.]$$

4. Prove that if ϕ is a solution of the integral equation

$$y(t) = e^{it} + \alpha \int_t^\infty \sin(t - s) \frac{y(s)}{s^2} \, ds$$

(assuming the existence of the integral), then ϕ satisfies the differential equation $y'' + (1 + \alpha/t^2)y = 0$.

Returning to the main question of proving the existence of solutions of (3.3) (and thereby of (3.1), (3.2)), we outline a plausible method of attacking this problem. We start by using the constant function $\phi_0(t) = y_0$ as an approximation to a solution. We substitute this approximation into the right side of (3.3) and use the result

$$\phi_1(t) = y_0 + \int_{t_0}^t f(s, \phi_0(s)) \, ds$$

as a next approximation to a solution. Then we substitute this approximation $\phi_1(t)$ into the right side of (3.3) to obtain what we hope is a still better approximation $\phi_2(t)$ given by

$$\phi_2(t) = y_0 + \int_{t_0}^t f(s, \phi_1(s)) \, ds$$

and we continue the process. Our goal is to find a function ϕ with the property that when it is substituted in the right side of (3.3), the result is the same function ϕ. If we continue our approximation procedure, we may hope that the sequence of functions $\{\phi_k(t)\}$, called **successive approximations**, converges to a limit function that has this property. Under suitable hypotheses this is the case, and precisely this approach will be used to prove the existence of a solution of the integral equation (3.3).

- **EXERCISE**

5. Construct the successive approximations to the solution ϕ of the differential equation $y' = -y$ that satisfies $\phi(0) = 2$. Do these successive approximations converge to a familiar function, and if so, is this function a solution of the problem?

We will consider the problem (3.1), (3.2) first with f and $\partial f/\partial y$ continuous on a closed rectangle $R = \{(t, y) | \, |t - t_0| \le a, |y - y_0| \le b\}$ centered at (t_0, y_0). Thus the functions f and $\partial f/\partial y$ are bounded on R, and there exist constants

$M > 0$, $K > 0$ such that

$$|f(t, y)| \leq M, \qquad \left|\frac{\partial f}{\partial y}(t, y)\right| \leq K \tag{3.6}$$

for all points (t, y) in R. If (t, y_1) and (t, y_2) are two points in R, then by the mean value theorem, there exists a number η between y_1 and y_2 such that

$$f(t, y_2) - f(t, y_1) = \frac{\partial f}{\partial y}(t, \eta)(y_2 - y_1)$$

Since the point (t, η) is also in R, $|\partial f/\partial y(t, \eta)| \leq K$, and we obtain

$$|f(t, y_2) - f(t_1, y_1)| \leq K|y_2 - y_1| \tag{3.7}$$

valid whenever (t, y_1) and (t, y_2) are in R.

Definition. *A function f that satisfies an inequality of the form* (3.7) *for all* (t, y_1), (t, y_2) *in a region D is said to satisfy a Lipschitz condition in D.*

The above argument shows that if f and $\partial f/\partial y$ are continuous on R, then f satisfies a Lipschitz condition in R. It is possible for f to satisfy a Lipschitz condition in a region without having a continuous partial derivative with respect to y there, for example, $f(t, y) = t\,|y|$ defined in any region containing $(0, 0)$. In this chapter, we assume the continuity of $\partial f/\partial y$ for simplicity, but we could instead assume that f satisfies a Lipschitz condition without substantial changes in the proofs.

Example 1. If $f(t, y) = y^{1/3}$ in the rectangle $R = \{(t, y) \mid |t| \leq 1, |y| \leq 2\}$, then f does not satisfy a Lipschitz condition in R.

To establish this, we need only to produce a suitable pair of points for which (3.7) fails to hold with any constant K. Consider the points

$$(t, y_1), \quad (t, 0), \qquad \text{with} \qquad -1 \leq t \leq 1, \quad y_1 > 0$$

Then

$$\frac{f(t, y_1) - f(t, 0)}{y_1 - 0} = \frac{y_1^{1/3}}{y_1} = y_1^{-2/3}$$

Now, choosing $y_1 > 0$ sufficiently small, it is clear that $K = y_1^{-2/3}$ can be made larger than any preassigned constant. Therefore (3.7) fails to hold for any K.

• EXERCISES

6. Compute a Lipschitz constant K as in (3.7), and then show that each of the following functions f satisfies the Lipschitz condition in the regions indicated.
   (a) $f(t, y) = t^2 + y^4$, $\{(t, y) | \, |t| \le 1, |y| \le 3\}$.
   (b) $f(t, y) = p(t) \cos y + q(t) \sin y$, $\{(t, y) | \, |t| \le 100, |y| < \infty\}$, where p, q are continuous functions on $-100 \le t \le 100$.
   (c) $f(t, y) = t \exp(-y^2)$, $\{(t, y) | \, |t| \le 1, |y| < \infty\}$.
7. Show that $f(t, y) = t|y|$ satisfies a Lipschitz condition in the region $\{(t, y) | \, |t| \le 1, |y| < \infty\}$.

We have already indicated that we will use an approximation procedure to establish the existence of solutions. Now let us define the **successive approximations** in the general case by the equations

$$\phi_0(t) = y_0$$

$$\phi_{j+1}(t) = y_0 + \int_{t_0}^{t} f(s, \phi_j(s))\, ds \qquad (j = 0, 1, 2, \ldots) \tag{3.8}$$

Before we can do anything with these successive approximations, we must show that they are defined properly. This means that in order to define ϕ_{j+1} on some interval I, we must first know that the point $(s, \phi_j(s))$ remains in the rectangle R for every s in I.

Lemma 3.2. *Define α to be the smaller of the positive numbers a and b/M. Then the successive approximations ϕ_j given by* (3.8) *are defined on the interval $I = \{t | \, |t - t_0| \le \alpha\}$, and on this interval*

$$|\phi_j(t) - y_0| \le M|t - t_0| \le b, \qquad (j = 0, 1, 2, \ldots) \tag{3.9}$$

Proof. The proof is by induction. It is obvious that $\phi_0(t)$ is defined on I and satisfies (3.9) with $j = 0$ on I. Now assume that for any $j = n \ge 1$, ϕ_n is defined and satisfies (3.9) on I (then, of course, the point $(t, \phi_n(t))$ remains in R for $t \in I$). Then by (3.8) ϕ_{n+1} is defined on I. To complete the proof we need to show that for $t \in I$, $\phi_{n+1}(t)$ remains in R, or analytically that ϕ_{n+1} satisfies (3.9) with $j = n + 1$. But from (3.8), the induction hypothesis, and (3.6) we have

$$|\phi_{n+1}(t) - y_0| = \left| \int_{t_0}^{t} f(s, \phi_n(s))\, ds \right| \le \left| \int_{t_0}^{t} |f(s, \phi_n(s))|\, ds \right|$$
$$\le M|t - t_0| \le M\alpha \le b$$

This establishes the lemma. ∎

In order to explain the choice of α in Lemma 3.2, we observe that the condition $|f(t, y)| \le M$ implies that a solution ϕ of (3.1), (3.2) cannot cross the lines of slope M and $-M$ through the initial point (t_0, y_0). The relation (3.9) established in the above lemma says that the successive approximations ϕ_j do not cross these lines either. The length of the interval I depends on where these lines meet the rectangle R. If they meet the vertical sides of the rectangle (Figure 3.1), then we define $\alpha = a$, while if they meet the top and bottom of the rectangle (Figure 3.2), then we define $\alpha = b/M$. In either case, all the successive approximations remain in the triangles indicated in the figures.

We can now state and prove the fundamental local existence theorem.

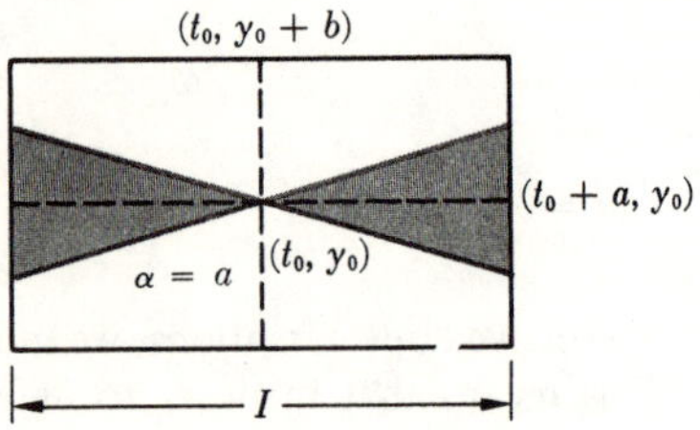

Figure 3.1

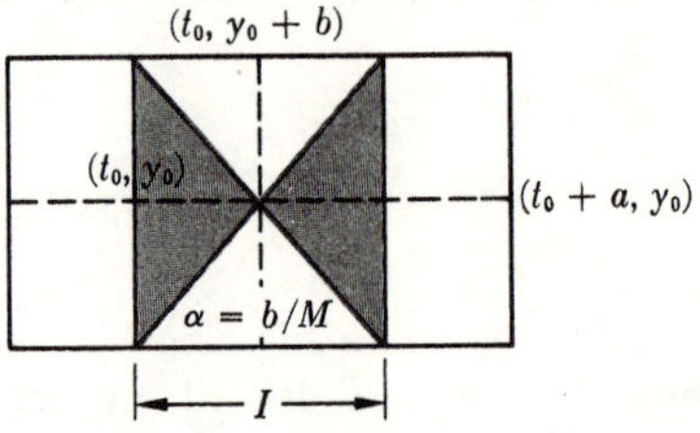

Figure 3.2

Theorem 3.1. *Suppose f and $\partial f/\partial y$ are continuous on the closed rectangle R and satisfy the bounds* (3.6). *Then the successive approximations ϕ_j, given by* (3.8), *converge (uniformly) on the interval $I = \{t \,|\, |t - t_0| \le \alpha\}$, to a solution ϕ of the differential equation* (3.1) *that satisfies the initial conditions* (3.2).

Proof. Lemma 3.2 shows that the successive approximations ϕ_j are defined on the interval I. To prove the convergence of the sequence $\{\phi_j\}$ on I we write the obvious identity

$$\phi_j(t) = \phi_0(t) + [\phi_1(t) - \phi_0(t)] + \cdots + [\phi_j(t) - \phi_{j-1}(t)]$$

$$= \phi_0(t) + \sum_{m=1}^{j-1} [\phi_{m+1}(t) - \phi_m(t)]$$

The next step is to estimate the difference between ϕ_j and ϕ_{j+1}. We work on the interval $t_0 \le t \le t_0 + \alpha$ to the right of t_0, but the argument can easily be modified to give the result on the interval $t_0 - \alpha \le t \le t_0$. We define

$$r_j(t) = |\phi_{j+1}(t) - \phi_j(t)|, \qquad (j = 0, 1, 2, \ldots)$$

Then, using the definition (3.8) and the Lipschitz condition (3.7), we have

$$r_j(t) = |\phi_{j+1}(t) - \phi_j(t)| = \left| \int_{t_0}^{t} [f(s, \phi_j(s)) - f(s, \phi_{j-1}(s))] \, ds \right|$$

$$\leq \int_{t_0}^{t} |f(s, \phi_j(s)) - f(s, \phi_{j-1}(s))| \, ds$$

$$\leq K \int_{t_0}^{t} |\phi_j(s) - \phi_{j-1}(s)| \, ds$$

$$= K \int_{t_0}^{t} r_{j-1}(s) \, ds, \qquad (j = 1, 2, \ldots) \tag{3.10}$$

The case $j = 0$ is slightly different. We have, from (3.6)

$$r_0(t) = |\phi_1(t) - \phi_0(t)| = \left| \int_{t_0}^{t} f(s, \phi_0(s)) \, ds \right|$$

$$\leq \int_{t_0}^{t} |f(s, \phi_0(s))| \, ds \leq M(t - t_0) \tag{3.11}$$

From (3.10) and (3.11) we will prove by induction that

$$r_j(t) \leq \frac{MK^j(t - t_0)^{j+1}}{(j+1)!}, \qquad (j = 0, 1, 2, \ldots ; t_0 \leq t \leq t_0 + \alpha) \tag{3.12}$$

The case $j = 0$ of (3.12) is already established. Assume that (3.12) is true for $j = p - 1$ for some integer $p > 1$; then (3.10) gives, on using the induction hypothesis,

$$r_p(t) \leq K \int_{t_0}^{t} r_{p-1}(s) \, ds \leq K \int_{t_0}^{t} \frac{MK^{p-1}(s - t_0)^p}{p!} \, ds$$

$$= \frac{MK^p(t - t_0)^{p+1}}{(p+1)!}, \qquad (t_0 \leq t \leq t_0 + \alpha)$$

which is (3.12) for $j = p$. This proves (3.12).

• **EXERCISE**

8. Prove the analogue of the inequality (3.12) for the interval $t_0 - \alpha \leq t \leq t_0$.

Combining (3.12) with the result of Exercise 8, we have

$$r_j(t) \le \frac{MK^j|t-t_0|^{j+1}}{(j+1)!} = \frac{M[K|t-t_0|]^{j+1}}{K(j+1)!} \le \frac{M(K\alpha)^{j+1}}{K(j+1)!} \qquad (3.13)$$

$$(j = 0, 1, 2, \dots; |t-t_0| \le \alpha)$$

It follows from (3.13) that the series $\sum_{j=0}^{\infty} r_j(t)$ is dominated on the interval $|t-t_0| \le \alpha$ by the series of positive constants $M/K \sum_{j=0}^{\infty} (K\alpha)^{j+1}/(j+1)!$, which converges to $(M/K)e^{K\alpha}$. By the comparison test, the series $\sum_{j=0}^{\infty} r_j(t)$ converges (in fact, uniformly) on the interval $I = \{t \,|\, |t-t_0| \le \alpha\}$. In view of the definition of the r_j, this implies the absolute (and uniform) convergence on $|t-t_0| \le \alpha$ of the series $\sum_{j=0}^{\infty}[\phi_{j+1}(t) - \phi_j(t)]$.

Since $\phi_j(t) = \phi_0(t) + \sum_{m=0}^{j-1} [\phi_{m+1}(t) - \phi_m(t)]$, this also proves the convergence of the sequence $\{\phi_j(t)\}$ for every t in the interval I to some function of t, which we call $\phi(t)$. We will show that the function ϕ is continuous and satisfies the integral equation (3.3) on I.

From the definition of $\phi(t)$,

$$\phi(t) = \phi_0(t) + \sum_{n=0}^{\infty} (\phi_{n+1}(t) - \phi_n(t))$$

therefore, using $\phi_j = \phi_0 + \sum_{m=1}^{j-1} [\phi_{m+1} - \phi_m]$,

$$\phi(t) - \phi_j(t) = \sum_{n=j}^{\infty} (\phi_{n+1}(t) - \phi_n(t)).$$

Now from (3.13)

$$|\phi(t) - \phi_j(t)| \le \sum_{n=j}^{\infty} |\phi_{n+1}(t) - \phi_n(t)| \le \sum_{n=j}^{\infty} r_n(t)$$

$$\le \frac{M}{K} \sum_{n=j}^{\infty} \frac{(K\alpha)^{n+1}}{(n+1)!} \le \frac{M}{K} \frac{(K\alpha)^{j+1}}{(j+1)!} \sum_{n=0} \frac{(K\alpha)^n}{n!} \qquad (3.14)$$

$$= \frac{M}{K} \frac{(K\alpha)^{j+1}}{(j+1)!} e^{K\alpha}, \qquad (t \in I)$$

It is an elementary exercise to see that

$$\varepsilon_j = \frac{(K\alpha)^{j+1}}{(j+1)!} \to 0 \quad \text{as} \quad j \to \infty$$

To prove the continuity of $\phi(t)$ on I, let $\varepsilon > 0$ be given. We have $\phi(t+h) - \phi(t) = \phi(t+h) - \phi_j(t+h) + \phi_j(t+h) - \phi_j(t) + \phi_j(t) - \phi(t)$, and thus

$$|\phi(t+h) - \phi(t)| \le |\phi(t+h) - \phi_j(t+h)| + |\phi_j(t+h) - \phi_j(t)| + |\phi_j(t) - \phi(t)| \le 2\varepsilon_j + |\phi_j(t+h) - \phi_j(t)|$$

by the above estimate. Choosing j sufficiently large and $|h|$ sufficiently small, and using $\lim_{j\to\infty} \varepsilon_j = 0$ and the continuity of the $\phi_j(t)$, we can make

$$|\phi(t+h) - \phi(t)| < \varepsilon$$

We now wish to show that the limit function $\phi(t)$ satisfies the integral equation (3.3). We will do this by letting $j \to \infty$ in the definition (3.8) of the successive approximations and by showing that

$$\lim_{j\to\infty} \int_{t_0}^{t} f(s, \phi_j(s))\, ds = \int_{t_0}^{t} f(s, \phi(s))\, ds \tag{3.15}$$

Once this is done the proof of the theorem is completed by applying Lemma 3.1. To prove (3.15), we have, using the Lipschitz condition (3.7) and the estimate (3.14),

$$\left| \int_{t_0}^{t} [f(s, \phi(s)) - f(s, \phi_j(s))]\, ds \right| \le K \left| \int_{t_0}^{t} |\phi(s) - \phi_j(s)|\, ds \right| \le \varepsilon_j \frac{M}{K} e^{K\alpha} \cdot K\alpha$$

and this approaches zero as $j \to \infty$ for every t on I. This establishes (3.15) and completes the proof. ∎

Incidentally, we have also established the following useful consequence in the course of the proof of (3.14).

Corollary. The error committed by stopping with the jth *approximation* $\phi_j(t)$ *satisfies the estimate*

$$|\phi(t) - \phi_j(t)| \le \frac{M}{K} \frac{(K\alpha)^{j+1}}{(j+1)!} e^{K\alpha}$$

for every t on I.

• EXERCISES

9. Construct the successive approximations to the solution ϕ of the differential equation $y' = y$, that satisfies $\phi(0) = 1$.

10. Construct the successive approximations to the solution ϕ of the problem in Exercise 9, but using $\phi_0(t) = \cos t$ instead of $\phi_0(t) = 1$. Do these successive approximations converge, and if so, what is their limit?

11. Construct the successive approximations $\phi_0, \phi_1, \phi_2, \phi_3$ to the solution ϕ of the differential equation $y' = \cos y$ that satisfies $\phi(0) = 0$.

12. Consider the integral equation

$$y(t) = y_0 + z_0 t - \int_0^t (t-s) g(s, y(s))\, ds \tag{3.4}$$

of Exercise 2, where $g(t, y)$, $\partial g/\partial y\ (t, y)$ are continuous on the rectangle $R = \{(t, y) \mid |t| \leq a, |y - y_0| \leq b\}$. (Thus they are automatically bounded on R.) Let $|g(t, y)| \leq M$, $|\partial g/\partial y(t, y)| \leq K$ for all $(t, y) \in R$. Define

$$\phi_0(t) = y_0$$

$$\phi_n(t) = y_0 + z_0 t - \int_0^t (t-s) g(s, \phi_{n-1}(s))\, ds \qquad (n = 1, 2, \ldots)$$

Show that (a) the ϕ_n are well defined for $|t| \leq \alpha$, where

$$\alpha = \min(a, b/\tilde{M}) \qquad \tilde{M} = |z_0| + M\frac{a}{2}$$

(b) $\{\phi_n\}$ converges to a solution of the integral equation (3.4) on $|t| \leq \alpha$. This together with Exercise 2 establishes the existence of solutions of the initial value problem $y'' + g(t, y) = 0$, $y(0) = y_0$, $y'(0) = z_0$.

13. Consider the integral equation

$$y(t) = e^{it} + \alpha \int_t^\infty \sin(t-s) \frac{y(s)}{s^2}\, ds$$

of Exercise 4. Define the successive approximations

$$\begin{cases} \phi_0(t) = 0 \\ \phi_n(t) = e^{it} + \alpha \displaystyle\int_t^\infty \sin(t-s) \frac{\phi_{n-1}(s)}{s^2}\, ds \qquad (1 \leq t < \infty) \end{cases}$$

(a) Show by induction that

$$|\phi_n(t) - \phi_{n-1}(t)| \leq \frac{|\alpha|^{n-1}}{(n-1)!\, t^{n-1}} \qquad (1 \leq t < \infty;\ n = 1, 2, \ldots)$$

Since $\phi_n(t) = \phi_0(t) + (\phi_1 - \phi_0) + \cdots + (\phi_n(t) - \phi_{n-1}(t))$ this shows that the ϕ_n are well defined for $1 \leq t < \infty$, and $\{\phi_n\}$ converge uniformly for $1 \leq t < \infty$ to a continuous limit function ϕ.

(b) Show that the limit function satisfies the given integral equation.

(c) Using

$$|\phi_n(t) \le |\phi_1(t) - \phi_0(t)| + \cdots + |\phi_n(t) - \phi_{n-1}(t)|$$

and the above estimate for $|\phi_n(t) - \phi_{n-1}(t)|$, show that the limit function satisfies the estimate

$$|\phi(t) \le e^{|\alpha|} \qquad 1 \le t < \infty$$

Theorem 3.1 is not the best possible result of its type. Under the hypotheses of Theorem 3.1, we can establish uniqueness of solutions of (3.1), (3.2), as we shall prove in Section 3.3. However, we may have existence of solutions without uniqueness. In fact, the following result is true.

Theorem 3.2. *Suppose f is continuous on the rectangle R, and suppose $|f(t, y)| \le M$ for all points (t, y) in R. Let α be the smaller of the positive numbers a* and *b/M. Then there is a solution ϕ of the differential equation* (3.1) *that satisfies the initial condition* (3.2) *existing on the interval $|t - t_0| < \alpha$.*

We shall make no attempt to prove Theorem 3.2, as its proof is considerably more difficult than the proof of Theorem 3.1. It cannot be proved by the method of successive approximations, as the successive approximations may not converge under the hypotheses of Theorem 3.2. A proof may be found in [4], Chapter 1, Theorem 1.2. That the hypotheses of Theorem 3.2 do not guarantee uniqueness is shown by the following example.

Example 2. Consider the equation $y' = 3y^{2/3}$, with $f(t, y) = 3y^{2/3}$, $(\partial f/\partial y(t, y)) = 2y^{-1/3}$. Since $\partial f/\partial y$ is not continuous for $y = 0$, we cannot apply Theorem 3.1 to deduce the existence of a solution of $y' = f(t, y)$ through the point $(0, 0)$. By the method of Example 1 (p. 112) we see that f does not satisfy a Lipschitz condition either. Since f is continuous in the whole (t, y) plane, we can apply Theorem 3.2 to this problem. In fact, there is an infinite number of solutions through $(0, 0)$. For each constant $c \ge 0$, the function ϕ_c defined by

$$\phi_c(t) = \begin{cases} 0 & (-\infty < t \le c) \\ (t - c)^3 & (c \le t < \infty) \end{cases}$$

is a solution of $y' = 3y^{2/3}$ through $(0, 0)$ (see Figure 3.3). In addition, the identically zero function is a solution of this initial value problem. Of course, for every initial point (t_0, y_0) with $y_0 \neq 0$ we have existence by Theorem 3.1 (and uniqueness by Theorem 3.4).

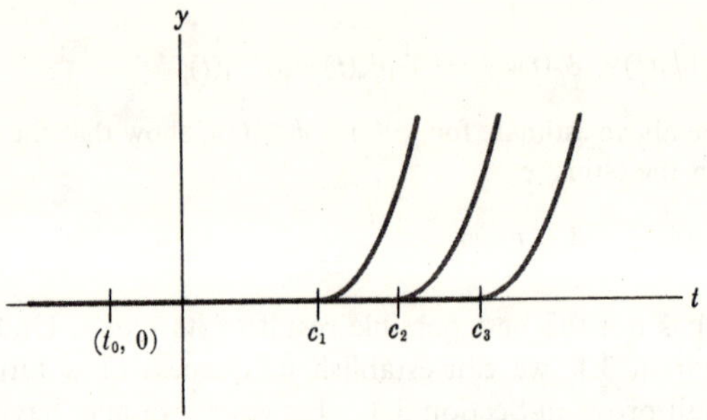

Figure 3.3

• EXERCISES

14. Do the successive approximations for solutions ϕ of $y' = 3y^{2/3}$ with $\phi(0) = 0$ converge to a solution?

15. Do the successive approximations for solutions of the problem considered in Exercise 14, but using

$$\phi_0(t) = \begin{cases} 0 & (0 \le t \le 1) \\ t - 1 & (1 \le t < \infty) \end{cases}$$

instead of $\phi_0(t) = 0$, converge to a solution?

If f and $\partial f/\partial y$ are continuous on a region D, not necessarily a rectangle, then given any point (t_0, y_0) in D we can construct a rectangle R lying entirely in D with center at (t_0, y_0). The hypotheses of Theorem 3.1 are then satisfied in R, and Theorem 3.1 gives us the existence of a solution $\phi(t)$ of $y' = f(t, y)$ through the point (t_0, y_0) on some interval about t_0. In fact, this solution may exist on a larger interval than the one constructed in the proof of Theorem 3.1. We shall return to this problem in Section 3.4.

Example 3. Consider the function f defined in the region D in the (t, y) plane, where D is given by $-\infty < t < 1$, $-\infty < y < \infty$, by

$$f(t, y) = \begin{cases} 0 & (-\infty < t \le 0, \ -\infty < y < \infty) \\ 2t & (0 < t < 1, \ -\infty < y < 0) \\ 2t - \dfrac{4y}{t} & (0 < t < 1, \ 0 \le y \le t^2) \\ -2t & (0 < t < 1, \ t^2 < y < \infty) \end{cases}$$

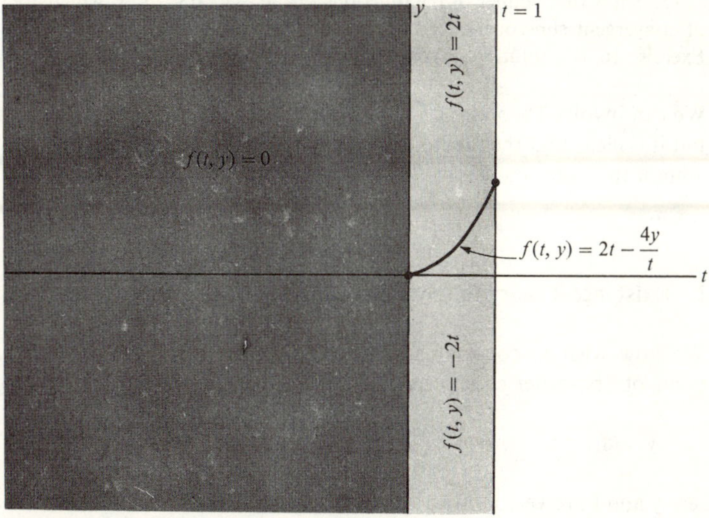

Figure 3.4

This function f is continuous and bounded by the constant 2 on D. The successive approximations to the solution ϕ of $y' = f(t, y)$ through the initial point (0, 0) are given by

$$\phi_0(t) = 0$$
$$\phi_{2k-1}(t) = t^2$$
$$\phi_{2k}(t) = -t^2 \qquad (0 \le t \le 1; k = 1, 2, \ldots)$$

Thus the successive approximations alternate between t^2 and $-t^2$ and do not converge.

• **EXERCISES**

16. Show by a direct computation that the successive approximations

$$\begin{cases} \phi_0(t) = 0 \\ \phi_n(t) = \int_0^t f(s, \phi_{n-1}(s))\, ds \end{cases}$$

where f is the function defined in Example 3, become

$$\phi_{2k-1}(t) = t^2$$
$$\phi_{2k}(t) = -t^2 \qquad (k = 0, 1, 2, \ldots; \quad 0 \le t \le 1)$$

17. Show that neither of the functions t^2 and $-t^2$, $0 \le t \le 1$, which are limits of convergent subsequences of successive approximations in Example 3 and in Exercise 16, is a solution of the problem $y' = f(t, y)$ $(0 \le t \le 1)$.

We can invoke Theorem 3.2 to give the existence of a solution in Example 3, but it is clear that the method of successive approximations cannot be used to obtain this solution.

3.2 Existence Theory for Systems of First-Order Equations

We now wish to consider the extension of the results of Section 3.1 to systems of first-order equations of the form

$$\mathbf{y}' = \mathbf{f}(t, \mathbf{y}) \tag{3.16}$$

where $\mathbf{y}$ and $\mathbf{f}$ are vectors with n components and where t is a scalar. Before proceeding, we remind the reader that because of the equivalence of single scalar differential equations of nth order and systems of first-order equations (established in Section 1.3, p. 14), every result that is established for (3.16) has an immediate interpretation for an nth-order scalar equation, or for that matter a system of such equations of any order.

In what follows D will represent a region in $n + 1$ dimensions (see Section 1.6, p. 24). Let $\mathbf{f}$ be continuously differentiable with respect to t and with respect to the components of $\mathbf{y}$ at all points of D (for short, we write $\mathbf{f} \in C'(D)$), and suppose that there exists a constant $K > 0$ such that the norms of $\partial\mathbf{f}/\partial y_j$ satisfy

$$\left|\frac{\partial \mathbf{f}}{\partial y_j}(t, \mathbf{y})\right| \le K \qquad (j = 1, \ldots, n) \tag{3.17}$$

for all $(t, \mathbf{y})$ in D. Such an inequality is automatically satisfied if $\mathbf{f} \in C'$, for example, on the closed "box" $B = \{(t, \mathbf{y}) \mid |t - t_0| \le a, |\mathbf{y} - \boldsymbol{\eta}| \le b\}$ for some fixed positive numbers a and b, or on any closed, bounded set in $(n + 1)$-dimensional space. It then follows that for any points $(t, \mathbf{y})$, $(t, \mathbf{z})$ in D we have the inequality

$$|\mathbf{f}(t, \mathbf{y}) - \mathbf{f}(t, \mathbf{z})| \le K|\mathbf{y} - \mathbf{z}| \tag{3.18}$$

This may be seen by applying the mean value theorem to each variable separately and then using (3.17), or by the following argument. Define the function $\mathbf{G}$ by

$$\mathbf{G}(\sigma) = \mathbf{f}(t, \mathbf{z} + \sigma(\mathbf{y} - \mathbf{z})) \qquad (0 \leq \sigma \leq 1)$$

and consider $\mathbf{f}(t, \mathbf{y}) - \mathbf{f}(t, \mathbf{z})$. We have

$$\mathbf{f}(t, \mathbf{y}) - \mathbf{f}(t, \mathbf{z}) = \mathbf{G}(1) - \mathbf{G}(0) = \int_0^1 \mathbf{G}'(\sigma)\, d\sigma.$$

By the chain rule, letting $\mathbf{f}_{y_j} = \partial \mathbf{f}/\partial y_j\ (j = 1, \ldots, n)$, we have

$$\mathbf{G}'(\sigma) = \mathbf{f}_{y_1}(t, \mathbf{z} + \sigma(\mathbf{y} - \mathbf{z}))(y_1 - z_1) + \mathbf{f}_{y_2}(t, \mathbf{z} + \sigma(\mathbf{y} - \mathbf{z}))(y_2 - z_2)$$
$$+ \cdots + \mathbf{f}_{yn}(t, \mathbf{z} + \sigma(\mathbf{y} - \mathbf{z}))(y_n - z_n)$$

Using the bound (3.17), we find

$$|\mathbf{f}(t, \mathbf{y}) - \mathbf{f}(t, \mathbf{z})| \leq \int_0^1 |\mathbf{G}'(\sigma)|\, d\sigma$$
$$\leq K\{|y_1 - z_1| + |y_2 - z_2| + |y_n - z_n|\} = K\,|\mathbf{y} - \mathbf{z}|$$

which is (3.18). A function $\mathbf{f}$ satisfying an inequality of the form (3.18) for any points $(t, \mathbf{y})$, $(t, \mathbf{z})$ in D is said to satisfy a **Lipschitz condition in** D with Lipschitz constant K. A function $\mathbf{f}$ satisfying (3.18) need not, of course, be of the class C' and all the remarks made in the simple case of scalar functions apply here.

The analogue of Theorem 3.1 is the following result.

Theorem 3.3. *Let* $\mathbf{f}$ *and* $\partial \mathbf{f}/\partial y_j$ $(j = 1, \ldots, n)$ *be continuous on the box* $B = \{(t, \mathbf{y}) \mid |t - t_0| \leq a, |\mathbf{y} - \boldsymbol{\eta}| \leq b\}$, *where a and b are positive numbers, and satisfying the bounds*

$$|\mathbf{f}(t, \mathbf{y})| \leq M, \qquad \left|\frac{\partial \mathbf{f}(t, \mathbf{y})}{\partial y_j}\right| \leq K \qquad (j = 1, \ldots, n) \tag{3.19}$$

for $(t, \mathbf{y})$ *in* B. *Let* α *be the smaller of the numbers a and* b/M *and define the successive approximations*

$$\begin{cases} \boldsymbol{\phi}_0(t) = \boldsymbol{\eta} \\ \boldsymbol{\phi}_n(t) = \boldsymbol{\eta} + \displaystyle\int_{t_0}^{t} \mathbf{f}(s, \boldsymbol{\phi}_{n-1}(s))\, ds \end{cases} \tag{3.20}$$

Then the sequence $\{\boldsymbol{\phi}_j\}$ of successive approximations converges (uniformly) on the interval $|t - t_0| \leq \alpha$ to a solution $\boldsymbol{\phi}(t)$ of (3.16), that satisfies the initial condition $\boldsymbol{\phi}(t_0) = \boldsymbol{\eta}$.

The choice of α is suggested by the same reasoning as in Theorem 3.1.

The proof is step by step, line by line the same as the proof of Theorem 3.1, with the scalars f, ϕ, y_0 replaced by the vectors $\mathbf{f}$, $\boldsymbol{\phi}$, $\boldsymbol{\eta}$, and in obvious places the absolute value is replaced by the norm. We remind the reader that the first step is to establish the equivalence of the initial value problem with the integral equation

$$\mathbf{y}(t) = \boldsymbol{\eta} + \int_{t_0}^{t} \mathbf{f}(s, \boldsymbol{\phi}(s))\, ds \tag{3.21}$$

and then work with (3.21). This is the analogue of Lemma 3.1.

• EXERCISES

1. Give a detailed proof of Theorem 3.3. (The reader is urged to carry out this proof with care, in order to appreciate the usefulness of introducing vectors.)
2. By writing the scalar equation $y^{(n)} = g(t, y, y', \ldots, y^{(n-1)})$ as a system of n first-order equations (see Section 1.3, p. 14), apply Theorem 3.3 to deduce an existence theorem for this scalar equation.
3. Given the system

$$y_1' = y_1^2 + y_2^2 + 1$$
$$y_2' = y_1^2 - y_2^2 - 1$$

Let $y = \begin{pmatrix} y_1 \\ y_2 \end{pmatrix}$ and let B be the "box" $\{(t, \mathbf{y}) \,\big|\, |t| \leq 1, \; |\mathbf{y}| \leq 2\}$. Determine the bounds M, K in (3.19) for $\mathbf{f}$ and $\partial \mathbf{f}/\partial y_j$ for this case. Determine α of Theorem 3.3. Compute the first three successive approximations of the solution $\boldsymbol{\phi}(t)$ satisfying the initial condition $\boldsymbol{\phi}(0) = \mathbf{0}$, $\boldsymbol{\phi} = \begin{pmatrix} \phi_1 \\ \phi_2 \end{pmatrix}$.

3.3 Uniqueness of Solutions

Our next goal is to prove that under suitable hypotheses there is only one solution $\boldsymbol{\phi}$ of the system of differential equations

$$\mathbf{y}' = \mathbf{f}(t, \mathbf{y}) \tag{3.22}$$

that satisfies the initial condition

$$\boldsymbol{\phi}(t_0) = \boldsymbol{\eta} \tag{3.23}$$

We have seen by examples in the scalar case that the assumption of continuity of $\mathbf{f}$ is not enough to guarantee uniqueness. On the other hand, we assert that the hypotheses of Theorem 3.3 are enough to guarantee uniqueness. The principal tool in the proof is the Gronwall inequality (Theorem 1.4, p. 31).

Theorem 3.4. *Suppose* $\mathbf{f}$ *and* $\partial\mathbf{f}/\partial y_j$ $(j = 1, \ldots, n)$ *are continuous on the "box"*

$$B = \{(t, \mathbf{y}) \,\big|\, |t - t_0| \le a, |\mathbf{y} - \boldsymbol{\eta}| \le b\}$$

Then there exists at most one solution of (3.22) *satisfying the initial condition* (3.23).

We recall that under the hypotheses of Theorem 3.4 we have already established the existence of at least one solution $\boldsymbol{\phi}$ of (3.22), (3.23) existing on the interval $|t - t_0| < \alpha$, where α is defined as in Theorem 3.3. We also recall that the hypotheses of Theorem 3.4 imply the inequality (3.18) (the Lipschitz condition).

Proof of Theorem 3.4. Suppose that $\boldsymbol{\phi}_1$ and $\boldsymbol{\phi}_2$ are two solutions of (3.22), (3.23) which both exist on some common interval J containing t_0. Since every solution of (3.22), (3.23) also satisfies the integral equation (3.21), we have

$$\boldsymbol{\phi}_1(t) = \boldsymbol{\eta} + \int_{t_0}^{t} \mathbf{f}(s, \boldsymbol{\phi}_1(s))\, ds$$

$$\boldsymbol{\phi}_2(t) = \boldsymbol{\eta} + \int_{t_0}^{t} \mathbf{f}(s, \boldsymbol{\phi}_2(s))\, ds$$

for every t in J. Subtracting these two equations, we obtain

$$\boldsymbol{\phi}_2(t) - \boldsymbol{\phi}_1(t) = \int_{t_0}^{t} [\mathbf{f}(s, \boldsymbol{\phi}_2(s)) - \mathbf{f}(s, \boldsymbol{\phi}_1(s))]\, ds$$

Taking norms and using (3.18), we have

$$\begin{aligned} |\boldsymbol{\phi}_2(t) - \boldsymbol{\phi}_1(t)| &\le \left| \int_{t_0}^{t} |\mathbf{f}(s, \boldsymbol{\phi}_2(s)) - \mathbf{f}(s, \boldsymbol{\phi}_1(s))|\, ds \right| \\ &\le K \left| \int_{t_0}^{t} |\boldsymbol{\phi}_2(s) - \boldsymbol{\phi}_1(s)|\, ds \right| \qquad (\text{for } t \in J) \end{aligned}$$

where $|\partial \mathbf{f}/\partial y_j(t, \mathbf{y}) \le K$ $(j = 1, \ldots, n)$, for all $(t, \mathbf{y}) \in R$. Taking first the case $t \ge t_0$ and then $t \le t_0$, we find that the Gronwall inequality (p. 31) implies for both cases that $|\boldsymbol{\phi}_2(t) - \boldsymbol{\phi}_1(t)| \le 0$. Since $|\boldsymbol{\phi}_2(t) - \boldsymbol{\phi}_1(t)|$ is nonnegative, we have $|\boldsymbol{\phi}_2(t) - \boldsymbol{\phi}_1(t)| = 0$ for all t in J, or $\boldsymbol{\phi}_2(t) = \boldsymbol{\phi}_1(t)$ for t in J. Thus there cannot be two distinct solutions of (3.22), (3.23) on J, and this proves uniqueness. ∎

It is not necessary to assume as much as continuity of $\partial \mathbf{f}/\partial y_j$ $(j = 1, \ldots, n)$ to insure uniqueness. It is clear from the proof of Theorem 3.4 that the Lipschitz condition (3.18), which here follows automatically from the continuity of $\partial \mathbf{f}/\partial y_j$ $(j = 1, \ldots, n)$, could be used in the hypothesis of Theorem 3.4 instead of the continuity of $\partial \mathbf{f}/\partial y_j$ $(j = 1, \ldots, n)$ without changing the proof. It is possible to prove uniqueness of solutions under considerably weaker hypotheses, but in most problems Theorem 3.4 is applicable and such more refined results are not needed.

• EXERCISE

1. State and prove a uniqueness theorem for solutions of the initial value problem

$$y'' + g(t, y) = 0, \qquad y(0) = y_0, \qquad y'(0) = z_0$$

where g is a given function defined on a rectangle $R: |t| \le a$, $|y - y_0| \le b$. Refer to Exercise 12, Section 3.1. [*Hint:* Under appropriate hypotheses if ϕ_1 and ϕ_2 are both solutions of the initial value problem existing on some interval $|t| \le \alpha$, we have

$$\phi_k(t) = y_0 + z_0 t - \int_0^t (t - s) g(s, \phi_k(s))\, ds \qquad (k = 1, 2)$$

for $|t| \le \alpha$; subtract the two equations corresponding to $k = 1$ and $k = 2$, then use the Lipschitz condition (3.7), and finally appeal to the Gronwall inequality in the form given in Exercise 3, Section 1.7, p. 32.]

We will prove one other uniqueness theorem for a scalar equation, because its proof illustrates the type of reasoning used in proving general uniqueness theorems and because it is related to Example 3 of Section 3.1.

Theorem 3.5. *Suppose f is continuous on the rectangle $R = \{(t, y) \mid |t - t_0| \le a, |y - y_0| \le b\}$, and monotone nonincreasing in y for each fixed t on the rectangle R. Then the initial value problem*

$$y' = f(t, y) \tag{3.1}$$

$$\phi(t_0) = y_0 \tag{3.2}$$

has at most one solution on any interval J with t_0 as left end point.

Proof. Suppose ϕ_1 and ϕ_2 are two solutions of (3.1), (3.2), which may agree on some interval to the right of t_0, but which differ somewhere in $t_0 \leq t < t_0 + \alpha_1$ for some $\alpha_1 > 0$. We may assume that $\phi_2 > \phi_1$ on some interval $t_1 < t < t_1 + h < t_0 + \alpha$ while $\phi_2 = \phi_1$ on $t_0 \leq t \leq t_1$ (here t_1 may be the same as t_0, but need not be). More precisely, t_1 is the greatest lower bound of the set E of values of t on which $\phi_2 > \phi_1$. This greatest lower bound exists because the set E is bounded below, at any rate, by t_0.

This implies $f(t, \phi_1(t)) \geq f(t, \phi_2(t))$ on $t_1 < t < t_1 + h$. Since ϕ_1 and ϕ_2 are both solutions of (3.1), we have $\phi_1' \geq \phi_2'$ on $t_1 < t < t_1 + h$. The function $u = \phi_2 - \phi_1$, then, is by hypothesis strictly positive on $t_1 < t < t_1 + h$. Since $u' = \phi_2' - \phi_1' \leq 0$, u has a nonpositive derivative on $t_1 < t < t_1 + h$, and satisfies $u(t_1) = 0$. This is impossible, and the supposition $\phi_2 > \phi_1$ on $t_1 < t < t_1 + h$ must be false. Thus the solutions ϕ_1 and ϕ_2 are actually identical on $t_0 \leq t < t_0 + \alpha_1$. This completes the proof of the theorem. ▮

- **EXERCISE**

2. Use Theorem 3.5 to prove uniqueness of the solution for $t \geq 0$ in Example 3 of Section 3.1.

Exercise 2 shows that in Example 3, Section 3.1, we have uniqueness. We have already shown that the successive approximations for that example do not converge. Thus uniqueness does not imply the convergence of successive approximations. On the other hand, Example 2 and Exercise 14, Section 3.1 show that the convergence of successive approximations does not imply uniqueness.

3.4 Continuation of Solutions

The existence theorems of Sections 3.1 and 3.2 say that under suitable hypotheses there is a solution of a differential equation or system of differential equations that exists on some, possibly small, interval. The question to be studied in this section is whether this solution in fact exists on a larger interval.

Example 1. Consider the scalar initial value problem

$$y' = y^2, \quad y(0) = 1$$

whose solution ϕ can be written explicitly as $\phi(t) = 1/(1-t)$ (see Example 1, Section 1.3, p. 11). Clearly the solution exists on $-\infty < t < 1$. Suppose we try to determine the interval of validity as given by Theorem 3.1. Here R is the rectangle $R = \{(t, y) \mid |t| \le a, |y-1| \le b\}$ and α and M of Theorem 3.1 are given by $M = \max_R (y^2) = (1+b)^2$, $\alpha = \min(a, b/M)$. The largest value of the positive number $b/(1+b)^2$ is $1/4$.

• EXERCISE

1. Using calculus or otherwise, show that

$$\max_{0 \le b < \infty} \left(\frac{b}{(1+b)^2} \right) = \frac{1}{4}$$

Therefore, independent of the choice of a, $\alpha \le 1/4$. Thus, Theorem 3.1 gives existence of a solution on $|t| \le 1/4$. From the explicit formula for the solution it is clear that the solution actually exists on a much larger interval.

Suppose that $\mathbf{f}$ and $\partial \mathbf{f}/\partial y_j$ $(j = 1, \ldots, n)$ are continuous in a given region D in $(n+1)$-dimensional space. Let $(t_0, \boldsymbol{\eta})$ be a given point in D. Consider the initial value problem

$$\mathbf{y}' = \mathbf{f}(t, \mathbf{y}) \tag{3.22}$$

$$\mathbf{y}(t_0) = \boldsymbol{\eta} \tag{3.23}$$

By the definition of a region (p. 24), there exist numbers $a > 0$, $b > 0$ such that the "box" $B = \{(t, \mathbf{y}) \mid |t - t_0| \le a, |\mathbf{y} - \boldsymbol{\eta}| \le b\}$, centered at $(t_0, \boldsymbol{\eta})$ is contained in D. By Theorem 3.3 there exists a number $\alpha > 0$ and a unique solution $\boldsymbol{\phi}$ of (3.22), (3.23) existing on the interval $|t - t_0| \le \alpha$ (see Figure 3.5).

Now consider the point $(t_0 + \alpha, \boldsymbol{\phi}(t_0 + \alpha))$ as a new initial point. Since this point is in D (definitely not on the boundary of D), there exist numbers $a_1 > 0$, $b_1 > 0$ such that the box

$$B_1 = \{(t, \mathbf{y}) \mid |t - (t_0 + \alpha)| \le a_1, |\mathbf{y} - \boldsymbol{\phi}(t_0 + \alpha)| \le b_1\},$$

centered at $(t_0 + \alpha, \boldsymbol{\phi}(t_0 + \alpha))$ is contained in D (see Figure 3.6). Consider the system of differential equations (3.22), subject to the initial condition

$$\mathbf{y}(t_0 + \alpha) = \boldsymbol{\phi}(t_0 + \alpha) \tag{3.24}$$

By Theorem 3.3, there exists a number $\alpha_1 > 0$ and a unique solution $\boldsymbol{\psi}$ of (3.22), (3.24) existing on the interval $|t - (t_0 + \alpha)| \le \alpha_1$ (see Figure 3.6).

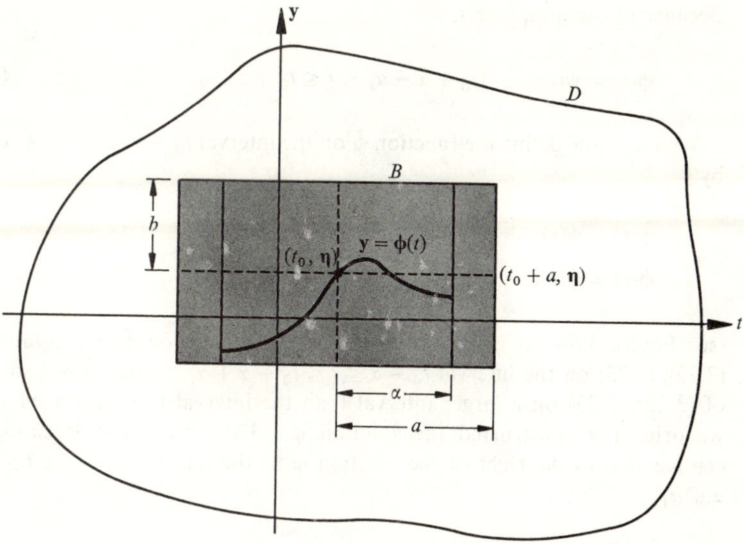

Figure 3.5

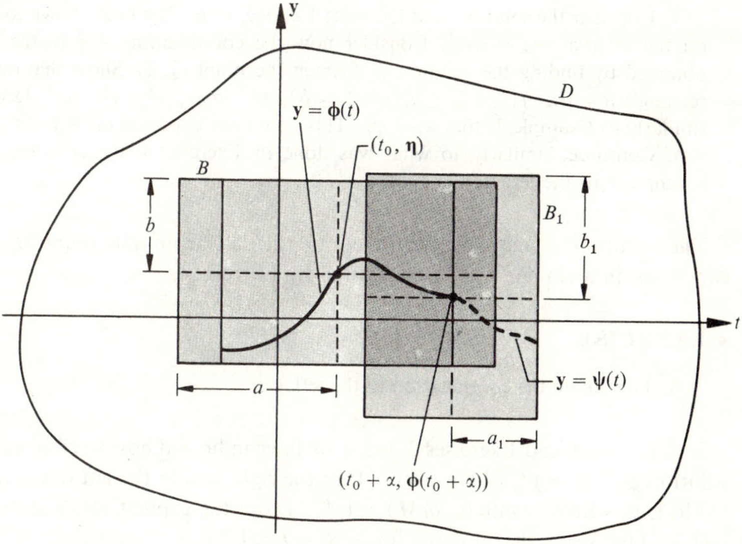

Figure 3.6

Because of the uniqueness,

$$\phi(t) = \psi(t), \qquad (t_0 + \alpha - \alpha_1 \leq t \leq t_0 + \alpha) \tag{3.25}$$

We may now define the function $\hat{\phi}$ on the interval $t_0 - \alpha < t < t_0 + \alpha + \alpha_1$ by

$$\hat{\phi}(t) = \phi(t), \qquad (t_0 - \alpha \leq t \leq t_0 + \alpha)$$
$$\hat{\phi}(t) = \psi(t), \qquad (t_0 + \alpha \leq t \leq t_0 + \alpha + \alpha_1)$$

(see Figure 3.6). It is easy to verify that this function $\hat{\phi}$ is a solution of (3.22), (3.23) on the interval $t_0 - \alpha \leq t \leq t_0 + \alpha + \alpha_1$. Thus $\hat{\phi}$ is a solution of (3.22), (3.23) on a larger interval than the interval $|t - t_0| \leq \alpha$ on which we originally constructed the solution ϕ. This solution $\hat{\phi}$ is called the **continuation to the right** of the solution ϕ to the interval $t_0 - \alpha \leq t \leq t_0 + \alpha + \alpha_1$.

• **EXERCISES**

2. Show that the function $\hat{\phi}$ satisfies the integral equation (3.21) on the interval $t_0 - \alpha \leq t \leq t_0 + \alpha + \alpha_1$.
3. Consider the solution ϕ of Example 1 above, which has been shown to exist on the interval $-\frac{1}{4} \leq t \leq \frac{1}{4}$. Consider now the continuation of ϕ to the right obtained by finding the solution ψ through the point $(\frac{1}{4}, \frac{4}{3})$. Show that on any rectangle $R = \{(t, y) \mid |t - \frac{1}{4}| \leq a, |y - \frac{4}{3}| \leq b\}$, $M = \max_R y^2 = (\frac{4}{3} + b)^2$. Deduce, similarly to Example 1, that $\alpha_1 = \frac{3}{16}$. This now gives existence on $-\frac{1}{4} \leq t \leq \frac{7}{16}$.
4. Continue, similarly to what was done in Exercise 3, the solution ϕ of Example 1 to the left of the point $(-\frac{1}{4}, \frac{4}{5})$.

The solution ϕ may be continued to the left from the point $(t_0 - \alpha, \phi(t_0 - \alpha))$ in a similar way, as suggested by Exercise 4.

• **EXERCISE**

5. Formulate the continuation to the left.

In Example 1 and Exercises 3 and 4 we have indicated how to continue the solution ϕ of $y' = y^2$, where $\phi(0) = 1$, to the right and to the left of the interval $|t| \leq \alpha$, with $\alpha = \min(a, b/M) = 1/4$. From the explicit solution $\phi(t) = 1/(1 - t)$ we know that it exists for $-\infty < t < 1$.

In order to establish this fact, and a general result on how far the solution can be continued, we need the following auxiliary result.

Lemma 3.3. *Suppose* $\mathbf{f}, \partial\mathbf{f}/\partial y_j$ $(j = 1, \ldots, n)$ *are continuous in a domain D and suppose* $|\mathbf{f}|$ *is bounded in D. Let* $\boldsymbol{\phi}$ *be a solution of*

$$\mathbf{y}' = \mathbf{f}(t, \mathbf{y}) \tag{3.22}$$

$$\boldsymbol{\phi}(t_0) = \boldsymbol{\eta} \tag{3.23}$$

existing on a finite interval $\gamma < t < \delta$. *Then* $\lim_{t\to\delta^-} \boldsymbol{\phi}(t)$ *and* $\lim_{t\to\gamma^+} \boldsymbol{\phi}(t)$ *exist.*

Proof. Let t_1 and t_2 be any two points on the interval $\gamma < t < \delta$ with $t_1 < t_2$. Then, since $\boldsymbol{\phi}$ satisfies the integral equation (3.21),

$$\boldsymbol{\phi}(t_1) = \boldsymbol{\eta} + \int_{t_0}^{t_1} \mathbf{f}(s, \boldsymbol{\phi}(s))\, ds$$

$$\boldsymbol{\phi}(t_2) = \boldsymbol{\eta} + \int_{t_0}^{t_2} \mathbf{f}(s, \boldsymbol{\phi}(s))\, ds$$

Subtraction gives

$$\boldsymbol{\phi}(t_2) - \boldsymbol{\phi}(t_1) = \int_{t_1}^{t_2} \mathbf{f}(s, \boldsymbol{\phi}(s))\, ds$$

and the assumption $|\mathbf{f}(t, \mathbf{y})| \le M$ for $(t, \mathbf{y}) \in D$ now gives

$$|\boldsymbol{\phi}(t_2) - \boldsymbol{\phi}(t_1)| \le M\,|t_2 - t_1|$$

Since the right side tends to zero as t_1 and t_2 both tend to δ from below, the Cauchy convergence criterion shows that $\boldsymbol{\phi}(t)$ tends to a limit as t tends to δ from below. We obtain the proof that $\boldsymbol{\phi}(t)$ tends to a limit as t tends to γ from above in an analogous way by letting t_1 and t_2 tend to γ from above. ∎

Now we can define

$$\boldsymbol{\phi}(\delta) = \lim_{t\to\delta^-} \boldsymbol{\phi}(t), \quad \boldsymbol{\phi}(\gamma) = \lim_{t\to\gamma^+} \boldsymbol{\phi}(t)$$

and we have the solution $\boldsymbol{\phi}$ defined on the closed interval $\gamma \le t \le \delta$. The continuation process described above can be repeated provided the graph of the solution remains in a region in which $\mathbf{f}$ and $\partial\mathbf{f}/\partial y_j$ $(j = 1, \ldots, n)$ are continuous, and this continuation process is carried out by means of boxes remaining in D, precisely as B and B_1 were used above.

Theorem 3.6. *Suppose that* $\mathbf{f}$ *and* $\partial \mathbf{f}/\partial y_j$ $(j = 1, \ldots, n)$ *are continuous in a given region* D *and suppose* $\mathbf{f}$ *is bounded on* D. *Let* $(t_0, \boldsymbol{\eta})$ *be a given point of* D. *Then the unique solution* $\boldsymbol{\phi}$ *of the system* $\mathbf{y}' = \mathbf{f}(t, \mathbf{y})$ *passing through the point* $(t_0, \boldsymbol{\eta})$ *can be extended until its graph meets the boundary of* D.

Proof. Suppose the solution $\boldsymbol{\phi}$ cannot be extended up to the boundary of D, but can be extended to the right only to an interval $t_0 \leq t < \hat{t}$. By Lemma 3.3, $\boldsymbol{\phi}(\hat{t}) = \lim_{t \to \hat{t}^-} \boldsymbol{\phi}(t)$ exists. If $(\hat{t}, \boldsymbol{\phi}(\hat{t}))$ is a boundary point of D, we are done. If not, there is a box centered at $(\hat{t}, \boldsymbol{\phi}(\hat{t}))$ lying in D. But now we can extend the solution $\boldsymbol{\phi}$ to the right of $\hat{t}$ by the method given above, resulting in a contradiction. This shows that $\boldsymbol{\phi}$ can be extended up to the boundary of D. An analogous argument shows that $\boldsymbol{\phi}$ can also be extended to the left up to the boundary of D, and this completes the proof. ∎

The reader should note that in the proof of Theorem 3.6, it is possible that $\hat{t} = +\infty$. If $\hat{t} = +\infty$, then it is possible that $\lim_{t \to \hat{t}^-} |\boldsymbol{\phi}(t)| = \infty$. If $\hat{t} < \infty$, since $\mathbf{f}$ is assumed to be bounded on D, we must have $\lim_{t \to \hat{t}^-} |\boldsymbol{\phi}(t)| < \infty$.

In many problems the hypothesis that $\mathbf{f}$ is bounded on D made in Theorem 3.6 is not satisfied if D is taken as the maximal possible domain of definition of $\mathbf{f}$. For the equation $y' = y^2$ considered in Example 1, if D is taken to be the whole (t, y) plane, the function $f(t, y) = y^2$ is not bounded. We may handle this problem as follows.

Example 2. Consider the solution ϕ of the equation $y' = y^2$ such that $\phi(0) = 1$. Here D is the whole plane. If we consider the equation on the region $D_A = \{(t, y) \mid -\infty < t < \infty, |y| < A\}$ instead of D, then $f(t, y) = y^2$ is bounded on D_A by A^2 and Theorem 3.6 is applicable. It shows that the solution $\phi(t) = 1/(1 - t)$ can be continued for those values of t such that $|\phi(t)| = 1/(1 - t) < A$, or $-\infty < t < 1 - 1/A$. However, since this result holds for an arbitrary $A > 0$, $\phi(t)$ can be continued to the interval $-\infty < t < 1$, but not to the interval $-\infty < t \leq 1$.

In Exercises 3 and 4 above, the solution ϕ was "built up" on a union (or sum) of contiguous (adjacent) closed intervals. If such a continuation can be accomplished in a finite number of steps, the resulting solution will be valid on a closed interval. If, however, an infinite number of steps is required, as is in fact the case for the equation in Example 2, then the resulting solution may be valid only on an open interval. This can happen because an infinite union of closed intervals need not be closed. For example, the union of the intervals $1 - (1/n) \leq t \leq 1 - 1/(n + 1)$ $(n = 1, 2, \ldots)$ is the interval $0 < t < 1$.

In applications a very common example is the case where the domain D is the whole $(t, \mathbf{y})$ space. We start with an arbitrary infinite box $B_A = \{(t, \mathbf{y}) \mid -\infty < t < \infty, |\mathbf{y}| < A\}$ containing the initial point $(t_0, \boldsymbol{\eta})$ and employ the argument of Example 2 above. By Theorem 3.6, the solution $\boldsymbol{\phi}$ can be continued to the boundary of B_A. Clearly, either this gives the existence of the solution $\boldsymbol{\phi}$ on $-\infty < t < \infty$ or there exists a $\hat{t}$ such that $\lim_{t \to \hat{t}} |\boldsymbol{\phi}(t)| = A_j$ in the second alternative we enlarge the box B_A by increasing A. It may happen that by choosing A finite but large enough we obtain existence of the solution $\boldsymbol{\phi}$ on $-\infty < t < \infty$. In this case, $\boldsymbol{\phi}$ is also bounded on $-\infty < t < \infty$, since $|\boldsymbol{\phi}(t)| < A$ on $-\infty < t < \infty$. If, however, the solution $\boldsymbol{\phi}$ reaches the portion $|\mathbf{y}| = A$ of the boundary of B_A for $A > 0$, there are two possibilities. Either $\boldsymbol{\phi}$ exists on $-\infty < t < \infty$ (but is not bounded), or there exists a finite number T such that $\lim_{t \to T} |\boldsymbol{\phi}(t)| = \infty$. We can now summarize.

Corollary to Theorem 3.6. If D is the entire $(t, \mathbf{y})$ *space and if* $\mathbf{f}$ *and* $\partial \mathbf{f}/\partial y_j$ $(j = 1, \ldots, n)$ *are continuous on D, then the solution* $\boldsymbol{\phi}$ *of* $\mathbf{y}' = \mathbf{f}(t, \mathbf{y})$ *can be continued uniquely in both directions for as long as* $|\boldsymbol{\phi}(t)|$ *remains finite.*

A similar result holds for the case where D is the infinite strip $a < t < b$, $|\mathbf{y}| < \infty$, and $\mathbf{f}$, $\partial \mathbf{f}/\partial y_j$ $(j = 1, \ldots, n)$ are continuous on D.

This corollary does not say that if D is the entire $(t, \mathbf{y})$ space, then every solution can be extended to the interval $-\infty < t < \infty$.

• EXERCISES

6. Show that the solution ϕ of $y' = -y^2$, where $\phi(1) = 1$ exists for $0 < t < \infty$ but cannot be continued to the left beyond $t = 0$.
7. Show that no solution, other than $\phi(t) \equiv 0$, of the equation $y' = -y^2$ can be extended to the interval $-\infty < t < \infty$.
8. Formulate, as another corollary to Theorem 3.6, the result on continuation if $\mathbf{f}$ is defined on the infinite strip $D = \{(t, \mathbf{y}) \mid a < t < b, |\mathbf{y}| < \infty\}$.

For the existence of solutions of the linear system

$$\mathbf{y}' = A(t)\mathbf{y} + \mathbf{g}(t) \tag{3.26}$$

where $A(t)$ is a continuous n-by-n matrix and $\mathbf{g}(t)$ is a continuous vector on some interval I, we refer the reader to Theorem 2.1, p. 37, where existence is established for the case $I = \{t \mid a \le t \le b\}$.* The case of existence of solutions of the system (3.26) on an arbitrary interval follows from Theorem 3.6, the

* We note that Theorem 2.1 depends on Theorem 1.1, which has now been completely justified.

corollary, and Exercise 8, using the same estimates that were obtained in the proof of Theorem 2.1.

There are special cases where we can prove a global result directly from successive approximations, as is illustrated by the following result.

• EXERCISES

9. Let f be a continuous scalar function defined on the domain $D = \{(t, y) \mid |t - t_0| \leq \alpha, |y| < \infty\}$ where α is an arbitrary positive real number. Suppose $\partial f/\partial y$ is bounded on D, that is, there exists a constant $K = K(\alpha) > 0$ depending on α but not on y, such that

$$\left|\frac{\partial f}{\partial y}(t, y)\right| \leq K(\alpha)$$

Show that the successive approximations

$$\phi_0(t) = y_0$$

$$\phi_n(t) = y_0 + \int_{t_0}^{t} f(s, \phi_{n-1}(s))\, ds \qquad (n = 1, 2, \ldots)$$

approach a (unique) solution ϕ of $y' = f(t, y)$ satisfying the initial condition $\phi(t_0) = y_0$. Also show

$$|\phi(t)| \leq |y_0| + \frac{M}{K}(e^{\alpha K} - 1) \qquad \text{for} \quad |t - t_0| \leq \alpha$$

where $K = K(\alpha)$ and $M = \max_{|t - t_0| \leq \alpha} |f(t, y_0)|$. [*Hint:* The trick now is to get around the fact that $f(t, y)$ itself is not necessarily bounded on D, even though $f(t, y_0)$ is bounded.

(a) Induction easily shows that each ϕ_n $(n = 0, 1, \ldots)$ is well defined for $|t - t_0| \leq \alpha$. Now

$$|\phi_1(t) - \phi_0(t)| \leq \left|\int_{t_0}^{t} |f(s, y_0)|\, ds\right| \leq M|t - t_0|$$

(b) Show by induction that

$$|\phi_n(t) - \phi_{n-1}(t)| \leq \frac{MK^{n-1}|t - t_0|^n}{n!}$$

(c) Deduce the uniform convergence of $\{\phi_n\}$ to a limit function ϕ on $|t - t_0| \leq \alpha$. Then, for $|t - t_0| \leq \alpha$,

$$|\phi_n(t) - y_0| = \left|\sum_{k=1}^{n} \phi_n(t) - \phi_{n-1}(t)\right| \leq \sum_{k=1}^{n} |\phi_n(t) - \phi_{n-1}(t)|$$

$$\leq \sum_{k=1}^{\infty} \frac{MK^{n-1}|t - t_0|^n}{n!} = \frac{M}{K}(e^{K\alpha} - 1)$$

from which, on taking the limit, we obtain the asserted bound for $|\phi(t)|$.

(d) Show that the limit function $\phi(t)$ is a solution of the problem.]

10. Let $f(t, y)$ be a continuous scalar function on the whole (t, y) plane. Suppose that $\partial f/\partial y$ is also continuous and for any $\alpha > 0$ suppose

$$\left|\frac{\partial f}{\partial y}(t, y)\right| \leq K = K(\alpha) \qquad (|t| \leq \alpha, \quad |y| < \infty)$$

Show that given any (t_0, y_0) the equation $y' = f(t, y)$ has a unique solution $\phi(t)$ on $-\infty < t < \infty$, such that $\phi(t_0) = y_0$. [*Hint:* By Exercise 9, $|\phi(t)|$ is bounded on every interval $|t - t_0| \leq \alpha$; now apply the corollary to Theorem 3.6.]

11. Show that the equation

$$y' = p(t) \cos y + q(t) \sin y$$

where p, q are continuous functions on $-\infty < t < \infty$ has a (unique) solution ϕ on $-\infty < t < \infty$, satisfying the arbitrary initial condition $\phi(t_0) = y_0$. [*Hint:* Use Exercise 10.]

3.5 Dependence on Initial Conditions and Parameters

A solution $\boldsymbol{\phi}$ of the system of differential equations

$$\mathbf{y}' = \mathbf{f}(t, \mathbf{y}) \tag{3.27}$$

passing through the point $(t_0, \boldsymbol{\eta})$ depends not only on t, but also on the initial point $(t_0, \boldsymbol{\eta})$. When we wish to emphasize this dependence, we will write the solution as $\boldsymbol{\phi}(t, t_0, \boldsymbol{\eta})$. We will show that under suitable hypotheses $\boldsymbol{\phi}$ depends continuously on the initial values, and in fact that $\boldsymbol{\phi}$ is a continuous function of the "triple" $(t, t_0, \boldsymbol{\eta})$. As in the previous sections of this chapter, we make no attempt to prove the most refined result of this type.

Theorem 3.7. *Suppose* $\mathbf{f}$ *and* $\partial \mathbf{f}/\partial y_j$ $(j = 1, \ldots, n)$ *are continuous and bounded in a given region* D. *We assume that the bounds* (3.19) *are satisfied on* D (rather than on the box B). *Let* $\boldsymbol{\phi}$ *be the solution of the system* (3.27) *passing through the point* $(t_0, \boldsymbol{\eta})$ *and let* $\boldsymbol{\psi}$ *be the solution of* (3.27) *passing through the point* $(\hat{t}_0, \hat{\boldsymbol{\eta}})$. *Suppose that* $\boldsymbol{\phi}$ *and* $\boldsymbol{\psi}$ *both exist on some interval* $\alpha < t < \beta$. *Then to each* $\varepsilon > 0$ *there corresponds* $\delta > 0$ *such that if* $|t - \hat{t}| < \delta$, *and* $|\boldsymbol{\eta} - \hat{\boldsymbol{\eta}}| < \delta$, *then*

$$|\boldsymbol{\phi}(t) - \boldsymbol{\psi}(\hat{t})| < \varepsilon \qquad (\alpha < t < \beta, \quad \alpha < \hat{t} < \beta) \tag{3.28}$$

Proof. Since $\boldsymbol{\phi}$ is the solution of (3.27) through the point $(t_0, \boldsymbol{\eta})$, we have for every t, $\alpha < t < \beta$,

$$\boldsymbol{\phi}(t) = \boldsymbol{\eta} + \int_{t_0}^{t} \mathbf{f}(s, \boldsymbol{\phi}(s))\, ds \tag{3.29}$$

Since $\boldsymbol{\psi}$ is the solution of (3.27) through the point $(\hat{t}_0, \hat{\boldsymbol{\eta}})$, we have for every $\hat{t}$, $\alpha < \hat{t} < \beta$,

$$\boldsymbol{\psi}(t) = \hat{\boldsymbol{\eta}} + \int_{\hat{t}_0}^{t} \mathbf{f}(s, \boldsymbol{\psi}(s))\, ds \tag{3.30}$$

Since

$$\int_{t_0}^{t} \mathbf{f}(s, \boldsymbol{\phi}(s))\, ds = \int_{\hat{t}_0}^{t} \mathbf{f}(s, \boldsymbol{\phi}(s))\, ds + \int_{t_0}^{\hat{t}_0} \mathbf{f}(s, \boldsymbol{\phi}(s))\, ds$$

subtraction of (3.30) from (3.29) gives

$$\boldsymbol{\phi}(t) - \boldsymbol{\psi}(t) = \boldsymbol{\eta} - \hat{\boldsymbol{\eta}} + \int_{\hat{t}_0}^{t} [\mathbf{f}(s, \boldsymbol{\phi}(s)) - \mathbf{f}(s, \boldsymbol{\psi}(s))]\, ds + \int_{t_0}^{\hat{t}_0} \mathbf{f}(s, \boldsymbol{\phi}(s))\, ds$$

and therefore

$$|\boldsymbol{\phi}(t) - \boldsymbol{\psi}(t)| \le |\boldsymbol{\eta} - \hat{\boldsymbol{\eta}}| + \left| \int_{\hat{t}_0}^{t} |\mathbf{f}(s, \boldsymbol{\phi}(s)) - \mathbf{f}(s, \boldsymbol{\psi}(s))|\, ds \right| + \left| \int_{t_0}^{\hat{t}_0} |\mathbf{f}(s, \boldsymbol{\phi}(s))|\, ds \right| \tag{3.31}$$

Using (3.19) to estimate the right-hand side of (3.31), we obtain

$$|\boldsymbol{\phi}(t) - \boldsymbol{\psi}(t)| \le |\boldsymbol{\eta} - \hat{\boldsymbol{\eta}}| + K \left| \int_{\hat{t}_0}^{t} |\boldsymbol{\phi}(s) - \boldsymbol{\psi}(s)|\, ds \right| + M |\hat{t}_0 - t_0|$$

If $|t_0 - \hat{t}_0| < \delta$, $|\boldsymbol{\eta} - \hat{\boldsymbol{\eta}}| < \delta$, then we have

$$|\boldsymbol{\phi}(t) - \boldsymbol{\psi}(t)| \le \delta + K \left| \int_{\hat{t}_0}^{t} |\boldsymbol{\phi}(s) - \boldsymbol{\psi}(s)|\, ds \right| + M\delta \tag{3.32}$$

The Gronwall inequality (Theorem 1.4, p. 31) applied to (3.32) gives

$$|\boldsymbol{\phi}(t) - \boldsymbol{\psi}(t)| \le \delta(1 + M) \exp(K|t - \hat{t}_0|) \le \delta(1 + M) \exp[K(\beta - \alpha)]$$

using the fact that $|t - \hat{t}_0| < \beta - \alpha$. Since $|\boldsymbol{\psi}(t) - \boldsymbol{\psi}(\hat{t})| < \left|\int_{\hat{t}}^{t} |\mathbf{f}(s, \boldsymbol{\psi}(s))| ds\right| \le M|t - \hat{t}| \le M\delta$ if $|t - \hat{t}| < \delta$, we have

$$|\boldsymbol{\phi}(t) - \boldsymbol{\psi}(\hat{t})| \le |\boldsymbol{\phi}(t) - \boldsymbol{\psi}(t)| + |\boldsymbol{\psi}(t) - \boldsymbol{\psi}(\hat{t})| \le \delta(1 + M)e^{K(\beta - \alpha)} + \delta M$$

Now, given $\varepsilon > 0$, we need only choose $\delta < \varepsilon/[M + (1 + M)e^{K(\beta - \alpha)}]$ to obtain (3.28) and thus complete the proof. ∎

Theorem 3.7 shows that the solution $\boldsymbol{\phi}(t, t_0, \boldsymbol{\eta})$ of (3.27) passing through the point $(t_0, \boldsymbol{\eta})$ is a continuous function of the "triple" $(t, t_0, \boldsymbol{\eta})$. It is possible to show that if the initial point $(\hat{t}_0, \hat{\boldsymbol{\eta}})$ of $\boldsymbol{\psi}$ is sufficiently close to the graph of the solution $\boldsymbol{\phi}$, then there is a common interval on which both solutions $\boldsymbol{\phi}$ and $\boldsymbol{\psi}$ exist (see [4]. Ch. 1, Theorem 7.1).

We also remark that in practice the solutions $\boldsymbol{\phi}$ and $\boldsymbol{\psi}$ are usually known to remain in a closed bounded subset of D, so that the hypothesis that $\mathbf{f}$ and $\partial\mathbf{f}/\partial y_j$ are bounded is automatically satisfied on this subset (which is all that is needed in the above proof), even though it may not hold on D.

Continuous dependence on initial conditions is true under considerably weaker hypotheses than those of Theorem 3.7. In fact, it can be shown that uniqueness of solutions by itself implies the continuous dependence of solutions on initial conditions. For a proof, see [4, Ch. 2, Theorem 4.1].

The technique used to prove Theorem 3.7 can be applied to establish the following result.

Theorem 3.8. *Let* $\mathbf{f}$, $\mathbf{g}$ *be defined in a domain* D *and satisfy the hypotheses of Theorem* 3.7. *Let* $\boldsymbol{\phi}$ *and* $\boldsymbol{\psi}$ *be solutions of* $\mathbf{y}' = \mathbf{f}(t, \mathbf{y})$, $\mathbf{y}' = \mathbf{g}(t, \mathbf{y})$ *respectively, such that* $\boldsymbol{\phi}(t_0) = \boldsymbol{\eta}$, $\boldsymbol{\psi}(t_0) = \hat{\boldsymbol{\eta}}$, *existing on a common interval* $\alpha < t < \beta$. *Suppose*

$$|\mathbf{f}(t, \mathbf{y}) - \mathbf{g}(t, \mathbf{y})| \le \varepsilon$$

for $(t, \mathbf{y})$ *in* D. *Then the solutions* $\boldsymbol{\phi}$, $\boldsymbol{\psi}$ *satisfy the estimate*

$$|\boldsymbol{\phi}(t) - \boldsymbol{\psi}(t)| \le |\boldsymbol{\eta} - \hat{\boldsymbol{\eta}}| \exp(K|t - t_0|) + \varepsilon(\beta - \alpha)\exp(K|t - t_0|)$$

for all t *in* $\alpha < t < \beta$.

• EXERCISE

1. Prove Theorem 3.8. [*Hint:* Write the integral equations satisfied by $\boldsymbol{\phi}$ and $\boldsymbol{\psi}$, and subtract to obtain

$$\boldsymbol{\phi}(t) - \boldsymbol{\psi}(t) = \boldsymbol{\eta} - \hat{\boldsymbol{\eta}} + \int_{t_0}^{t} \{\mathbf{f}(s, \boldsymbol{\phi}(s)) - \mathbf{f}(s, \boldsymbol{\psi}(s))\}\, ds$$

$$+ \int_{t_0}^{t} \{\mathbf{f}(s, \boldsymbol{\psi}(s)) - \mathbf{g}(s, \boldsymbol{\psi}(s))\}\, ds$$

Then take norms, and use (3.19), the hypotheses, and the Gronwall inequality to obtain the result.]

Theorem 3.8 says roughly that if two differential equations have their right-hand sides "close together," their solutions cannot differ by very much.

One immediate consequence of Theorem 3.8 is another proof of uniqueness (Theorem 3.4). We simply take $\mathbf{g}(t, \mathbf{y}) = \mathbf{f}(t, \mathbf{y})$, $\boldsymbol{\eta} = \hat{\boldsymbol{\eta}}$. Then $\varepsilon = 0$, and Theorem 3.8 gives $\boldsymbol{\phi}(t) \equiv \boldsymbol{\psi}(t)$.

Theorem 3.8 also gives a different type of continuity property, as indicated below.

• EXERCISE

2. Let $\{\mathbf{f}_k(t, \mathbf{y})\}$ be a sequence of vector functions that converges to $\mathbf{f}(t, \mathbf{y})$ in the sense that $|\mathbf{f}(t, \mathbf{y}) - \mathbf{f}_k(t, y)| \le \varepsilon_k$ for all $(t, \mathbf{y})$ in D, with $\varepsilon_k \to 0$ as $k \to \infty$ and let $\mathbf{f}$ and $\mathbf{f}_k(\mathrm{k} = 1, 2, \ldots)$ satisfy the hypotheses of Theorem 3.8. Let $\{\hat{\mathbf{y}}_k\}$ be a sequence of constants converging to $\boldsymbol{\eta}$. Let $\boldsymbol{\psi}_k(t)$ be the solution of $\mathbf{y}' = \mathbf{f}_k(t, \mathbf{y})$ with $\boldsymbol{\psi}_k(t_0) = \hat{\mathbf{y}}_k$ $(k = 1, 2, \ldots)$, existing on $\alpha < t < \beta$, and let $\boldsymbol{\phi}(t)$ be the solution of $\mathbf{y}' = \mathbf{f}(t, \mathbf{y})$ with $\boldsymbol{\phi}(t_0) = \boldsymbol{\eta}$, existing on $\alpha < t < \beta$. Show that $\lim_{k \to \infty} \boldsymbol{\psi}_k(t) = \boldsymbol{\phi}(t)$ for $\alpha < t < \beta$.

• MISCELLANEOUS EXERCISES

1. Consider the differential equation $y' = y$, with the solution $\phi(t) = e^t$ satisfying the initial condition $\phi(0) = 1$. Let $p(t)$ be *any* polynomial in t and define

$$\phi_0(t) = p(t)$$

$$\phi_k(t) = 1 + \int_0^t \phi_{k-1}(s)\, ds \qquad (k = 1, 2, \ldots)$$

(a) Is $\lim_{k \to \infty} \phi_k(t) = \phi(t)$? Prove your answer.

(b) What general statement can you make concerning the initial "guess" $\phi_0(t)$ and the convergence of the successive approximations? What happens if the 1 in the definition of $\phi_k(t)$ for $k \ge 1$ is replaced by some other function, say $p(t)$?

2. Consider the scalar differential equation

$$y' = f(t, y)$$

where f and $\partial f/\partial y$ are continuous in a region D in the (t, y) plane, and let (t_0, y_0) be a point in D. Let G be a **bounded** subregion of D containing (t_0, y_0) and let $\bar{G}$ be the closure of G. Define

$$M = \max_{\bar{G}} |f(t, y)|$$

Through (t_0, y_0) construct the lines AB and CD of slope M and $-M$, respectively, as shown in Figure 3.7. Now construct vertical lines HI, JK intersecting the t axis at α and β, respectively, so that the isosceles triangles HOI, JOK are contained in G. Let $\phi_0(t)$ be any continuous function defined on $\alpha \le t \le \beta$ such that the set of points $\{(t, \phi_0(t)) \mid \alpha \le t \le \beta\}$ is contained in $\bar{G}$. Define

$$\phi_1(t) = y_0 + \int_{t_0}^t f(s, \phi_0(s))\, ds \qquad (\alpha \le t \le \beta)$$

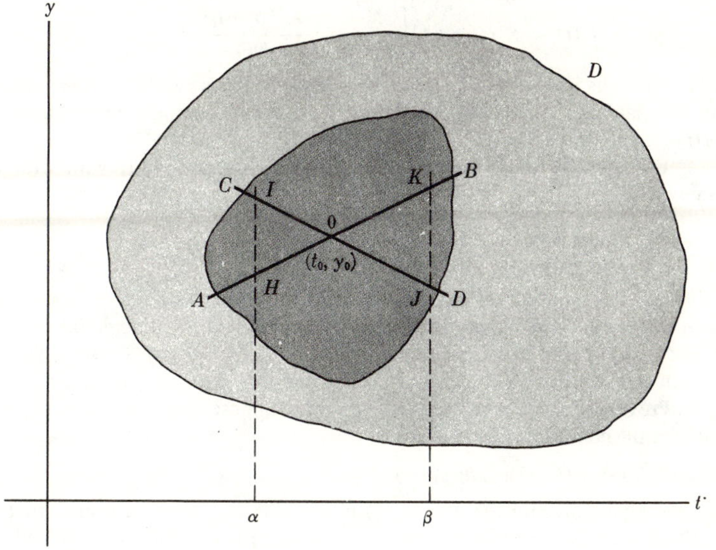

Figure 3.7

and generally

$$\phi_k(t) = y_0 + \int_{t_0}^{t} f(s, \phi_{k-1}(s))\, ds \qquad (k = 1, 2, 3, \ldots; \alpha \leq t \leq \beta)$$

Prove that the successive approximations $\{\phi_k(t)\}$ converge (uniformly) to the (unique) solution $\phi(t)$ of $y' = f(t, y)$ with $\phi(t_0) = y_0$ on $\alpha \leq t \leq \beta$, and estimate the error $|\phi(t) - \phi_m(t)|$ on $\alpha \leq t \leq \beta$ in terms of

$$K_1 = \max_{\alpha \leq t < \beta} |\phi_0(t) + \phi_1(t)| \qquad \text{for } m = 0$$

$$K_2 = \max_{\bar{G}} \left| \frac{\partial f}{\partial y}(t, y) \right| \qquad \text{for } m \geq 1$$

[*Hint:* Follow the proof of Theorem 3.1.]

(REMARK. The same result holds if f satisfies a Lipschitz condition.)

3. In the notation of Exercise 2, suppose that a function $\phi_0(t)$ has been found satisfying the hypotheses of Exercise 2, and also constants $K > 0$, $\delta \geq 0$ are known such that

$$|\phi_1(t) - \phi_0(t)| \leq K\, |t - t_0|^{\delta} \qquad (\alpha \leq t \leq \beta)$$

Show that

$$|\phi(t) - \phi_1(t)| \leq K|t - t_0|^{\delta+1}\left[\frac{K_2}{\delta+1} + \frac{K_2{}^2|t - t_0|}{(\delta+1)(\delta+2)} + \cdots\right] \qquad (\alpha \leq t \leq \beta)$$

where K_2 is as in Exercise 2. Can you generalize this result? [*Hint:* Assume an estimate for $|\phi_m(t) - \phi_{m-1}(t)|$ on $\alpha \leq t \leq \beta$ and compute an estimate for $|\phi(t) - \phi_m(t)|$ on $\alpha \leq t \leq \beta$.]

4. Let $\phi(t)$ be the solution of $y' = t^2 + y^2$ on $0 \leq t \leq 1$, with $\phi(0) = 0$. Show that

$$\left|\phi(t) - \left(\frac{1}{3}t^3 + \frac{1}{63}t^7\right)\right| \leq 0.00155t^8 \qquad (0 \leq t \leq 1)$$

[*Hint:* In the notation of Exercise 3, let $\phi_0(t) = t^3/3$, and compute $\phi_2(t)$ and $|\phi_1(t) - \phi_0(t)|$. Then apply the result of Exercise 3 to the differential equation $y' = t^2 + y^2$ on the closed rectangle $\{(t, y) \mid 0 \leq t \leq 1, |y| \leq A\}$ for some suitably chosen $A > 0$. Such a choice is $A = 0.36$, and this gives $K_2 = 0.72$.

5. Prove **the Osgood uniqueness theorem**: Suppose that the function $f(t, y)$ satisfies the condition

$$|f(t, y_2) - f(t, y_1)| \leq h(|y_2 - y_1|)$$

for every pair of points (t, y_1), (t, y_2) in a region D. Here we assume that the function $h(u)$ is continuous for $0 < u < \alpha$ for some $\alpha > 0$, that $h(u) > 0$, and that

$$\lim_{\varepsilon \to 0+} \int_\varepsilon^\alpha \frac{du}{h(u)} = \infty$$

Then through each point (t_0, y_0) in D there is at most one solution of the equation $y' = f(t, y)$. [*Hint:* Suppose ϕ_1 and ϕ_2 are two solutions with $\phi_1(t_0) = \phi_2(t_0) = y_0$. Define $\psi(t) = \phi_2(t) - \phi_1(t)$ and suppose $\psi(t) \not\equiv 0$. Then show that $|\psi'(t)| \leq h(|\psi(t)|) \leq 2h(|\psi(t)|)$. Suppose $\psi(t_1) \neq 0$, for some $t_1 > t_0$, and let $u(t)$ be the solution of $u' = 2h(u)$ satisfying the initial condition $u(t_1) = |\psi(t_1)|$; $u(t)$ is strictly positive for $t_0 \leq t \leq t_1$. Show that $\psi'(t_1) < u'(t_1)$ and therefore $\psi(t) > u(t)$ on some interval to the left of t_1. Then show that $\psi(t) > u(t)$ for $t_0 \leq t \leq t_1$ and obtain a contradiction.]

6. Show that the functions $h(u) = Ku^\alpha$ $(\alpha \geq 1)$, $h(u) = Ku \log |u|$, $h(u) = Ku \log |u \,|\log|\log |u||$, and so on, satisfy the Osgood condition of Exercise 5. (The case $h(u) = Ku$ is, of course, the Lipschitz condition.)

7. Let $f(t, y)$ and $g(t, y)$ be continuous and satisfy a Lipschitz condition with respect to y in a region D. Suppose $|f(t, y) - g(t, y)| < \varepsilon$ in D for some $\varepsilon > 0$. Let $\phi_1(t)$ be a solution of $y' = f(t, y)$ and let $\phi_2(t)$ be a solution of $y' = g(t, y)$ such that $|\phi_2(t_0) - \phi_1(t_0)| < \delta$ for some t_0 and some $\delta > 0$. Show that for all t for which $\phi_1(t)$ and $\phi_2(t)$ both exist,

$$|\phi_2(t) - \phi_1(t)| \leq \delta \exp(K|t - t_0| + \frac{\varepsilon}{K}(\exp(K|t - t_0| - 1)$$

where K is the Lipschitz constant. [*Hint:* Show that $|\phi_2(t) - \phi_1(t)| \leq \delta + \int_{t_0}^t [K|\phi_2(s) - \phi_1(s)| + \varepsilon]\, ds$ and complete the argument by a slight generalization of the Gronwall inequality (see Exercise 3, Section 1.7, p. 32).]

8. Obtain the analogues of Exercises 2, 5, and 7 for systems of the form $\mathbf{y}' = \mathbf{f}(t, \mathbf{y})$, where $\mathbf{f}$ and $\mathbf{y}$ are vectors and where $\mathbf{f}$ satisfies suitable hypotheses.

9. Let $q(t)$ be a differentiable function on $0 \le t \le \pi$. Show that the differential equation

$$-y'' + q(t)y = \lambda y$$

has a solution $\phi_1(t, \lambda)$ on $0 \le t \le \pi$ with $\phi_1(0, \lambda) = 0$, $\phi_1'(0, \lambda) = 1$, such that

$$\phi_1(t, \lambda) = \sin \frac{\sqrt{\lambda} t}{\sqrt{\lambda}} + M_1$$

$$\phi_1'(t, \lambda) = \cos \sqrt{\lambda} t + M_2$$

where $|M_1| \le K/\lambda$, $|M_2| \le K/\sqrt{\lambda}$ on $0 \le t \le \pi$ for some constant K. [*Hint:* Consider the integral equation

$$\phi(t, \lambda) = \frac{\sin \sqrt{\lambda} t}{\sqrt{\lambda}} + \frac{1}{\sqrt{\lambda}} \int_0^t \sin \sqrt{\lambda}(t-s) q(s) \phi(s, \lambda)\, ds$$

and use successive approximations as in Exercise 13, Section 3.1, to prove the existence of a bounded solution $\phi_1(t, \lambda)$. Then estimate

$$M_1 = \frac{1}{\sqrt{\lambda}} \int_0^t \sin \sqrt{\lambda}(t-s) q(s) \phi_1(s, \lambda)\, ds.$$

Differentiate the integral equation to obtain the desired estimate for $\phi_1'(t, \lambda)$.]

10. Show that the differential equation considered in Exercise 9 has a solution $\phi_2(t, \lambda)$ on $0 \le t \le \pi$, with $\phi_2(0, \lambda) = 1$, $\phi_2'(0, \lambda) = 0$, such that

$$\phi_2(t, \lambda) = \cos \sqrt{\lambda} t + M_3$$

$$\phi_2'(t, \lambda) = -\sqrt{\lambda} \sin \sqrt{\lambda} t + \tfrac{1}{2} \cos \sqrt{\lambda} t \int_0^t q(s)\, ds = M_4$$

where $|M_3| \le K/\sqrt{\lambda}$, $|M_4| \le K/\sqrt{\lambda}$ on $0 \le t \le \pi$ for some constant K. [*Hint:* Proceed as in Exercise 9, using an appropriate integral equation.]

REMARK. Exercises 9 and 10 above, coupled with the Liouville transformation [2, Exercise 5, p. 234] are the starting points for the study of the asymptotic behavior of eigenvalues and eigenfunctions of general Sturm–Liouville boundary value problems; see, for example, [26].

11. Let $\mathbf{f}$, $\partial \mathbf{f}/\partial y_j$ $(j = 1, \ldots, n)$ be continuous in a domain D. Let $\boldsymbol{\phi}(t, t_0, \boldsymbol{\eta})$ be the solution of $\mathbf{y}' = \mathbf{f}(t, \mathbf{y})$ for which $\boldsymbol{\phi}(t_0, t_0, \boldsymbol{\eta}) = \boldsymbol{\eta}$ and suppose that $\boldsymbol{\phi}$ is differentiable with respect to each of its $n + 2$ variables. Show that $\partial \boldsymbol{\phi}/\partial \eta_j (t, t_0, \boldsymbol{\eta})$ $(j = 1, \ldots, n)$ is that solution of the linear system

$$\mathbf{w}' = \mathbf{f}_{\mathbf{y}}(t, \boldsymbol{\phi})\mathbf{w} \tag{3.3}$$

for which $\partial\boldsymbol{\phi}/\partial\eta_j(t_0, t_0, \boldsymbol{\eta}) = \mathbf{e}_j$, the familiar unit vector in the jth coordinate direction where $\mathbf{f_y}(t, \boldsymbol{\phi})$ denotes the matrix $(\partial f_i/\partial y_j(t, \boldsymbol{\phi}))$.

12. Under the hypotheses of Exercise 11, show that $\partial\boldsymbol{\phi}/\partial t_0\ (t, t_0, \boldsymbol{\eta})$ is the solution of (3.33) for which

$$\frac{\partial\boldsymbol{\phi}}{\partial t_0}(t_0, t_0, \boldsymbol{\eta}) = -f(t_0, \boldsymbol{\eta})$$

13. Under the hypotheses of Exercise 11, show that

$$\frac{\partial\boldsymbol{\phi}}{\partial t_0}(t, t_0, \boldsymbol{\eta}) = -\sum_{j=1}^{n} \frac{\partial\boldsymbol{\phi}}{\partial\eta_j}(t, t_0, \boldsymbol{\eta}) f_j(t_0, \boldsymbol{\eta})$$

Chapter 4 STABILITY OF LINEAR AND ALMOST LINEAR SYSTEMS

4.1 Introduction

Ideally one would like to compute explicitly all solutions of every differential equation, or system of differential equations. However, as we have seen, there are actually very few equations (beyond linear equations with constant coefficients, and even here there are difficulties if the order of the equation or system is high) for which we can do this. In this chapter we begin to study qualitative properties of solutions of differential equations, **without solving the equations explicitly**. This marks the beginning of the modern theory of differential equations, which was pioneered primarily by the independent work of two mathematicians, A. M. Lyapunov and H. Poincaré, at the turn of the century; their ideas continue to stimulate current research in the subject. The reader might well ask why we do not simply calculate the solutions numerically on a high-speed computer and use whatever information can be obtained from the numerical solution. This is precisely what we can do if we desire information about one specific solution. However, in many problems, for example, those of design of complex systems, automatic controls, and so forth, we want certain information of a qualitative nature about **all solutions.**

Moreover, we often want to know whether a certain property of these solutions remains unchanged if the system is subjected to various types of changes (usually called perturbations). We shall be much more specific later; however, the point is that for such studies the computer and the calculation of a few specific solutions do not provide a satisfactory answer. These qualitative studies are also important from the practical point of view because in most problems (this is already true for the simplest mass-spring or pendulum system) both the differential equations and the measurement of initial values and various other data involve approximations. Indeed, in almost every mathematical model of a physical problem a number of effects have been neglected. It is therefore important to study how sensitive the particular model is to small perturbations or changes of initial conditions and of various parameters. Another drawback in the use of numerical approximations is that often it is of interest to show that a solution of a differential equation tends to zero as $t \to \infty$. While a numerical approximation method may **suggest** that this is true, it **cannot** be used to prove it.

One qualitative phenomenon of great practical interest is the notion of **stability** of a certain state or solution of a system of differential equations. This chapter and Chapter 5 are devoted primarily to the study of this property and conditions under which a solution is stable. This concept will be motivated and defined precisely in the next section. Then stability of linear systems (which represent the simplest case) will be considered briefly and followed by an investigation of **"almost linear" systems** using analytical methods. Such nonlinear systems are important because no physical system is truly linear—linearization is just another idealization of reality. The aim of this investigation is to find out under what conditions the addition of nonlinear terms does **not** drastically alter the behavior of the linear system; or, to put it another way: Under what conditions does the perturbed system behave more or less like the linearized system? We shall also look briefly at the concept of asymptotic equivalence, which is important for certain problems to which the other techniques cannot be applied. Finally, we shall apply some of the techniques developed here to the study and stability of periodic solutions of nonlinear periodic systems. Several important cases of nonlinear systems that are not necessarily almost linear will be studied in Chapter 5.

We remark that our objective is not to prove the most general or most sophisticated results on stability; rather, we wish to convey a simple and intuitive view of this important subject. We will also see from several examples in this and the subsequent chapters that some nonlinear problems exhibit phenomena that cannot be duplicated by linear systems, and in certain applications, such as the theory of automatic controls studied in Chapter 6, we therefore need to build nonlinearities into the system in order to achieve a desired effect.

4.2 Definitions of Stability

To be able to present the ideas in a simple geometrical setting, let us consider a physical system whose equations of motion are given by the

autonomous system

$$\mathbf{y}' = \mathbf{f}(\mathbf{y}) \tag{4.1}$$

where the real continuous vector-valued function $\mathbf{f}$ with n components is defined in some region D in real n-dimensional Euclidean space. We assume throughout that $\mathbf{f} \in C^1(D)$, which guarantees existence and uniqueness of solutions of the initial value problem for (4.1) (Theorems 3.3, p. 123, and 3.4, p. 125). Then, as we have seen in Section 2.8 (p. 184), the solutions of (4.1) can be conveniently pictured as curves in the phase space. The behavior of solutions of (4.1) is indicated by these curves. Let $\mathbf{y} = \mathbf{y}_0$ be an isolated **critical point** of (4.1) (that is, $\mathbf{f}(\mathbf{y}_0) = \mathbf{0}$), so that $\boldsymbol{\phi}(t) \equiv \mathbf{y}_0$—a constant vector—is a solution of (4.1). A critical point of (4.1) is an **equilibrium** or **rest point** of the associated physical system. In phase space the equilibrium solution $\boldsymbol{\phi}(t) \equiv \mathbf{y}_0$ is represented as a single point. Now suppose the system is displaced slightly from this equilibrium state; that is, suppose we consider a solution $\boldsymbol{\psi}(t)$ of (4.1) that passes through the point $\boldsymbol{\eta}$ at a time t_0 (see Figure 4.1), where the Euclidean distance $\|\boldsymbol{\eta} - \mathbf{y}_0\|$ is small. (We remind the reader

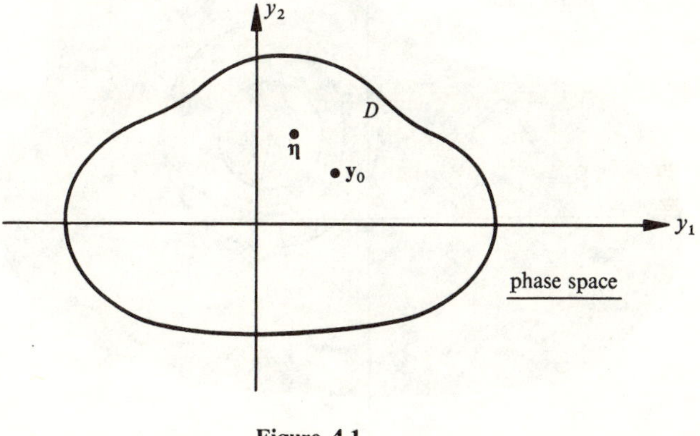

Figure 4.1

that because of errors in measurements it would be impossible to start the physical system exactly at $\mathbf{y}_0$.) Let us now inquire what happens to the system when we start the motion at a point $\boldsymbol{\eta}$ different from $\mathbf{y}_0$, but near $\mathbf{y}_0$. Will the resulting motion $\boldsymbol{\psi}$ remain close to the equilibrium state for $t \geq t_0$? If it does, this, roughly speaking, is stability of the equilibrium solution $\mathbf{y}_0$. (The reader may find it helpful to think of an undamped simple pendulum displaced slightly from the equilibrium at $\theta = 0, \theta' = 0$.) If the motion (solution ψ) does remain near the equilibrium position $\mathbf{y}_0$ and if in addition the solution ψ tends to return to the equilibrium position as time t increases to infinity, then, roughly speaking, this is **asymptotic stability of the equilibrium solution $\mathbf{y}_0$**. (The reader might think of a damped pendulum.) If, on the other hand, the solution ψ leaves every small neighborhood of $\mathbf{y}_0$, this, roughly speaking, is **instability of the equilibrium solution $\mathbf{y}_0$**. (The undamped pendulum displaced slightly from the equilibrium at $\theta = \pi, \theta' = 0$.) More precisely, we have the following definitions.

Definition 1. (See Figure 4.2.) *The equilibrium solution* $\mathbf{y}_0$ *of* (4.1) *is said to be stable if for each number* $\varepsilon > 0$ *we can find a number* $\delta > 0$ (depending *on* ε) *such that if* $\psi(t)$ *is any solution of* (4.1) *having* $\|\psi(t_0) - \mathbf{y}_0\| < \delta$, *then the solution* $\psi(t)$ *exists for all* $t \geq t_0$ *and* $\|\psi(t) - \mathbf{y}_0\| < \varepsilon$ *for* $t \geq t_0$ (where for convenience the norm is the Euclidean distance that makes neighborhoods spherical).

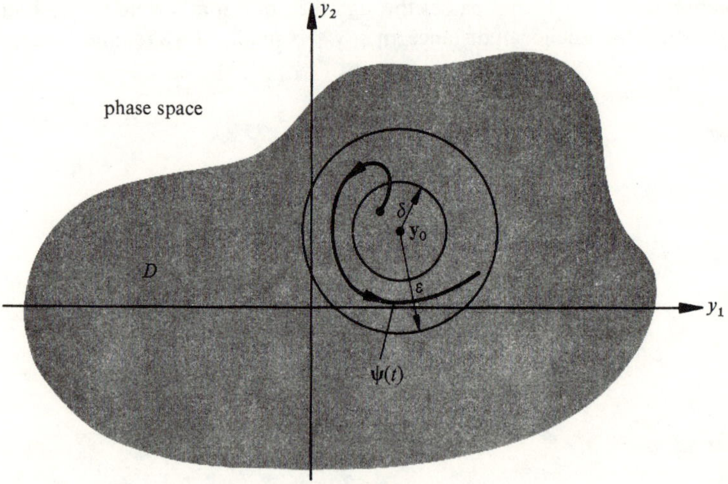

Figure 4.2

Definition 2. (See Figure 4.3.) *The equilibrium solution* $\mathbf{y}_0$ *is said to be asymptotically stable if it is stable and if there exists a number* $\delta_0 > 0$ *such that if* $\psi(t)$ *is any solution of* (4.1) *having* $\|\psi(t_0) - \mathbf{y}_0\| < \delta_0$, *then* $\lim_{t \to +\infty} \psi(t) = \mathbf{y}_0$.

The equilibrium solution $\mathbf{y}_0$ is said to be **unstable** if it is not stable.

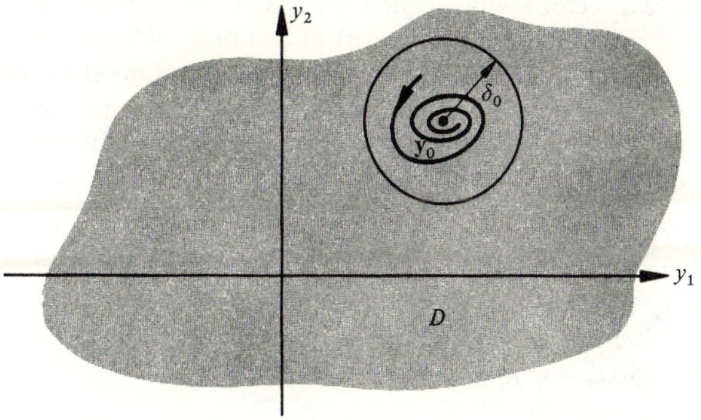

Figure 4.3

• **EXERCISE**

1. Decide for which of the two-dimensional linear systems considered in Section 2.8, p. 86, the origin $\mathbf{y} = 0$ is a stable, asymptotically stable, or unstable equilibrium point.

We stress the fact that the system (4.1) can have all solutions (motions) approaching a critical point $\mathbf{y}_0$ without the critical point $\mathbf{y}_0$ being asymptotically stable. An example of this type of behavior is given by the two-dimensional system

$$x' = \frac{x^2(y - x) + y^5}{(x^2 + y^2)\{1 + (x^2 + y^2)^2\}}, \qquad y' = \frac{y^2(y - 2x)}{(x^2 + y^2)\{1 + (x^2 + y^2)^2\}}$$

(See [8, p.191].) It should be observed that because (4.1) is autonomous, and hence (see Section 2.8, p. 84) invariant under translations of time, the numbers δ, δ_0 in Definitions 1 and 2 are independent of t_0. In this case the stability and asymptotic stability are **uniform**. We shall not pursue the concepts of uniform stability and uniform asymptotic stability here. Rather, we refer the reader to more advanced books, such as [3, 5, 8, 9, 15, 21, 25].

We now generalize the concept of stability to that of an arbitrary solution of the nonautonomous system

$$\mathbf{y}' = \mathbf{f}(t, \mathbf{y}) \tag{4.2}$$

where the real vector $\mathbf{f}$ with n components is defined and continuous in some region $D = \{(t, \mathbf{y}) \mid 0 \le t \le \infty, |\mathbf{y}| < a\}$ of real $(n + 1)$-dimensional Euclidean space, where $a \ge 0$ is some constant. We shall assume throughout that $\mathbf{f}$, $\partial\mathbf{f}/\partial y_j$ $(j = 1, \ldots, n)$ are continuous in D. Let $\boldsymbol{\phi}(t)$ be some solution of (4.2) existing for $0 \le t < \infty$. We shall denote by $\boldsymbol{\psi}(t, t_0, \mathbf{y}_0)$, where $t_0 \ge 0$, a solution of (4.2) for which $\boldsymbol{\psi}(t_0, t_0, \mathbf{y}_0) = \mathbf{y}_0$.

Definition 3. *A solution $\boldsymbol{\phi}(t)$ of (4.2) is said to be stable if for every $\varepsilon > 0$ and every $t_0 \ge 0$ there exists a $\delta > 0$ (δ now depends on both ε and possibly t_0) such that whenever $|\boldsymbol{\phi}(t_0) - \mathbf{y}_0| < \delta$, the solution $\boldsymbol{\psi}(t, t_0, \mathbf{y}_0)$ exists for $t > t_0$ and satisfies $|\boldsymbol{\phi}(t) - \boldsymbol{\psi}(t, t_0, \mathbf{y}_0)| < \varepsilon$ for $t \ge t_0$.*

We can picture this concept geometrically for the case $n = 1$ as in Figure 4.4.

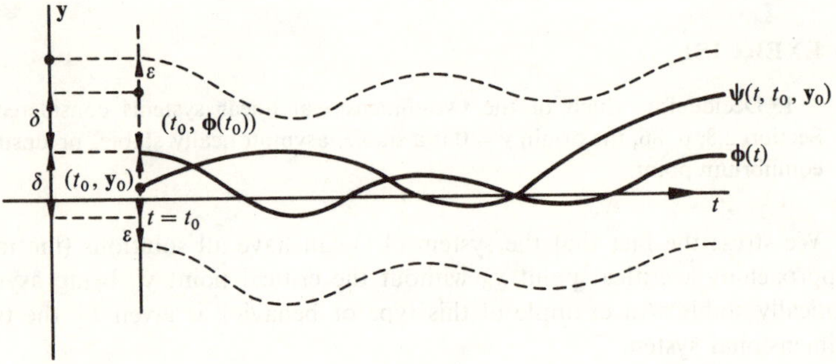

Figure 4.4

It may be shown, by an argument involving uniqueness of solutions and continuous dependence on initial conditions, that if a solution is stable in the above sense for a given t_0, then it is also stable for any other initial time $t_1 \ge t_0$. For this reason, in verifying the stability of a solution, we may work with a fixed t_0, such as $t_0 = 0$.

Definition 4. *The solution $\boldsymbol{\phi}(t)$ of (4.2) is said to be asymptotically stable if it is stable and if there exists $\delta_0 > 0$ such that whenever $|\boldsymbol{\phi}(t_0) - \mathbf{y}_0| < \delta_0$ the solution $\boldsymbol{\psi}(t, t_0, \mathbf{y}_0)$ approaches the solution $\boldsymbol{\phi}(t)$ as $t \to \infty$* (that is, $\lim_{t\to\infty} |\boldsymbol{\psi}(t, t_0, \mathbf{y}_0) - \boldsymbol{\phi}(t)| = 0$).

We note that δ_0 may also depend on t_0.

Example 1. The scalar equation $y' = -y/(1 + t)$ has $y \equiv 0$ as a solution. We see, by separation of variables, that $\psi(t, t_0, y_0) = y_0(1 + t_0)/(1 + t)$ is the solution through the point (t_0, y_0). Clearly, therefore, the zero solution is stable (Definition 3), even asymptotically stable (Definition 4). Note that since $|(1 + t_0)/(1 + t)| \leq 1$ for $t \geq t_0$ both δ and δ_0 are independent of t_0. We may take $\delta = \varepsilon$ in Definition 3 and δ_0 arbitrary in Definition 4.

• **EXERCISE**

2. Show by using Definitions 3 and 4 (or Definitions 1 and 2) with $\mathbf{y}_0 = 0$ that the zero solution of the scalar equation $y' = -\alpha y$ $(\alpha > 0)$ is stable and also asymptotically stable. Find suitable numbers δ, δ_0 and show that in this case they do **not** depend on t_0. Note that here the given equation is autonomous. Sketch the solutions in (t, y) space and also in phase space. (Note that the equation is autonomous and one-dimensional so that phase space is just a line!)

The reader should observe that even for a stable solution, the neighboring solution may behave badly for $t < t_0$. For example, the solution $y \equiv 0$ of the equation $y' = -y^3$ is asymptotically stable. By separation of variables, we find $\psi(t, t_0, y_0) = |y_0|\,[1 + 2y_0^2(t - t_0)]^{-1/2}$, which approaches zero as $t \to +\infty$, but it does not exist for all $t < t_0$; in particular, it becomes infinite as $t \to t_0 - 1/2y_0^2$.

The solutions in Example 1 and Exercise 2 have the property that for every t_0, y_0, not only for small $|y_0|$, the corresponding solution $\psi(t, t_0, y_0)$ approaches zero as $t \to \infty$. (Note that the given equations are linear and we shall see in the next section that this is generally true for linear systems.) We say in these cases that the zero solution is **globally asymptotically stable**. That this property does not hold in general when the equation is nonlinear can be seen from the equation $y' = -y + y^2$.

• **EXERCISE**

3. Show that the zero solution of $y' = -y + y^2$ is asymptotically stable, **but not globally**; that is, **not all solutions tend to zero as** $t \to +\infty$. [*Hint:* Show by separation of variables that $\psi(t, t_0, y_0) = [((1/y_0) - 1)\exp(-(t - t_0)) + 1]^{-1}$ and note that both $y = 0$ and $y = 1$ are critical points of the equation.] Sketch all solutions in the (t,y) plane; you may take $t_0 = 0$. Also, sketch all solutions in phase space. (The system is autonomous and one-dimensional; thus phase space is just a line!) What can you conclude about the solution $y \equiv 1$?

The equation considered in Exercise 3 suggests an important problem. If a solution $\boldsymbol{\phi}(t)$ is asymptotically stable, but not globally (that is, if not all solutions starting anywhere approach $\boldsymbol{\phi}$ as $t \to \infty$), what is the **region of asymptotic stability**? In other words, for a given t_0, describe the set of initial values $\mathbf{y}_0$ such that $\lim_{t\to\infty} |\boldsymbol{\psi}(t, t_0, \mathbf{y}_0) - \boldsymbol{\phi}(t)| = 0$. This, in general, is a difficult (and largely unsolved) but important practical problem. However, the method of Lyapunov developed in Chapter 5 gives a possible technique for obtaining an estimate for such a region, and this is done in Section 5.5.

We remark that for the purpose of studying the stability of a given solution $\boldsymbol{\phi}(t)$ of the system (4.1) or (4.2), it is convenient to make the change of variable

$$\mathbf{y} = \mathbf{x} + \boldsymbol{\phi}(t)$$

Then $\mathbf{y}' = \mathbf{x}' + \boldsymbol{\phi}'(t) = \mathbf{f}(t, \mathbf{x} + \boldsymbol{\phi}(t))$ and since $\boldsymbol{\phi}'(t) = \mathbf{f}(t, \boldsymbol{\phi}(t))$ we have

$$\mathbf{x}' = \mathbf{f}(t, \mathbf{x} + \boldsymbol{\phi}(t)) - \mathbf{f}(t, \boldsymbol{\phi}(t))$$

Observe that $\mathbf{x} \equiv 0$ is a solution of the transformed system. Thus if we define $\hat{\mathbf{f}}(t, \mathbf{x}) = \mathbf{f}(t, \mathbf{x} + \boldsymbol{\phi}(t)) - \mathbf{f}(t, \boldsymbol{\phi}(t))$, then the solution $\boldsymbol{\phi}(t)$ of (4.2) corresponds to the identically zero solution of $\mathbf{x}' = \hat{\mathbf{f}}(t, \mathbf{x})$. We can therefore always assume—without any loss of generality—that $\mathbf{y} \equiv \mathbf{0}$ is a solution of the given system (4.1) or (4.2) and we can limit our study of stability to that of the zero solution. This will be done in the remainder of this chapter unless otherwise noted. The reader should note that for an autonomous system the above change of variable may yield a nonautonomous system, unless the solution $\boldsymbol{\phi}$ whose stability is being investigated is a constant.

• EXERCISES

4. For each of the following scalar differential equations, decide whether the solution given is asymptotically stable, stable but not asymptotically stable, or unstable.

(a) $y' = 0$, $\phi(t) \equiv 1$. (b) $y' = y$, $\phi(t) \equiv 0$.
(c) $y' = y$, $\phi(t) = e^t$. (d) $y' = -y$, $\phi(t) \equiv 0$.
(e) $y' = -y$, $\phi(t) = e^{-t}$. (f) $y' = -y^2$, $\phi(t) \equiv 0$.

5. Formulate and interpret Definitions 1–4 for the case when zero is a solution of the given equation and the stability or asymptotic stability of this zero solution is to be defined.

6. By writing the scalar equation

$$u^{(n)} = f(t, u, u', \ldots, u^{(n-1)})$$

as an equivalent system, define the concept of stability and asymptotic stability of a solution $u = \phi(t)$ defined for $t \geq t_0$. Here it is assumed that the function $f(t, z_1, z_2, \ldots, z_n)$ is defined on some $(n+1)$-dimensional region, say $0 \leq t < \infty$, $|z_1| < a$, $|z_2| < a, \ldots, |z_n| < a$ for some constant $a > 0$. [*Hint:* Let $\psi(t, t_0, u_0)$ be the solution with $\psi(t_0, t_0, u_0) = u_0$, $\psi'(t_0, t_0, u_0) = u_0'$, ..., $\psi^{(n-1)}(t_0, t_0, u_0) = u_0^{(n-1)}$ existing on $t_0 \le t < \infty$. Now look at Definitions 3 and 4 and apply them to the given equation.]

7. If in Exercise 6, $n = 2$, give a physical interpretation of stability and asymptotic stability of the **zero solution** (that is, assume that $f(t, 0, 0) \equiv 0$) if u represents displacement from equilibrium and u' the velocity.

8. Show that for the linear system $\mathbf{y}' = A(t)\mathbf{y}$, where $A(t)$ is continuous for $0 \le t < \infty$, stability or asymptotic stability of the zero solution implies that of every other solution. [*Hint:* Let $\boldsymbol{\phi}(t)$ be the solution whose stability is to be tested and consider the solution $\mathbf{y} = \boldsymbol{\phi} + \mathbf{z}$ where $\mathbf{z}$ is "small." Show that $\mathbf{z}'(t) = A(t)\mathbf{z}(t)$ and now use the fact that 0 is a stable or asymptotically stable solution of this equation.]

9. For each of the following scalar differential equations, decide whether the solution given is asymptotically stable, stable but not asymptotically stable, or unstable.

(a) $y'' = 0$, $\phi(t) \equiv 0$.
(b) $y'' = 0$, $\phi(t) = t$.
(c) $y'' + 4y = 0$, $\phi(t) \equiv 0$.
(d) $y'' + 4y = 0$, $\phi(t) = \cos 2t$.
(e) $y'' + 4y' + 4y = 0$, $\phi(t) \equiv 0$.
(f) $y'' + 4y' + 4y = 0$, $\phi(t) = e^{-2t}$.
(g) $y'' - 4y = 0$, $\phi(t) \equiv 0$.
(h) $y'' - 4y = 0$, $\phi(t) = e^{-2t}$.

4.3 Linear Systems

The simplest general system for which stability questions are easily and completely decided is the linear system

$$\mathbf{y}' = A\mathbf{y} \tag{4.3}$$

where A **is a real constant** $n \times n$ **matrix**. As we shall see, a simple criterion for the zero solution of (4.3) can be given in terms of the eigenvalues of A. It should be noted that even this case can present serious practical computational difficulties, especially if the real parts of some of the eigenvalues are near zero, and the accuracy of the calculation is in doubt.

By simply looking at Theorem 2.10 (p. 80) and its corollary (p. 81), we immediately establish the following basic result.

Theorem 4.1. *If all eigenvalues of A have nonpositive real parts and all those eigenvalues with zero real parts are simple, then the solution* $\mathbf{y} = \mathbf{0}$ *of* (4.3) *is stable. If* (and only if) *all eigenvalues of A have negative real parts, the zero solution of* (4.3) *is asymptotically stable. In fact, in this case if* $\Psi(t, t_0)$ *denotes the fundamental matrix of* (4.3) *which is the identity at* $t = t_0$, $\Psi(t, t_0) = \exp((t - t_0)A)$ *and there exist constants* $K > 0$, $\sigma > 0$ *such that*

$$|\Psi(t, t_0)| \le K \exp(-\sigma(t - t_0)) \qquad (t_0 \le t < \infty) \tag{4.4}$$

with $\sigma > 0$ *in the case that all eigenvalues of* A *have negative real parts and* $\sigma = 0$ *if there are simple eigenvalues with zero real part. If one or more eigenvalues of* A *have a positive real part, the zero solution of* (4.3) *is unstable.*

The case with some eigenvalues with zero real part **not** simple, assuming the remaining eigenvalues have negative real parts, requires special investigation which we shall not undertake in general (see Example 3 below). We note that the estimate (4.4) is just the aforementioned corollary to Theorem 2.10, with $t = 0$ replaced by $t = t_0$. The number σ is any real number such that $-\sigma$ is larger than the real part of every eigenvalue of A. If all eigenvalues with the largest real part are simple, then $-\sigma$ may be taken equal to this largest real part. To prove the stability of the zero solution in this case we note that every solution of (4.3) has the form $\psi(t) = \Psi(t, t_0)\psi(t_0)$. Then using (4.4) with $\sigma = 0$, we have $|\psi(t)| \le K|\psi(t_0)|$ for $t_0 \le t < \infty$. Thus using Definition 1 or 3, we take $\delta < \varepsilon/K$ and we have $|\psi(t)| < \varepsilon$ for $t_0 \le t < \infty$ provided $|\psi(t_0)| < \varepsilon/K$.

- **EXERCISES**

1. Give a rigorous proof of asymptotic stability for the case $\mathscr{R}\lambda_j < 0$ ($j = 1, 2, \ldots, n$) and of instability in the case $\mathscr{R}\lambda_j > 0$, for one or more λ_j where $\lambda_1, \ldots, \lambda_n$ are the eigenvalues of A.
2. State Theorem 4.1 for the scalar equation $u^{(n)} + a_1 u^{(n-1)} + \cdots + a_{n-1}u' + a_n u = 0$ where $a_1, a_2, \ldots, a_n$ are constants. [*Hint:* See Exercise 6, Section 4.2. Write as an equivalent system. Compute the characteristic polynomial and use it to state the result.]

The reader should observe that actually much more than Theorem 4.1 is true in the case of the system (4.3). Namely, the stability properties are global. In particular, if all $\mathscr{R}\lambda_j < 0$ ($j = 1, \ldots, n$), all solutions, not only those with $|\psi(0)|$ small, approach zero as $t \to \infty$. That stability properties are global for all linear systems, **not merely for those with constant coefficients,** follows from the fact that for a linear system we have the representation theorem (Theorem 2.2, p. 43, and the remarks following it, p. 45) for **every** solution in the form $\Phi(t, t_0)\mathbf{c}$ where Φ is, say, the fundamental matrix that is the identity at $t = t_0$ and $\mathbf{c}$ is a constant vector.

Example 1. For the scalar equation $u'' - u = 0$ we obtain the equivalent system of the form (4.3) with

$$A = \begin{pmatrix} 0 & 1 \\ 1 & 0 \end{pmatrix}$$

and with eigenvalues ± 1. Therefore by Theorem 4.1 the zero solution of the system and hence of the scalar equation (see Exercise 2) is **unstable.**

Example 2. Proceeding in the same way to analyze the scalar equation $u'' + 2ku' + u = 0$, with $k > 0$, we obtain the matrix

$$A = \begin{pmatrix} 0 & 1 \\ -1 & -2k \end{pmatrix}$$

with eigenvalues $\lambda = -k \pm (k^2 - 1)^{1/2}$. Now if $k \geq 1$, both eigenvalues are real and negative, while if $0 < k < 1$, both eigenvalues have negative real parts. Thus in both cases by Theorem 4.1 and Exercise 2, the origin is asymptotically stable. Note that if $k < 0$, the zero solution is unstable. What happens if $k = 0$?

Example 3. For the equation $u^{(4)} + 2u'' + u = 0$ the matrix of the equivalent system

$$A = \begin{pmatrix} 0 & 1 & 0 & 0 \\ 0 & 0 & 1 & 0 \\ 0 & 0 & 0 & 1 \\ -1 & 0 & -2 & 0 \end{pmatrix}$$

has eigenvalues $+i$ and $-i$ each of multiplicity two, real part zero. It is easily seen that the zero solution is not stable, because $\cos t$, $\sin t$, $t \cos t$, $t \sin t$ are linearly independent solutions of the original equation, and the result follows from Exercise 6, Section 4.2, with $n = 4$ because every solution has the form $\psi(t) = c_1 \cos t + c_2 \sin t + c_3 t \cos t + c_4 t \sin t$; unless $c_3 = c_4 = 0$ such a solution and its derivatives are not bounded.

However, this situation is not always the one that prevails in the case of multiple eigenvalues with zero real part. For example, the system

$$\mathbf{y}' = A\mathbf{y}$$

where A is the zero matrix, clearly has the zero solution stable. (Prove this!) In general, it can be shown that in the case of eigenvalues with zero real part, one has stability if those with zero real part are simple roots of the minimal polynomial (see [7, Vol. I, p. 129]), while the remaining ones have real parts negative.

• EXERCISES

Determine for which of the following scalar equations or systems the zero solution is stable, asymptotically stable, or unstable.

3. $u'' + u = 0$.
4. $u'' + 5u' + 6u = 0$.
5. $u'' + 2u' + u = 0$.
6. $u^{(4)} - 2u'' + u = 0$.
7. $u'' + 2ku' + \alpha^2 u = 0$ ($k > 0$, $\alpha^2 > 0$ constants).
8. $u'' + u' - 6u = 0$.
9. $\mathbf{y}' = A\mathbf{y}$;

$$A = \begin{pmatrix} 1 & -1 & 1 & -1 \\ -3 & 3 & -5 & 4 \\ 8 & -4 & 4 & -4 \\ 15 & -10 & 11 & -11 \end{pmatrix}$$

10. $\mathbf{y}' = A\mathbf{y}$;

$$A = \begin{pmatrix} 0 & 0 & 0 & 0 \\ 0 & 0 & 0 & 0 \\ 0 & 0 & -1 & 3 \\ 0 & 0 & 0 & -1 \end{pmatrix}$$

11. Sketch the solution curves in Exercises 3, 4, 5, 7, 8 in the phase plane, and decide in which cases the origin is an attractor (see Section 2.8, p. 95).

12. Suppose that the solution $\mathbf{y} = 0$ of the system $\mathbf{y}' = A(t)\mathbf{y}$ is stable. Prove that **every** solution of the system is bounded.

For the scalar equation

$$u^{(n)} + a_1 u^{(n-1)} + \cdots + a_{n-1}u' + a_n u = 0$$

where $a_1, a_2, \ldots, a_n$ are real constants (see Exercise 2 above), it can be shown (see Routh–Hurwitz criterion [3, p. 21]) that the zero solution is asymptotically stable if and only if the determinants

$$D_1 = a_1 \quad D_2 = \det\begin{pmatrix} a_1 & a_3 \\ 1 & a_2 \end{pmatrix}$$

$$D_k = \det\begin{pmatrix} a_1 & a_3 & a_5 & \cdots & a_{2k-1} \\ 1 & a_2 & a_4 & \cdots & a_{2k-2} \\ 0 & a_1 & a_3 & \cdots & a_{2k-3} \\ \cdot & 1 & a_2 & \cdots & a_{2k-4} \\ \cdot & 0 & a_1 & \cdots & a_{2k-5} \\ \cdot & \cdot & 1 & \cdots & \cdot \\ \cdot & \cdot & 0 & \cdots & \cdot \\ \vdots & \vdots & \vdots & \ddots & \vdots \\ 0 & 0 & 0 \cdots 0 & 1 & a_k \end{pmatrix} \qquad k = 2, 3, \ldots, n$$

and with $a_j = 0$ for $j > n$, are strictly positive, and it is stable if all the determinants are nonnegative. The reader should verify the criterion for Exercises 4, 5, 6, 7, 8 above; however, for a proof we refer, for example, to [5]. We remark only that this stability criterion is simply the requirement that all roots of the characteristic polynomial

$$F(z) = z^n + a_1 z^{n-1} + \cdots + a_{n-1} z + a_n$$

lie in the left half of the complex z plane.

• **EXERCISE**

13. Use the Routh–Hurwitz criterion to derive a criterion for the asymptotic stability of the zero solution of each of the following scalar equations. Note that for a second-order equation, $D_1 = a_1$, $D_2 = \det\begin{pmatrix} a_1 & 0 \\ 1 & a_2 \end{pmatrix}$.

(a) $u'' + pu' + qu = 0$ (Answer: $p > 0, q > 0$);
(b) $u''' + pu'' + qu' + ru = 0$ (Answer: $p > 0, r > 0, pq > r$);
(c) $u^{(4)} + pu''' + qu'' + ru' + su = 0$ (Answer: $s > 0, q > 0, p > 0,$ $r(pq - r) > p^2 s$);

where p, q, r, s are real constants.

We now turn briefly to some of the simpler aspects of the much more difficult case of linear systems with variable coefficients; we consider first the system

$$\mathbf{y}' = (A + B(t))\mathbf{y} \tag{4.5}$$

where A is a constant matrix and $B(t)$ is continuous and "small" in the sense that $\lim_{t \to +\infty} B(t) = 0$. Thus for large t we would expect solutions of (4.5) to behave like those of $\mathbf{y}' = A\mathbf{y}$.

Example 4. Determine whether the zero solution of the scalar equation

$$y' = \left(-1 + \frac{1}{t+1}\right) y$$

is asymptotically stable. We would expect this to be the case, from comparison with the equation $y' = -y$. Indeed, by separation of variables we see that every solution of the given equation has the form

$$\phi(t) = y_0(t+1)e^{-t}$$

so that the solution $y = 0$ is asymptotically stable (in fact, globally).

This example generalizes to the following result.

Theorem 4.2. *Let all the eigenvalues of A have real parts negative and let $B(t)$ be continuous for $0 \le t < \infty$ with $\lim_{t\to\infty} B(t) = 0$. Then the zero solution of (4.5) is globally asymptotically stable.*

Proof. For any given $(t_0, \mathbf{y}_0)$, $t_0 > 0$, we have, from Theorem 2.1, p. 37, with $A(t) = A + B(t)$, that the (unique) solution $\boldsymbol{\psi}(t, t_0, \mathbf{y}_0)$, satisfying the initial condition $\boldsymbol{\psi}(t_0, t_0, \mathbf{y}_0) = \mathbf{y}_0$ exists for all $t \ge t_0$. Our problem is to show that this solution through the arbitrary point $(t_0, \mathbf{y}_0)$ satisfies the conditions of Definitions 3 and 4 (Section 4.2). By the variation of constants formula (Theorem 2.6, p. 53), using $B(t)\mathbf{y}$ as the "inhomogeneous term," we can express the solution by means of the equivalent integral equation

$$\boldsymbol{\psi}(t, t_0, \mathbf{y}_0) = \exp((t - t_0)A)\mathbf{y}_0 + \int_{t_0}^{t} \exp((t - s)A)B(s)\boldsymbol{\psi}(s, t_0, \mathbf{y}_0)\, ds \qquad (t_0 \le t < \infty) \qquad (4.6)$$

By the hypothesis on A, $\exp((t - t_0)A)$ satisfies (4.4) (p. 152) with σ, K as defined there. Since $\lim_{t\to\infty} B(t) = 0$, given any $\eta > 0$, there exists a number $T \ge t_0$ such that $|B(t)| < \eta$ for $t \ge T$. Similar to (4.6) and using $\boldsymbol{\psi}(T, t_0, \mathbf{y}_0)$ as initial value as well as uniqueness, we have

$$\boldsymbol{\psi}(t, t_0, \mathbf{y}_0) = e^{(t-T)A}\boldsymbol{\psi}(T, t_0, \mathbf{y}_0) + \int_{T}^{t} e^{(t-s)A}B(s)\boldsymbol{\psi}(s, t_0, \mathbf{y}_0)\, ds \qquad (T \le t < \infty)$$

Thus using (4.4) (with $t_0 = T$) and $|B(t)| < \eta$ for $t \ge T$, we obtain

$$|\boldsymbol{\psi}(t, t_0, \mathbf{y}_0)| \le Ke^{-\sigma(t-T)}|\boldsymbol{\psi}(T, t_0, \mathbf{y}_0)| + K\eta \int_{T}^{t} e^{-(t-s)\sigma}|\boldsymbol{\psi}(s, t_0, \mathbf{y}_0)|\, ds \qquad (T \le t < \infty)$$

Multiplying both sides of the inequality by $e^{\sigma t}$ and applying the Gronwall inequality (Theorem 1.4, p. 31) to the function $|\boldsymbol{\psi}(t, t_0, \mathbf{y}_0)|\, e^{\sigma t}$, we obtain

$$|\boldsymbol{\psi}(t, t_0, \mathbf{y}_0)| \le K|\boldsymbol{\psi}(T, t_0, \mathbf{y}_0)|\, e^{-(\sigma - K\eta)(t-T)} \qquad (T \le t < \infty) \qquad (4.7)$$

From this we can conclude that if $0 < \eta < \sigma/K$, the solution $\boldsymbol{\psi}(t, t_0, \mathbf{y}_0)$ will approach zero, in fact, exponentially. This does not yet prove that the zero solution of (4.5) is stable. To do this we compute a bound on $|\boldsymbol{\psi}(T, t_0, \mathbf{y}_0)|$. Returning to (4.6) and now restricting t to the interval $t_0 \le t \le T$, we have

$$|\psi(t, t_0, \mathbf{y}_0)| \le K \exp[-\sigma(t - t_0)]\,|\mathbf{y}_0| + K_1 K \int_{t_0}^{t} \exp[-\sigma(t - s)]\,|\psi(s, t_0, \mathbf{y}_0)|\, ds$$

where

$$K_1 = \max_{t_0 \le t \le T} |B(s)|$$

Multiplying by $e^{\sigma t}$ and applying the Gronwall inequality again, we obtain easily

$$\begin{aligned} |\psi(t, t_0, \mathbf{y}_0)| &\le K|\mathbf{y}_0| \exp[-\sigma(t - t_0) + K_1 K(t - t_0)] \\ &\le K|\mathbf{y}_0| \exp[K_1 K(t - t_0)] \qquad (t_0 \le t \le T) \end{aligned} \tag{4.8}$$

Therefore

$$|\psi(T, t_0, \mathbf{y}_0)| \le K|\mathbf{y}_0| \exp[K_1 K(T - t_0)] \qquad (t_0 \le T) \tag{4.9}$$

From this we can see that we can make $|\psi(T, t_0, \mathbf{y}_0)|$ small by choosing $|\mathbf{y}_0|$ sufficiently small. This together with (4.7) gives the stability. In detail, substituting (4.9) into (4.7), we obtain

$$|\psi(t, t_0, \mathbf{y}_0)| \le [K^2\,|\mathbf{y}_0| \exp(K_1 K(T - t_0))] \exp[-(\sigma - K\eta)(t - T)] \qquad (T \le t < \infty) \tag{4.10}$$

Let $K_2 = \max\,[K \exp[K_1 K(T - t_0)],\ K^2 \exp[K_1 K(T - t_0)]]$. Then from (4.8) and (4.10) we have

$$|\psi(t, t_0, \mathbf{y}_0)| \le \begin{cases} K_2\,|\mathbf{y}_0| & (t_0 \le t \le T) \\ K_2\,|\mathbf{y}_0|\, e^{-(\sigma - K\eta)(t - T)} & (T \le t < \infty) \end{cases} \tag{4.11}$$

Now for a given matrix A we can compute K and σ; we next pick any $0 < \eta < \sigma/K$ and then $T \ge t_0$ so that $|B(t)| < \eta$ for $t \ge T$. We then compute K_1, and finally also K_2. Now (see Definition 3, Section 4.2), given any $\varepsilon > 0$, choose $\delta < \varepsilon/K_2$. Then from (4.11), if $|\mathbf{y}_0| < \delta$, $|\psi(t, t_0, \mathbf{y}_0)| < \varepsilon$ for all $t \ge t_0$ so that the zero solution is stable. From (4.11) it is clear that the zero solution is globally asymptotically stable. ∎

The proof of Theorem 4.2 actually yields a slightly stronger result.

Corollary to Theorem 4.2. Let all eigenvalues of A have negative real parts, so that

$$|e^{At}| \leq Ke^{-\sigma t}$$

for some constants $K > 0$, $\sigma > 0$ *and all* $t \geq 0$. *Let* $B(t)$ *be continuous for* $0 \leq t < \infty$ *and let there exist* $T > 0$ *such that*

$$|B(t)| < \frac{\sigma}{K} \qquad [t \geq T]$$

Then the zero solution of (4.5) *is globally asymptotically stable.*

• EXERCISES

14. Prove the above corollary.

15. Prove the following result: Let all eigenvalues of A have real parts negative, and let $B(t)$ be continuous for $0 \leq t < \infty$ and such that $\int_0^\infty |B(s)|\, ds < \infty$. Then the zero solution of (4.5) is asymptotically stable. [*Hint:* Start with (4.6) and take norms; then use the Gronwall inequality.]

16. Show that if all solutions of $\mathbf{y}' = A\mathbf{y}$ are bounded and if $B(t)$ satisfies the hypothesis of Exercise 15, then all solutions of (4.5) are bounded on $t_0 < t < \infty$. Is the zero solution of (4.5) stable? (Obviously it need not be asymptotically stable.)

At this point one might be greatly tempted to conjecture that for the general case

$$\mathbf{y}' = A(t)\mathbf{y} \tag{4.12}$$

where $A(t)$ is continuous for $t_0 \leq t < \infty$, the following result should hold. If all the eigenvalues of $A(t)$ (these are now functions of t) have their real parts negative and bounded away from zero (that is, there exists a constant $\alpha > 0$ such that $\mathcal{R}\lambda_j(t) \leq -\alpha$, $j = 1, 2, \ldots, n$), then $\mathbf{y} = \mathbf{0}$ is an asymptotically stable solution of (4.12). **Unfortunately no such result holds.** For example (see [21, p. 494]), if

$$A(t) = \begin{pmatrix} -1 - 9\cos^2 6t + 12 \sin 6t \cos 6t & 12\cos^2 6t + 9 \sin 6t \cos 6t \\ -12 \sin^2 6t + 9 \sin 6t \cos 6t & -(1 + 9\sin^2 6t + 12 \sin 6t \cos 6t) \end{pmatrix}$$

the eigenvalues are $\lambda_1 = -1$, $\lambda_2 = -10$ (constants). However, as is easily verified,

$$\Phi(t) = \begin{pmatrix} e^{2t}(\cos 6t + 2\sin 6t) & e^{-13t}(\sin 6t - 2\cos 6t) \\ e^{2t}(\cos 6t - \sin 6t) & e^{-13t}(2\sin 6t + \cos 6t) \end{pmatrix}$$

is a fundamental matrix of (4.12) so that the zero solution is not even stable. It can be shown, however, that the strict negativity of all the eigenvalues of $A(t)$ for $t \geq T > 0$ plus, for example, the conditions (i) the eigenvalues of $A(\infty) = \lim_{t \to +\infty} A(t)$ all have real parts negative and (ii) the elements of $A(t)$ are continuous and have only a finite number of maxima and minima on the interval $T \leq t < \infty$, lead to the asymptotic stability of the zero solution of (4.12). Observe that neither (i) nor (ii) holds in the above example.

• EXERCISE

17. Show that for the system $\mathbf{y}' = A(t)\mathbf{y}$, where

$$A(t) = \begin{pmatrix} -1 & e^{2t} \\ 0 & -1 \end{pmatrix}$$

the solution $\mathbf{y} = \mathbf{0}$ is unstable even though the eigenvalues of $A(t)$ are $\lambda_1(t) = -1$, $\lambda_2(t) = -1$.

While there exist several general results about the asymptotic behavior of solutions of (4.12) in some special cases, little beyond what we have done can be proved at an elementary level. We remark that there is a complete theory for (4.12) if the matrix $A(t)$ is periodic, and some general results if $A(t)$ is almost periodic. Also useful for asymptotic behavior of solutions of (4.12) are the theories of "asymptotic equivalence" (Section 4.6) and the so-called "type numbers" (see, for example, [3, p. 50]). We choose not to pursue these specialized topics deeply in order to devote our attention to nonlinear systems. We should remark that one can also learn more about the asymptotic behavior of solutions of (4.12) by refining the concepts of stability and asymptotic stability. However, ordinarily it is very hard to check whether a given solution possesses this refined type of stability or not, except in the case of constant or periodic coefficients or whenever the system is autonomous, and we shall not pursue this topic further.

• EXERCISES

18. State and prove the analogue of Theorem 4.1 for the system

$$\mathbf{y}' = A(t)\mathbf{y}$$

where $A(t)$ is a continuous, periodic matrix of real period ω. [*Hint:* Use Theorem 2.12, p. 96, and the definition of multiplier or characteristic exponent.]

19. State and prove the analogue of Theorem 4.2 for the system

$$\mathbf{y}' = (A(t) + B(t))\mathbf{y}$$

where $A(t)$ is a continuous periodic matrix of real period ω.

4.4 Almost Linear Systems

Suppose that $\mathbf{y}_0$ is a critical point of the nonlinear autonomous system

$$\mathbf{y}' = \mathbf{F}(\mathbf{y}) \tag{4.13}$$

where $\mathbf{F}$ is continuous and has continuous first partial derivatives in a domain D in n-dimensional phase space. To test whether this equilibrium point is stable or not we consider (see Definitions 1 and 2, Section 4.2, and Figures 4.2 and 4.3) the solution $\boldsymbol{\psi}$ given by

$$\boldsymbol{\psi}(t) = \mathbf{y}_0 + \mathbf{z}(t) \tag{4.14}$$

Thus

$$\mathbf{z}'(t) = \mathbf{F}(\mathbf{y}_0 + \mathbf{z}(t))$$

but since $\mathbf{F}(\mathbf{y}_0) = \mathbf{0}$, $\mathbf{F}(\mathbf{y}_0 + \mathbf{z}(t)) = \mathbf{F}(\mathbf{y}_0 + \mathbf{z}(t)) - \mathbf{F}(\mathbf{y}_0)$. Applying the mean value theorem to this difference we obtain the almost linear system

$$\mathbf{z}'(t) = \mathbf{F}_\mathbf{y}(\mathbf{y}_0)\mathbf{z} + \mathbf{g}(\mathbf{z}) \tag{4.15}$$

where $\mathbf{g}(\mathbf{z})$ is continuous,

$$\lim_{|\mathbf{z}|\to 0} \frac{|\mathbf{g}(\mathbf{z})|}{|\mathbf{z}|} = 0$$

so that $\mathbf{g}(\mathbf{0}) = \mathbf{0}$, and where $\mathbf{F}_\mathbf{y}(\mathbf{y}_0)$ is **the constant matrix** whose element in the ith row and jth column is $\partial F_i/\partial y_j\,(\mathbf{y}_0)$. If the components of $\mathbf{F}(\mathbf{y})$ can be expanded in power series, $\mathbf{g}(\mathbf{z})$ would be a vector whose components are power series beginning with quadratic terms in the components of $\mathbf{z}$. Clearly, if $\mathbf{z} \equiv \mathbf{0}$ is a stable or asymptotically stable solution of (4.15), then the same will be true of the equilibrium point $\mathbf{y}_0$ of (4.13), as can be seen by applying the definition of stability or asymptotic stability to the solution $\mathbf{y} \equiv \mathbf{y}_0$ of (4.13). The above remarks suggest that we should study the stability properties of the zero solution of systems of the form

$$\mathbf{y}' = A\mathbf{y} + \mathbf{f}(t, \mathbf{y}) \tag{4.16}$$

where A is a constant matrix and where we shall assume that $\mathbf{f}$ is continuous in $(t, \mathbf{y})$ in the region $D = \{(t, \mathbf{y}) \mid 0 \le t < \infty, |\mathbf{y}| < k\}$, where $k > 0$ is some constant, $\mathbf{f}(t, \mathbf{0}) \equiv \mathbf{0}$, and $|\mathbf{f}|$ is small for small $|\mathbf{y}|$ in the sense of (4.18) below.

• **EXERCISE**

1. Show that if $\boldsymbol{\phi}(t)$ is a solution on $0 \le t < \infty$ of the system $\mathbf{y}' = \mathbf{F}(t, \mathbf{y})$, where $\mathbf{F}$ and $\mathbf{F_y}$ are continuous for $0 \le t < \infty$, $|\mathbf{y}| < k$, then to test the stability of the solution $\boldsymbol{\phi}$ (see Definitions 3 and 4, Section 4.2), it suffices to consider the stability of the zero solution of the almost linear system

$$\mathbf{z}' = A(t)\mathbf{z} + \mathbf{f}(t, \mathbf{z}) \tag{4.17}$$

where $A(t) = \mathbf{F_y}(t, \boldsymbol{\phi}(t))$ and $\mathbf{f}$ is a continuous vector satisfying

$$\lim_{|\mathbf{z}| \to 0} \frac{|\mathbf{f}(t, \mathbf{z})|}{|\mathbf{z}|} = 0$$

uniformly in t, so that $\mathbf{f}(t, \mathbf{0}) \equiv \mathbf{0}$. [*Hint:* Consider the solution $\boldsymbol{\psi}(t) = \boldsymbol{\phi}(t) + \mathbf{z}(t)$ of $\mathbf{y}' = \mathbf{F}(t, \mathbf{y})$ where $\mathbf{z}$ is "small." Now apply the same steps that led from (4.13) to (4.15).]

The basic idea is to compare solutions of (4.16) (more generally of (4.17)), called the **perturbed system,** to those of the linear system resulting from dropping all the nonlinear terms $\mathbf{f}$, called the unperturbed system. Concerning (4.16), we prove the following basic stability result, which is due to Poincaré and Perron.

Theorem 4.3. *Suppose all eigenvalues of A have negative real parts,* $\mathbf{f}(t, \mathbf{y})$ *and* $(\partial \mathbf{f}/\partial y_j)(t, \mathbf{y})$ $(j = 1, \ldots, n)$ *are continuous in* $(t, \mathbf{y})$ *for* $0 \le t < \infty$, $|\mathbf{y}| < k$ *where $k > 0$ is a constant, and* $\mathbf{f}$ *is small in the sense that*

$$\lim_{|\mathbf{y}| \to 0} \frac{|\mathbf{f}(t, \mathbf{y})|}{|\mathbf{y}|} = 0 \tag{4.18}$$

uniformly with respect to t on $0 \le t < \infty$. *Then the solution* $\mathbf{y} \equiv \mathbf{0}$ *of* (4.16) *is asymptotically stable.*

Thus, the addition of "small" (in the sense of (4.18)) nonlinear terms to the linear system $\mathbf{y}' = A\mathbf{y}$ does not affect the asymptotic stability of the zero solution of the system. However, the zero solution of (4.16) is not necessarily **globally** asymptotically stable. We shall see that the proof uses a variant of the method used to prove Theorem 4.2 (and is in fact easier than the proof of Theorem 4.2). We remark that the hypothesis of existence and continuity of $\partial \mathbf{f}/\partial y_j$ $(j = 1, \ldots, n)$ is actually unnecessary, as we could use an existence theorem (Theorem 3.2, p. 119) that does not require this hypothesis instead of Theorem 3.3 (p. 123).

Proof of Theorem 4.3. Given any point $(t_0, \mathbf{y}_0)$, $|\mathbf{y}_0| < k$, the equation (4.16) has a (local) solution $\boldsymbol{\psi}(t, t_0, \mathbf{y}_0)$ satisfying $\boldsymbol{\psi}(t_0, t_0, \mathbf{y}_0) = \mathbf{y}_0$, according to Theorem 3.3 (p. 123), existing on some, possibly small, interval to the right of t_0. For as long as this solution exists, we can express it by using the variation of constants formula (Theorem 2.6, p. 53) in the form of the integral equation

$$\boldsymbol{\psi}(t, t_0, \mathbf{y}_0) = \exp[(t - t_0)A]\mathbf{y}_0 + \int_{t_0}^{t} \exp[(t - s)A]\mathbf{f}(s, \boldsymbol{\psi}(s, t_0, \mathbf{y}_0))\, ds \tag{4.19}$$

which is equivalent to (4.16) and the initial condition $\boldsymbol{\psi}(t_0, t_0, \mathbf{y}_0) = \mathbf{y}_0$. To establish the stability of the zero solution of (4.16) we apply Definition 3, Section 4.2, and in order to do this we must show that for $|\mathbf{y}_0|$ small enough the solution $\boldsymbol{\psi}(t, t_0, \mathbf{y}_0)$ can be continued in such a way that $|\boldsymbol{\psi}|$ stays small. From the condition (4.18) we have that given any $\eta > 0$ there exists a number $\alpha > 0$, independent of t, such that $|\mathbf{f}(t, \mathbf{y})| < \eta |\mathbf{y}|$, provided $|\mathbf{y}| < \alpha$. We will naturally always take $\alpha \le k$, so that $\mathbf{f}$ will be defined. Let $|\mathbf{y}_0| < \alpha$. Because of the assumption regarding the eigenvalues of A, $\exp[(t - t_0)A]$ satisfies the estimate (4.4) and therefore, for as long as $|\boldsymbol{\psi}(t, t_0, \mathbf{y}_0)| < \alpha$, we have from (4.19)

$$|\boldsymbol{\psi}(t, t_0, \mathbf{y}_0)| \le K \exp[-\sigma(t - t_0)]\, |\mathbf{y}_0| + \int_{t_0}^{t} K\eta \exp[-\sigma(t - s)]\, |\boldsymbol{\psi}(s, t_0, \mathbf{y}_0)|\, ds \qquad (t \ge t_0)$$

Now, multiplying by $e^{\sigma t}$, we obtain from the Gronwall inequality (Theorem 1.4, p. 31)

$$\exp(\sigma t)\, |\boldsymbol{\psi}(t, t_0, \mathbf{y}_0)| \le K \exp(\sigma t_0)\, |\mathbf{y}_0| \exp[\eta K(t - t_0)]$$

and therefore

$$|\boldsymbol{\psi}(t, t_0, \mathbf{y}_0)| \le K |\mathbf{y}_0| \exp[-(\sigma - \eta K)(t - t_0)] \tag{4.20}$$

for all those $t \ge t_0$ for which $|\boldsymbol{\psi}(t)| < \alpha$. But now we choose a fixed η, $0 < \eta < \sigma/K$. Then for any given ε, $0 < \varepsilon \le \alpha$, we choose $\delta = \varepsilon/K \le \alpha$. Then if $|\mathbf{y}_0| < \delta$, it follows from (4.20) that $|\boldsymbol{\psi}(t, t_0, \mathbf{y}_0)| < \varepsilon \le \alpha$ for all $t \ge t_0$ and this bound is independent of t. Therefore, by Theorem 3.6 (p. 132) the local solution $\boldsymbol{\psi}$ can be continued to the boundary of D, that is, for all $t \ge t_0$ and (4.20) is satisfied. This means that

$$|\psi(t, t_0, \mathbf{y}_0)| \le K|\mathbf{y}_0| \exp[-(\sigma - \eta K)(t - t_0)] \quad (t_0 \le t < \infty) \quad (4.21)$$

But now we have shown much more than the stability of the zero solution of (4.16); namely, all solutions $\psi(t, t_0, \mathbf{y}_0)$ with $|\mathbf{y}_0| \le \varepsilon/K$, for any ε with $0 < \varepsilon \le \alpha$, approach zero (exponentially by (4.21)) and therefore the zero solution of (4.16) is asymptotically stable. It may appear at first glance that circular reasoning is involved. However, the reader should convince himself that the argument is not circular. ∎

The motion of a damped simple pendulum is governed by a differential equation of the form

$$\theta'' + \frac{k}{m}\theta' + \frac{g}{L}\sin\theta = 0$$

Often one replaces this by the simpler model

$$\theta'' + \frac{k}{m}\theta' + \frac{g}{L}\theta = 0$$

which can be solved explicitly (see, for example, [2, Section 3.4]). If we write these equations as systems, by setting $\theta = y_1$, $\theta' = y_2$, we obtain $\mathbf{y}' = A\mathbf{y} + \mathbf{f}(\mathbf{y})$ and $\mathbf{y}' = A\mathbf{y}$, respectively, for the two models with

$$A = \begin{pmatrix} 0 & 1 \\ -\dfrac{g}{L} & -\dfrac{k}{m} \end{pmatrix} \qquad \mathbf{f}(\mathbf{y}) = \begin{pmatrix} 0 \\ \dfrac{g}{L}(y_1 - \sin y_1) \end{pmatrix}$$

Since A has eigenvalues $-k/2m \pm (k^2/4m^2 - g/L)^{1/2}$, which both have negative real parts if k, m, g, and L are positive, and since

$$\left|\frac{g}{L}(\sin\theta - \theta)\right| = \frac{g}{L}\left|\frac{\theta^3}{3!} + \cdots\right| \le M|\theta|^3$$

for some constant M, we may apply Theorem 4.3. We see that the use of the more refined model, including the nonlinear term $\mathbf{f}(\mathbf{y})$, does not give a radically different behavior of the solution from that obtained from the simpler linear model for $|\mathbf{y}|$ sufficiently small as $t \to \infty$.

As a simple application of Theorem 4.3 we return for a moment to the two-dimensional autonomous systems studied in Section 2.8 (p. 95). There we claimed that if the origin is an **attractor** for the linear system $\mathbf{z}' = A\mathbf{z}$ where A is a 2×2 constant matrix, the same is true for the perturbed autonomous system $\mathbf{y}' = A\mathbf{y} + \mathbf{h}(\mathbf{y})$, where the perturbation terms satisfy (4.18). Because the origin is an attractor of the linear system, the matrix A has both eigenvalues with negative real parts. Thus Theorem 4.3 is applicable, and justifies the previous claim. The reader should note, however, that this does not mean that the configuration in the phase plane for the perturbed system in any way resembles the phase portrait of the linear system except in that the orbits approach the origin. It can be shown ([4, p. 382]), for example, that if the origin is a center for the linear system, it may be a center or a spiral point for the nonlinear system. This result requires quite a different analysis from what we can make there.

Theorem 4.3 has an immediate and rather obvious generalization to the system

$$\mathbf{y}' = (A + B(t))\mathbf{y} + \mathbf{f}(t, \mathbf{y}) \tag{4.22}$$

where A and $\mathbf{f}$ are exactly as in Theorem 4.3 and where the matrix $B(t)$ satisfies the hypothesis of Theorem 4.2. Several problems of considerable practical interest can be written in the form (4.22). Combining the techniques of Theorems 4.2 and 4.3, we easily prove the following result.

Theorem 4.4. *If A and $\mathbf{f}$ satisfy the hypothesis of Theorem 4.3 and if $B(t)$ is continuous for $0 \le t < \infty$ with $\lim_{t \to \infty} B(t) = 0$, then the zero solution of (4.22) is asymptotically stable.*

• **EXERCISES**

2. Prove Theorem 4.4. [*Hint:* Study the proofs of Theorems 4.2 and 4.3.]
3. Formulate and prove a corollary to Theorem 4.4 analogous to the corollary to Theorem 4.2.

In the equations for a damped simple pendulum, the linear part depends on the constants k, m, g, L, all of which are derived from experimental data and therefore subject to experimental error. This means that a complete discussion of these models should consider the effect of the addition of a linear term with small coefficients (it is to be hoped that the experimental errors are small). The content of Theorem 4.4 and its corollary (Exercise 3) is that such a term does not have a drastic effect on the solutions of the system.

Occasionally we encounter a system of the form

$$\mathbf{y}' = A\mathbf{y} + \mathbf{f}(t, \mathbf{y}) + \mathbf{h}(t, \mathbf{y}) \tag{4.23}$$

where A and $\mathbf{f}$ satisfy the hypothesis of Theorem 4.3 and where, for example, for small $|\mathbf{y}|$, say $|\mathbf{y}| < k$, $\mathbf{h}(t, \mathbf{y}) \to 0$ as $t \to +\infty$ uniformly with respect to $\mathbf{y}$

or $|\mathbf{h}(t, \mathbf{y})| \leq \lambda(t)$ where $\int_0^\infty \lambda(t)\,dt < \infty$. In such cases $\mathbf{y} \equiv \mathbf{0}$ need not be a solution of (4.23), because $\mathbf{h}(t, \mathbf{y})$ need not even depend on $\mathbf{y}$ (for example, we may have $\mathbf{h}(t, \mathbf{y}) = \mathbf{g}(t)$ where $\mathbf{g}(t) \to 0$ as $t \to +\infty$). Thus we cannot speak of the stability of the zero solution unless $\mathbf{h}(t, \mathbf{y}) \equiv 0$. Nevertheless, the techniques employed in the proofs of Theorems 4.3, 4.4 do apply and we may obtain a result on the asymptotic behavior of solutions of (4.23) as $t \to \infty$.

Theorem 4.5. *In equation* (4.23) *assume*

(i) *the eigenvalues of A all have negative real part;*
(ii) $\lim_{|\mathbf{y}| \to 0} |\mathbf{f}(t, \mathbf{y})|/|\mathbf{y}| = 0$ *uniformly in t on* $0 \leq t < \infty$;
(iii) $|\mathbf{h}(t, \mathbf{y})| \leq \lambda(t)$ *for* $0 \leq t < \infty$, $|\mathbf{y}| < k$ *for some* $k > 0$, *where λ is a continuous nonnegative function on* $0 \leq t < \infty$ *such that* $\Lambda(t) = \int_t^{t+1} \lambda(s)\,ds \to 0$ *as* $t \to \infty$.

There then exists $T_0 > 0$ such that every solution $\boldsymbol{\phi}$ of (4.23) *with $|\boldsymbol{\phi}(T)|$ small enough for any $T \geq T_0$ remains small for $t \geq T$, and* $\lim_{t \to \infty} \boldsymbol{\phi}(t) = \mathbf{0}$.

Proof. By the variation of constants formula (Theorem 2.6, p. 53), every solution $\boldsymbol{\phi}$ of (4.23) existing on some interval to the right of T satisfies the integral equation

$$\boldsymbol{\phi}(t) = e^{A(t-T)}\boldsymbol{\phi}(T) + \int_T^t e^{A(t-s)}\mathbf{f}(s, \boldsymbol{\phi}(s))\,ds + \int_T^t e^{A(t-s)}\mathbf{h}(s, \boldsymbol{\phi}(s))\,ds \tag{4.24}$$

By the hypothesis (ii), given any $\varepsilon > 0$ there exists $\delta > 0$ such that $|\mathbf{f}(s, \boldsymbol{\phi}(s))| \leq \varepsilon|\boldsymbol{\phi}(s)|$ for as long as $|\boldsymbol{\phi}(s)| \leq \delta$. Using this and (i), (iii) in (4.24), we obtain

$$|\boldsymbol{\phi}(t)| \leq Ke^{-\sigma(t-T)}|\boldsymbol{\phi}(T)| + \int_T^t \varepsilon K e^{-\sigma(t-s)}|\boldsymbol{\phi}(s)|\,ds + \int_T^t K\lambda(s)e^{-\sigma(t-s)}\,ds$$

for as long as $|\boldsymbol{\phi}(t)| \leq \delta$, for some constants $K > 0$, $\sigma > 0$ (recall (4.4), which follows from (i)). Multiplying this inequality by $e^{\sigma(t-T)}$ and applying a slight generalization of the Gronwall inequality (see Exercise 3, Section 1.7, p. 32), we obtain

$$|\boldsymbol{\phi}(t)| \leq K|\boldsymbol{\phi}(T)|\,e^{-(\sigma - K\varepsilon)(t-T)} + \int_T^t K\lambda(s)e^{-(\sigma - K\varepsilon)(t-s)}\,ds \tag{4.25}$$

for all $t \geq T$ for which $|\phi(t)| \leq \delta$. It will now be shown that if T is chosen sufficiently large, then $|\phi(t)| \leq \delta$ for all $t \geq T$. We will handle the two terms on the right side of (4.25) separately. To estimate the second one, we need the following lemma, which will be proved after the proof of Theorem 4.5 is completed.

Lemma 4.1. *If λ satisfies* (iii) *and if* $\omega > 0$, *then there exists T_0 such that*

$$\lim_{t \to \infty} \int_T^t e^{-\omega(t-s)} \lambda(s)\, ds = 0 \tag{4.26}$$

for all $T \geq T_0$.

We choose $\varepsilon < \sigma/K$, $\omega = \sigma - K\varepsilon$. By Lemma 4.1 there exists T_0 such that

$$\int_T^t K\lambda(s) e^{-(\sigma - K\varepsilon)(t-s)}\, ds \leq \frac{\delta}{2} \qquad (T \leq t < \infty)$$

for $T \geq T_0$. Then, by choosing $|\phi(T)| < \delta/2K$, we obtain $|\phi(t)| \leq \delta$ for $T \leq t < \infty$ from (4.25). Thus (4.25) is valid for $T \leq t < \infty$. From (4.25) we also see, using Lemma 4.1, that $\lim_{t \to \infty} |\phi(t)| = 0$. ∎

Note that without more information about λ we cannot say how rapidly $|\phi(t)|$ decays.

Proof of Lemma 4.1. It will first be shown that for every $T \geq 1$, $t \geq T$,

$$\int_T^t e^{\omega s} \lambda(s)\, ds \leq \int_{T-1}^t e^{\omega(s+1)} \Lambda(s)\, ds \tag{4.27}$$

To see this, we write

$$\begin{aligned} \int_{T-1}^t e^{\omega(s+1)} \Lambda(s)\, ds &= \int_{T-1}^t e^{\omega(s+1)} \left[\int_s^{s+1} \lambda(u)\, du \right] ds \\ &\geq \int_T^t \lambda(u) \left[\int_{u-1}^u e^{\omega(s+1)}\, ds \right] du \end{aligned}$$

as may be seen by interchanging the order of integration in the integral. The first iterated integral is extended over the region R in Figure 4.5. Note that in the last integral, we are integrating a nonnegative function over the smaller region in Figure 4.5. Evaluating the inner integral and using $e^{\omega} - 1 \geq \omega$

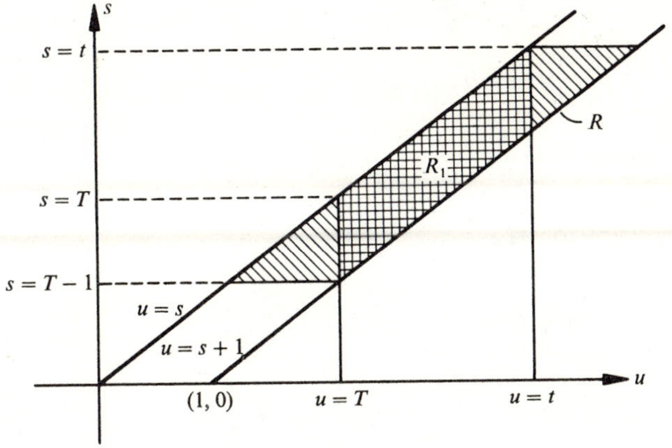

Figure 4.5

for $\omega \geq 0$, we obtain

$$\int_{T-1}^{t} e^{\omega(s+1)}\Lambda(s)\,ds \geq \int_{T}^{t} \lambda(u)e^{\omega u}\left(\frac{e^{\omega}-1}{\omega}\right)du \geq \int_{T}^{t} e^{\omega u}\lambda(u)\,du$$

which proves (4.27). To complete the proof of Lemma 4.1, given $\varepsilon > 0$, choose T_0 so that $\Lambda(s) < \varepsilon$ for all $s > T_0 - 1$. Then from (4.27)

$$\begin{aligned}\int_{T}^{t} e^{-\omega(t-s)}\lambda(s)\,ds &= e^{-\omega t}\int_{T}^{t} e^{\omega s}\lambda(s)\,ds\\ &\leq e^{-\omega t}\int_{T-1}^{t} e^{\omega(s+1)}\Lambda(s)\,ds\\ &\leq \varepsilon e^{-\omega t}\int_{T-1}^{t} e^{\omega(s+1)}\,ds < \frac{\varepsilon e^{\omega}}{\omega}\end{aligned}$$

Thus the integral can be made small by suitable choice of T_0, and every $T \geq T_0$. To prove (4.26) (T_0 having been fixed), given an arbitrary $\eta > 0$, $\tau > T \geq T_0$ so that $\Lambda(s) < \eta$ for $s > \tau - 1$, we write

$$\begin{aligned}\int_{T}^{t} e^{-\omega(t-s)}\lambda(s)\,ds &= \int_{T}^{\tau} e^{-\omega(t-s)}\lambda(s)\,ds + \int_{\tau}^{t} e^{-\omega(t-s)}\lambda(s)\,ds\\ &= e^{-\omega t}\int_{T}^{\tau} e^{\omega s}\lambda(s)\,ds + \int_{\tau}^{t} e^{-\omega(t-s)}\lambda(s)\,ds\end{aligned}$$

For the given η we may choose $t \geq T$ so large that

$$e^{-\omega t} \int_T^\tau e^{\omega s} \lambda(s)\, ds \leq \eta$$

By the argument used above,

$$\int_\tau^t e^{-\omega(t-s)} \lambda(s)\, ds \leq \eta \frac{e^\omega}{\omega}$$

for $t \geq \tau > T$. Therefore

$$\int_T^t e^{-\omega(t-s)} \lambda(s)\, ds \leq \eta \left(1 + \frac{e^\omega}{\omega}\right)$$

for t sufficiently large. Since η is arbitrary, this proves (4.26). ▮

• EXERCISES

4. Show that if $\lambda(t)$ is a continuous nonnegative function on $0 \leq t < \infty$ such that $\lim_{t\to\infty} \lambda(t) = 0$, then

$$\lim_{t\to\infty} \Lambda(t) = \lim_{t\to\infty} \int_t^{t+1} \lambda(s)\, ds = 0$$

5. Show that if $\lambda(t)$ is a continuous nonnegative function on $0 \leq t < \infty$ such that $\int_0^\infty \lambda(s)\, ds$ is finite, then

$$\lim_{t\to\infty} \Lambda(t) = \lim_{t\to\infty} \int_t^{t+1} \lambda(s)\, ds = 0$$

Exercises 4 and 5 give conditions under which hypothesis (iii) of Theorem 4.5 is satisfied.

Theorem 4.3 and its generalizations have several serious drawbacks from the practical point of view. In the first place, no hint is given as to how close to zero a solution must start in order to approach zero. No method is given in these theorems for estimating the region of asymptotic stability. We have already pointed out (see Exercise 3, Section 4.2) that when dealing with nonlinear problems we cannot expect the results to be global; thus it would be unreasonable to expect **all** solutions (with arbitrary initial conditions) in Theorem 4.3 to approach zero. If the region of stability or asymptotic stability is not the whole space, an estimate of it is of utmost importance. To illustrate why, suppose for example that the steady state (equilibrium) speed of some mechanical system is 17,000 mph and the best that can be done to estimate δ_0 in the definition of asymptotic stability is $\delta_0 = 0.5$; the result that the equilibrium solution is asymptotically stable is of little practical value, unless it can also be shown that perturbations of more than 0.5 from equilibrium at time $t = t_0$ lead to solutions that tend to move away from equilibrium. If such is the case, the given system is probably too sensitive to disturbances to be of any practical use.

In the second place, if the zero solution of the unperturbed linear system is stable but not asymptotically stable (see Theorem 4.1), Theorem 4.3 and its generalizations give no information about the behavior of the perturbed system. In particular, no information is obtained by this method if the matrix $A \equiv 0$. There is a very good reason for this. In such cases it is usually the nonlinear terms that determine the stability properties of the system. For example, consider the scalar equations $u' = -u^3$ and $u' = u^3$ (here A of (4.16) is the 1×1 zero matrix, so that all solutions of the unperturbed system in both cases are $u \equiv$ constant and the zero solution of the unperturbed system is stable but not asymptotically stable). By explicit solution (separation of variables) we see that for the first equation $u \equiv 0$ is globally asymptotically stable, while for the second $u \equiv 0$ is unstable. The methods employed so far are simply too crude to detect these differences. The Lyapunov method (Chapter 5) provides the only general technique available at present for handling such problems, although it is not always easy to apply in particular cases.*

In more advanced treatments, such as [3, 5, 21], the reader may find several other generalizations of Theorem 4.3 to the case where the unperturbed linear system has a variable coefficient matrix. These generalizations require the concept of uniform asymptotic stability, which we are not considering. To see that more than simple asymptotic stability of the unperturbed system is needed, consider the system

$$\begin{aligned} y_1' &= -ay_1 \\ y_2' &= (\sin \log t + \cos \log t - 2a)y_2 + {y_1}^2 \end{aligned} \tag{4.28}$$

which is of the form $\mathbf{y}' = A(t)\mathbf{y} + \mathbf{f}(t, \mathbf{y})$ with

$$A = \begin{pmatrix} -a & 0 \\ 0 & (\sin \log t + \cos \log t - 2a) \end{pmatrix} \qquad \mathbf{f}(t, \mathbf{y}) = \begin{pmatrix} 0 \\ {y_1}^2 \end{pmatrix}$$

---

* We do not mean to imply that the Lyapunov second method considered in Chapter 5 provides the answer to all stability questions. There are many important problems, such as the stability theory of Hamiltonian systems and stability theory of periodic solutions, in which it is of limited use.

Observe that $\mathbf{f}(t, \mathbf{y})$ does satisfy (4.18). Now the unperturbed system $\mathbf{z}' = A(t)\mathbf{z}$ has

$$\Psi(t) = \begin{pmatrix} e^{-at} & 0 \\ 0 & e^{t \sin \log t - 2at} \end{pmatrix}$$

as a fundamental matrix (by direct integration). Thus every solution of the unperturbed system is of the form $\Psi(t)\mathbf{c}$, where $\mathbf{c}$ is a constant vector, and this approaches zero for every $\mathbf{c}$ as $t \to \infty$, provided $1 < 2a$. Now if we look at the perturbed system, we see that its explicit solution $\boldsymbol{\phi}(t, t_0, \mathbf{y}_0)$ is (here it is convenient to write out the components, using $\mathbf{c} = (c_1, c_2) = \Psi^{-1}(t_0)\mathbf{y}_0$)

$$\begin{aligned} \phi_1(t, t_0, \mathbf{y}_0) &= c_1 e^{-at} \\ \phi_2(t, t_0, \mathbf{y}_0) &= e^{t \sin \log t - 2at}\left(c_2 + c_1{}^2 \int_0^t e^{-s \sin \log s}\, ds\right) \end{aligned} \tag{4.29}$$

as can be verified by the variation of constants formula in the second equation of (4.28).

• EXERCISE

6. Carry out the derivation of (4.29).

It can now be shown by a rather intricate analysis (see, for example, [3, p. 47]) that if we further restrict a to the interval $1 < 2a < 1 + e^{-\pi}/2$, then $\phi_2(t, t_0, \mathbf{y}_0) \to 0$ **only if** $c_1 = 0$. This shows that in the case of variable coefficients in the linear unperturbed equation, asymptotic stability of the zero solution of the unperturbed equation does **not** imply asymptotic stability of the perturbed equation, so that the analogue of Theorem 4.3 is not true without further restriction.

• EXERCISES

7. Show that if $c_1 \neq 0$, $|\phi_2(t, t_0, \mathbf{y}_0)|$ is unbounded, where ϕ_2 is defined in (4.29). This establishes the above conclusion. [*Hint:* Let $\{t_n\}_{n=1}^{\infty}$ be the sequence with $t_n = \exp[(2n + \frac{1}{2})\pi]$; then $\sin \log t_n = 1$, $n = 1, 2, \ldots$, and $-\sin \log s \geq \frac{1}{2}$ on the interval $\exp[(2n - \frac{1}{2})\pi] \leq s \leq \exp[(2n - \frac{1}{6})\pi]$. Show that

$$\int_0^t \exp(-s \sin \log s)\, ds \geq t_n\left[\exp\left(\frac{-2\pi}{3}\right) - \exp(-\pi)\right]\exp(\tfrac{1}{2}t_n e^{-\pi})$$

so that $|\phi_2(t, t_0, \mathbf{y}_0)| \geq |c_1|^2 \exp\{[1 - 2a + \frac{1}{2}e^{-\pi}]t_n\} + \alpha(n)$, where $\alpha(n)$ is a function that tends to 0 as $n \to \infty$. Since $1 - 2a + (e^{-\pi}/2) > 0$, this proves the result.]

8. State Theorem 4.3 and its generalizations for the scalar equation $u'' + a_1 u' + a_2 u = g(t, u, u')$ where a_1 and a_2 are real constants. Be sure to specify the condition that must be satisfied by a_1, a_2.

4.5 Conditional Stability

We now examine briefly what happens if the unperturbed system is unstable and we add perturbation terms of the type considered in Theorem 4.3. Up to now we have always started by assuming that the zero solution of the unperturbed system is asymptotically stable. As before, we assume that the unperturbed system has constant coefficients. Because the geometric interpretation is easy, we shall suppose that the coefficient matrix A is a 2×2 matrix with real coefficients having one eigenvalue positive and the other negative. We shall also assume that the perturbation terms do not depend on t explicitly, in order to permit the phase plane interpretation; however, this is not at all necessary. Thus we consider the autonomous two-dimensional system

$$\mathbf{y}' = A\mathbf{y} + \mathbf{g}(\mathbf{y}) \tag{4.30}$$

and by hypothesis the unperturbed system has a saddle point at the origin. (See Section 2.8, p. 92.) Without loss of generality we may suppose that a change of variable $\mathbf{y} = T\mathbf{z}$ has already been made so that A is in the simple canonical form

$$A = \begin{pmatrix} -\mu & 0 \\ 0 & \lambda \end{pmatrix} \qquad \lambda, \mu > 0 \tag{4.31}$$

Thus the unperturbed system $\mathbf{z}' = A\mathbf{z}$ has the phase portrait given in Figure 4.6. This has the important property that if a solution of $\mathbf{z}' = A\mathbf{z}$ starts on the z_1 axis, then it approaches the origin. If, on the other hand, a solution does not start on the z_1 axis, then it moves away (in fact, exponentially) from the origin as $t \to \infty$. The idea of our next result is to show that if the perturbation terms $\mathbf{g}(\mathbf{y})$ are suitably restricted, then there exists a curve C passing through the origin of $\mathbf{y}$-space such that if a solution of (4.30) starts away from C close enough to the origin, then it cannot approach the origin. (It in fact moves away from the origin.) This is the case of **the saddle point for the perturbed system.**

Theorem 4.6. *Let* $\mathbf{g}$, $\partial \mathbf{g}/\partial y_j$ $(j = 1, 2)$ *be continuous for* $|\mathbf{y}| < k$ *for some constant* $k > 0$ (k can be small), *and let* $\mathbf{g}(\mathbf{0}) = \mathbf{0}$ *and* $\lim_{|\mathbf{y}| \to 0} |\partial \mathbf{g}/\partial y_j| = 0 \, (j = 1, 2)$. *If the eigenvalues of* A *are* λ, $-\mu$, *with* λ, $\mu > 0$, *then there exists in* $\mathbf{y}$ *space a*

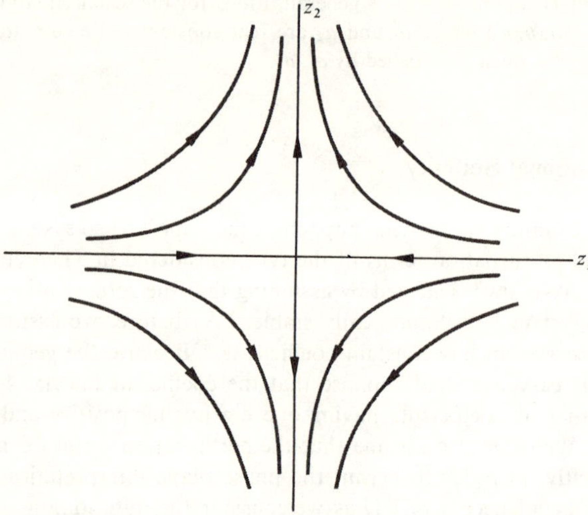

Figure 4.6

real curve C passing through the origin such that if ϕ is any solution of (4.30) *with* $\phi(0)$ (or $\phi(t_0)$) *on C and* $|\phi(0)|$ *small enough, then* $\phi(t) \to 0$ *as* $t \to \infty$. *Moreover, no solution* $\phi(t)$ *with* $|\phi(0)|$ *small enough, but not on C, can remain small for* $t \geq 0$; *in particular, the zero solution of* (4.30) *is unstable.*

For obvious reasons this type of behavior is sometimes called **conditional stability of the zero solution.** A similar behavior occurs in the general case with A an $n \times n$ matrix having k eigenvalues with negative real parts and $(n - k)$ eigenvalues with positive real parts. There are k linearly independent solutions of the unperturbed system that approach zero and $(n - k)$ linearly independent solutions of the unperturbed system that tend away from zero. When nonlinear terms satisfying suitable hypotheses are added, there exists a surface S of dimension k containing the origin of **y** space, and solutions of the full system with respect to S behave in the same way that those of (4.30) behave with respect to the curve C. For a precise statement and proof in the general (even nonautonomous) case see [4, p. 330].

Before proceeding with the proof, we remark again that we will not be able to consider the case when the matrix A has one or more eigenvalues with zero real part in this setting. This, as might be expected, is the really difficult case and except for isolated instances that may arise in the sequel (for example, Theorem 4.7; Example 3, Section 5.2, p. 197; and Sections 6.1, 6.2), we consider this case to be beyond the scope of this book.

Proof of Theorem 4.6. We remark that because (4.30) is autonomous we can assume that $t_0 = 0$ without loss of generality. Referring to (4.31), define the matrices

$$U_1(t) = \begin{pmatrix} e^{-\mu t} & 0 \\ 0 & 0 \end{pmatrix} \qquad U_2(t) = \begin{pmatrix} 0 & 0 \\ 0 & e^{\lambda t} \end{pmatrix} \tag{4.32}$$

so that

$$e^{tA} = U_1(t) + U_2(t)$$

Then

$$|U_1(t-s)| = e^{-\mu(t-s)}$$

and

$$|U_2(t-s)| = e^{\lambda(t-s)}$$

and therefore (we shall see in the proof why this is convenient) there exist constants $\alpha, \sigma > 0$ (just pick $0 < \alpha + \sigma \le \mu$ and $\sigma \le \lambda$) such that

$$|U_1(t-s)| \le e^{-(\alpha+\sigma)(t-s)} \qquad \text{if} \quad t \ge s \tag{4.33a}$$

and

$$|U_2(t-s)| \le e^{\sigma(t-s)} \qquad \text{if} \quad t \le s \tag{4.33b}$$

Using the hypothesis concerning $\mathbf{g}$, given $\varepsilon > 0$, we may choose $\delta > 0$ such that $0 < \delta < k$ and

$$|\mathbf{g}(\mathbf{y}) - \mathbf{g}(\mathbf{y}^*)| \le \varepsilon |\mathbf{y} - \mathbf{y}^*| \qquad (|\mathbf{y}|, |\mathbf{y}^*| \le \delta) \tag{4.34}$$

(Consider $\mathbf{g}(\mathbf{y}) - \mathbf{g}(\mathbf{y}^*)$, apply the mean value theorem component-wise, and then take norms.)

The proof is divided into several parts. We first assume that (4.30) has a solution $\mathbf{y}(t)$ with $|\mathbf{y}(t)|$ small (at any rate less than k), and this solution exists for $0 \le t < \infty$; we show that $\mathbf{y}(t)$ satisfies the integral equation

$$\mathbf{y}(t) = U_1(t)\mathbf{y}(0) + \int_0^t U_1(t-s)\mathbf{g}(\mathbf{y}(s))\, ds - \int_t^\infty U_2(t-s)\mathbf{g}(\mathbf{y}(s))\, ds \tag{4.35}$$

Define the function

$$\mathbf{z}(t) = \mathbf{y}(t) - U_1(t)\mathbf{y}(0) - \int_0^t U_1(t-s)\mathbf{g}(\mathbf{y}(s))\,ds + \int_t^\infty U_2(t-s)\mathbf{g}(\mathbf{y}(s))\,ds \tag{4.36}$$

It is an elementary exercise to see that $|\mathbf{z}(t)|$ is bounded on $0 \le t < \infty$. To prove this, we make use of (4.33), and the assumption (4.34) that implies that $|\mathbf{g}(\mathbf{y}(t))|$ is bounded whenever $|\mathbf{y}(t)| \le \delta$. From (4.36) we readily compute that $\mathbf{z}(t)$ satisfies

$$\mathbf{z}' = A\mathbf{z}. \tag{4.37}$$

• **EXERCISE**

1. Prove (4.37).

Now what initial conditions does $\mathbf{z}(t)$ satisfy? Clearly

$$\mathbf{z}(0) = \mathbf{y}(0) - U_1(0)\mathbf{y}(0) + \int_0^\infty U_2(-s)\mathbf{g}(\mathbf{y}(s))\,ds$$

and when written out, using $U_1(0)U_1(0) = U_1(0)$, $U_1(0)U_2(-s) = 0$, this yields $U_1(0)\mathbf{z}(0) = \mathbf{0}$. Therefore $z_1(0) = 0$, where $\mathbf{z} = (z_1, z_2)$. Now if $\mathbf{z}(0) \neq 0$, then $U_2(0)\mathbf{z}(0) = z_2(0) \neq 0$ because $U_1(0) + U_2(0)$ is the identity matrix. But then we have that $\mathbf{z}(t)$ given by (4.36) is a bounded solution of (4.37) with $z_1(0) = 0$, $z_2(0) \neq 0$; this means that $z_1(t) \equiv 0$, $z_2(t) = z_2(0)e^{\lambda t}$, $\lambda > 0$, and thus clearly $|\mathbf{z}(t)|$ cannot be bounded unless $z(t) \equiv 0$. This shows that $\mathbf{y}(t)$ satisfies (4.35).

We next show that (4.35) has a solution $\mathbf{w}(t)$ that stays small if $|\mathbf{y}(0)|$ is small enough (we hardly need to point out that $\mathbf{w}(t)$ also satisfies (4.30)). Define the successive approximations

$$\mathbf{w}_0(t) = \mathbf{0}$$

$$\mathbf{w}_{n+1}(t) = U_1(t)\mathbf{y}(0) + \int_0^t U_1(t-s)\mathbf{g}(\mathbf{w}_n(s))\,ds - \int_t^\infty U_2(t-s)\mathbf{g}(\mathbf{w}_n(s))\,ds$$

$$(n = 0, 1, \ldots) \tag{4.38}$$

We have, from (4.33), $\mathbf{g}(\mathbf{0}) = \mathbf{0}$, and (4.38),

$$|\mathbf{w}_1(t) - \mathbf{w}_0(t)| \le |U_1(t)|\,|\mathbf{y}(0)| < e^{-(\alpha+\sigma)t}|\mathbf{y}(0)| \le e^{-\alpha t}|\mathbf{y}(0)| \qquad (t \ge 0)$$

We now choose ε (see (4.34)) such that $\varepsilon/\sigma < \frac{1}{4}$; this determines a δ in (4.34). We then prove by induction (note that the case $k = 0$ is already verified)

$$|\mathbf{w}_{k+1}(t) - \mathbf{w}_k(t)| \leq \frac{|\mathbf{y}(0)|}{2^k} e^{-\alpha t} \qquad (t \geq 0) \tag{4.39}$$

• **EXERCISE**

2. Prove (4.39).

Define

$$\mathbf{w}(t) = \mathbf{w}_0(t) + (\mathbf{w}_1(t) - \mathbf{w}_0(t)) + \cdots + (\mathbf{w}_n(t) - \mathbf{w}_{n-1}(t)) + \cdots$$

Then

$$|\mathbf{w}(t)| \leq |\mathbf{y}(0)|\, e^{-\alpha t}\left(1 + \frac{1}{2} + \cdots + \frac{1}{2^k} + \cdots\right) = 2\,|\mathbf{y}(0)|\, e^{-\alpha t} \qquad (t \geq 0)$$

and clearly if $|\mathbf{y}(0)| < \delta/2$ (of (4.34)), then $\{\mathbf{w}_n(t)\}$ is a well-defined sequence that converges **uniformly** for all $t \geq 0$ and $\mathbf{w}(t)$ satisfies the estimate

$$|\mathbf{w}(t)| < \delta e^{-\alpha t} \qquad (t \geq 0) \tag{4.40}$$

so that it indeed remains small to $t \geq 0$. The reader should have no difficulty in verifying that $\mathbf{w}$ is the desired solution.

• **EXERCISE**

3. Show that the limit function $\mathbf{w}(t)$ satisfies the integral equation (4.35) and hence by equivalence also (4.30).

We next establish the uniqueness of solutions of (4.35) that are "small." (Note that it is obvious that under our hypotheses the differential equation (4.30) has a unique solution of the initial value problem. This, however, does not imply the result for (4.35) without proof.) For the same initial vector $\mathbf{y}(0)$ let $\mathbf{w}(t, \mathbf{y}(0))$, $\boldsymbol{\theta}(t, \mathbf{y}(0))$ be solutions of (4.35) such that $|\mathbf{w}(t, \mathbf{y}(0))|$, $|\boldsymbol{\theta}(t, \mathbf{y}(0))| \leq \delta$ for $t \geq 0$. Define $M = \underset{(t \geq 0)}{\text{lub}}\, |\mathbf{w}(t, \mathbf{y}(0)) - \boldsymbol{\theta}(t, \mathbf{y}(0))|$. Then from (4.35) we have, using (4.33), (4.34),

$$|\mathbf{w}(t, \mathbf{y}(0)) - \boldsymbol{\theta}(t, \mathbf{y}(0))| \leq \varepsilon e^{-\sigma t} \int_0^t e^{\sigma s} |\mathbf{w}(s, \mathbf{y}(0)) - \boldsymbol{\theta}(s, \mathbf{y}(0))|\, ds$$
$$+ \varepsilon e^{\sigma t} \int_t^{\infty} e^{-\sigma s} |\mathbf{w}(s, \mathbf{y}(0)) - \boldsymbol{\theta}(s, \mathbf{y}(0))|\, ds$$

Taking the least upper bound under the integral sign on the right and then integrating, we obtain

$$|\mathbf{w}(t, \mathbf{y}(0)) - \boldsymbol{\theta}(t, \mathbf{y}(0))| \leq \frac{\varepsilon}{\sigma} M(1 - e^{-\sigma t}) + \frac{\varepsilon}{\sigma} M \leq \frac{2\varepsilon M}{\sigma}$$

and thus also $M \leq (2\varepsilon/\sigma)M$. Since $\varepsilon < \sigma/4$ we get $M \leq M/2$, which is a contradiction, and this proves uniqueness of solutions of the integral equation (4.35).

We now return to the integral equation (4.35) to establish the existence of the curve C.

Define the vector $\mathbf{a} = (a_1, 0)$. From (4.35) we see that

$$\mathbf{w}(0) = U_1(0)\mathbf{y}(0) - \int_0^\infty U_2(-s)\mathbf{g}(\mathbf{w}(s))\, ds \tag{4.41}$$

but because of the special nature of U_1 and U_2 (see (4.32)) only the first component of $U_1(t)\mathbf{y}(0)$ (hence of $\mathbf{y}(0)$), and only the second component of the integral are relevant. Remembering (Theorem 3.7, p. 135) that the solution $\mathbf{w}$ is a continuous function of the initial conditions ($\mathbf{w} = \mathbf{w}(t, \mathbf{a})$), we put $\mathbf{y}(0) = \mathbf{a}$ and we have (writing out the components)

$$w_1(0, a) = a_1$$

$$w_2(0, a) = -\int_0^\infty U_2(-s)\mathbf{g}(\mathbf{w}(s, \mathbf{a}))\, ds$$

For small $|\mathbf{y}(0)|$, that is, for small $|\mathbf{a}|$, the equation

$$\psi(a_1) = w_2(0, a) = -\int_0^\infty U_2(-s)\mathbf{g}(\mathbf{w}(s, a))\, ds = 0$$

defines a curve C passing through the origin. It is clear from (4.40) that if the initial point $(y_1(0), y_2(0)) = (a_1, \psi(a_1))$ lies on C with $|a_1|$ small (so that $|a_1| + |\psi(a_1)| < \delta$), then the corresponding solution of (4.35) (hence of (4.30)) approaches zero.

It remains to be shown that if $\mathbf{r}(t)$ is any solution of (4.30) with $\mathbf{r}(0)$ **small but not on** C, then one cannot have $|\mathbf{r}(t)| \leq \delta$ for $t \geq 0$, where δ is defined by (4.34) and $\delta < k$. Suppose that $|\mathbf{r}(t)| < \delta$ for $t \geq 0$. Then integration of (4.30) and use of variation of constants gives

$$\mathbf{r}(t) = e^{tA}\mathbf{r}(0) + \int_0^t e^{(t-s)A}\mathbf{g}(\mathbf{r}(s))\, ds$$

Now using (4.32), we can write this in the form

$$\mathbf{r}(t) = U_1(t)\mathbf{r}(0) + U_2(t)\mathbf{r}(0) + \int_0^t U_1(t-s)\mathbf{g}(\mathbf{r}(s))\,ds$$
$$+ \int_0^\infty U_2(t-s)\mathbf{g}(\mathbf{r}(s))\,ds - \int_t^\infty U_2(t-s)\mathbf{g}(\mathbf{r}(s))\,ds \qquad (4.42)$$

where all integrals converge because of the bound (4.33) on $|U_2(t-s)|$, and because $|\mathbf{g}(\mathbf{r}(s))|$ is bounded when $|\mathbf{r}(s)| \le \delta$. Since (see the definition of U_2, (4.32))

$$\int_0^\infty U_2(t-s)\mathbf{g}(\mathbf{r}(s))\,ds = U_2(t)\int_0^\infty U_2(-s)\mathbf{g}(\mathbf{r}(s))\,ds$$

(4.42) can be written as

$$\mathbf{r}(t) = U_1(t)\mathbf{r}(0) + U_2(t)\mathbf{c} + \int_0^t U_1(t-s)\mathbf{g}(\mathbf{r}(s))\,ds$$
$$- \int_t^\infty U_2(t-s)\mathbf{g}(\mathbf{r}(s))\,ds \qquad (4.43)$$

where the constant vector $\mathbf{c}$ is

$$\mathbf{c} = \mathbf{r}(0) + \int_0^\infty U_2(-s)\mathbf{g}(\mathbf{r}(s))\,ds$$

Since $|\mathbf{r}(t)| \le \delta$, by hypothesis, the left-hand side is bounded for $0 \le t < \infty$; now, by the argument already used several times, each term on the right-hand side of (4.43) is also bounded, except possibly the term $U_2(t)\mathbf{c}$. But from (4.32) $U_2(t)\mathbf{c}$ is the vector $(0, e^{\lambda t}c_2)$, and $\lambda > 0$. Since the left-hand side of (4.43) is bounded, the right-hand side of (4.43) must also be bounded, and this implies that $c_2 = 0$. But if $c_2 = 0$, $U_2(t)\mathbf{c} = \mathbf{0}$, and therefore $\mathbf{r}(t)$ is a solution of (4.35) (compare (4.43) and (4.35) with $U_2(t)\mathbf{c} = \mathbf{0}$). But by uniqueness of "small" solutions of (4.35) already established, since $|\mathbf{r}(t)| < \delta$, $\mathbf{r}(0)$ must be a point on the curve C and therefore we have contradicted the assumption that $\mathbf{r}(0)$ is not on C. This completes the proof of Theorem 4.6. ▌

• **EXERCISE**

4. Show that the simple undamped pendulum, governed by the equation

$$\theta'' + \frac{g}{L}\sin\theta = 0$$

has an unstable (actually conditionally stable) equilibrium point at $\theta = \pi$. [*Hint:* (a) Write as a system

$$y_1' = y_2$$

$$y_2' = -\frac{g}{L}\sin y_1$$

(b) Make the change of variable $v_1 = y_1 - \pi$, $v_2 = y_2$ and show that the corresponding linear unperturbed system has a saddle point at $v_1 = 0$, $v_2 = 0$. Then apply Theorem 4.6.]

4.6 Asymptotic Equivalence

We shall now make a few remarks concerning the concept of **asymptotic equivalence** of two systems. This notion has already been touched on indirectly; it can sometimes be useful in studying asymptotic behavior. Consider the two systems of n equations

$$\mathbf{x}' = A(t)\mathbf{x} \tag{4.44}$$

$$\mathbf{y}' = A(t)\mathbf{y} + \mathbf{f}(t, \mathbf{y}) \tag{4.45}$$

where $A(t)$ is continuous for $t_0 \le t < \infty$ ($t_0 \ge 0$), and $\mathbf{f}(t, \mathbf{y})$ is continuous for $t_0 \le t < \infty$ and $|\mathbf{y}| < k$, where $k > 0$ is a constant.

Definition. *We say that the systems* (4.44) *and* (4.45) *are asymptotically equivalent if to each solution* $\mathbf{x}(t)$ *of* (4.44) *with* $|\mathbf{x}(t_0)|$ *sufficiently small there corresponds a solution* $\mathbf{y}(t)$ *of* (4.45) *such that*

$$\lim_{t\to\infty} |\mathbf{y}(t) - \mathbf{x}(t)| = 0$$

and if to each solution $\hat{\mathbf{y}}(t)$ *of* (4.45) *with* $|\hat{\mathbf{y}}(t_0)|$ *sufficiently small there corresponds a solution* $\hat{\mathbf{x}}(t)$ *of* (4.44) *such that*

$$\lim_{t\to\infty} |\hat{\mathbf{y}}(t) - \hat{\mathbf{x}}(t)| = 0$$

We remark that the solutions $\mathbf{x}(t)$ and $\mathbf{y}(t)$ (or $\hat{\mathbf{x}}(t)$ and $\hat{\mathbf{y}}(t)$) need not both satisfy the same initial conditions, and that they need only be defined for sufficiently large t.

Example 1. The scalar equations

$$x' = 1 \qquad y' = 1 + \frac{1}{t^2}$$

are asymptotically equivalent. Every solution of the first equation has the form $x(t) = t + c_1$ for some constant c_1, and every solution of the second equation has the form $y(t) = t - 1/t + c_2$ for some constant c_2. The equivalence in both directions is obtained by taking $c_1 = c_2$. Note that $c_1 = x(1) - 1$ but $c_2 = y(1)$.

If $A(t)$ is a constant matrix whose eigenvalues all have negative real parts and if $\mathbf{f}$ satisfies the hypothesis of Theorem 4.3, then (4.44) and (4.45) are asymptotically equivalent (in fact much more is true!).

Example 2. The scalar equations

$$x' = -x \qquad y' = -y + y^2$$

are asymptotically equivalent because every solution of each equation **with sufficiently small initial value** tends to zero as $t \to \infty$. Note, however, that there is no solution $x(t)$ of the first equation corresponding to the solution $y(t) \equiv 1$ of the second equation such that $\lim_{t\to\infty} |x(t) - y(t)| = 0$. This explains why the definition of asymptotic equivalence specified small initial values.

Example 3. We have shown in Exercise 4, Section 3.1 (p. 111) and Exercise 13, Section 3.1 (p. 118) that the integral equation

$$y(t) = e^{it} + \alpha \int_t^\infty \sin(t - s) \frac{y(s)}{s^2}\, ds$$

where α is a given constant, has a bounded solution $\phi(t)$, and that $\phi(t)$ is a solution of the differential equation

$$y'' + \left(1 + \frac{\alpha}{t^2}\right) y = 0 \tag{4.46}$$

By the same argument, we could establish the existence of a bounded solution $\phi(t)$, $|\phi(t)| \leq K$, of the integral equation

$$y(t) = c_1 e^{it} + c_2 e^{-it} + \alpha \int_t^\infty \sin(t - s) \frac{y(s)}{s^2}\, ds \tag{4.47}$$

for every c_1, c_2, and show that $\phi(t)$ is a solution of (4.46). Thus

$$|\phi(t) - (c_1 e^{it} + c_2 e^{-it})| \le |\alpha| \int_t^\infty \frac{|\phi(s)|}{s^2}\, ds \le K|\alpha| \int_t^\infty \frac{ds}{s^2} = \frac{K|\alpha|}{t}$$

and $\lim_{t\to\infty} |\phi(t) - (c_1 e^{it} + c_2 e^{-it})| = 0$. This shows that corresponding to a solution $c_1 e^{it} + c_2 e^{-it}$ of the equation $x'' + x = 0$ there is a solution $\phi(t)$ of (4.46) such that $\lim_{t\to\infty} |\phi(t) - (c_1 e^{it} + c_2 e^{-it})| = 0$. Since a solution of (4.46) satisfies (4.47) for some choice of c_1, c_2, to every solution $\phi(t)$ of (4.46) corresponds a solution $\hat{c}_1 e^{it} + \hat{c}_2 e^{-it}$ of $x'' + x = 0$ such that

$$\lim_{t\to\infty} |\hat{\phi}(t) - (\hat{c}_1 e^{it} + \hat{c}_2 e^{-it})| = 0.$$

Therefore the second-order equations $x'' + x = 0$ and (4.46) are asymptotically equivalent. As (4.46) can be obtained from the Bessel equation by a change of dependent variable (see, for example, [2, p. 171]), this asymptotic equivalence is of great value in studying the behavior of the Bessel functions $J_p(t)$ and $Y_p(t)$ for large t.

This last example suggests that asymptotic equivalence might be useful where some stronger concept, such as asymptotic stability, might not be obtainable. In particular, suppose $A(t) = A$ (a constant matrix) and we know that all solutions of (4.44) are bounded, but do not necessarily tend to zero. This happens if A has at least one eigenvalue with zero real part while the remaining ones have negative real parts. Then (4.44) has a periodic solution, $\mathbf{p}(t)$. If (4.44) and (4.45) are asymptotically equivalent, (4.45) must have a solution $\mathbf{y}(t)$ that behaves, asymptotically as $t \to +\infty$, like the periodic solution $\mathbf{p}(t)$. **This situation is not covered by any of the theorems discussed so far.**

A simple nontrivial result of this type on asymptotic equivalence is the following concerning linear systems, due to N. Levinson (1946).

Theorem 4.7. *Let A be a constant matrix such that all solutions of*

$$\mathbf{x}' = A\mathbf{x} \tag{4.48}$$

are bounded on $0 \le t < \infty$. Let $B(t)$ be a continuous matrix such that

$$\int_0^\infty |B(s)|\, ds < \infty \tag{4.49}$$

Then (4.48) *and the system*

$$\mathbf{y}' = (A + B(t))\mathbf{y} \tag{4.50}$$

are asymptotically equivalent.

Sketch of Proof. It is first shown that under the hypothesis all solutions of (4.50) are also bounded on $0 \le t < \infty$. (This is done using the variation of constants formula; see Exercise 16, Section 4.3.) Without loss of generality we next assume that a linear change of variable $\mathbf{x} = T\mathbf{z}$ has been made in (4.48) and T chosen so that $T^{-1}AT$ is of the form

$$\begin{pmatrix} A_1 & 0 \\ 0 & A_2 \end{pmatrix}$$

where A_1 is a matrix having all those eigenvalues of A with real parts negative and A_2 has all those eigenvalues of A with real parts zero. We assume that A is already in this form and, as in the proof of Theorem 4.6, we define the matrices

$$U_1(t) = \begin{pmatrix} e^{tA_1} & 0 \\ 0 & 0 \end{pmatrix} \qquad U_2(t) = \begin{pmatrix} 0 & 0 \\ 0 & e^{tA_2} \end{pmatrix}$$

It now follows from Theorem 2.10 (p. 80) and the hypothesis that there exist constant K, $\sigma > 0$, such that

$$\begin{aligned} |U_1(t-s)| &\le Ke^{-\sigma(t-s)} \qquad (0 \le s < t) \\ |U_2(t-s)| &\le K \qquad (s \ge t \ge 0) \end{aligned} \tag{4.51}$$

Now every solution $\mathbf{y}(t)$ of (4.50) can be written in the form

$$\mathbf{y}(t) = e^{tA}\mathbf{c} + \int_0^t U_1(t-s)B(s)\mathbf{y}(s)\,ds - \int_t^\infty U_2(t-s)B(s)\mathbf{y}(s)\,ds \tag{4.52}$$

where $e^{tA}\mathbf{c}$ is some solution of (4.48). This is easily established from the variation of constants formula, as was done in the proof of Theorem 4.6. For this solution $\mathbf{y}(t)$, we have, from (4.51), (4.52),

$$|\mathbf{y}(t) - e^{tA}\mathbf{c}| \le K\int_0^t e^{-\sigma(t-s)}|B(s)||\mathbf{y}(s)|\,ds + \int_t^\infty K|B(s)||\mathbf{y}(s)|\,ds$$

Since every solution is bounded (see Exercise 16, p. 158), we have $|\mathbf{y}(t)| \le K_1$, $(0 \le t < \infty)$, and using (4.49), we find that both integrals on the right-hand side approach zero (see Exercise 2 below). Therefore

$$\lim_{t \to \infty} |y(t) - e^{tA}\mathbf{c}| = 0$$

The justification of the above steps is contained in the following exercises.

• EXERCISES

1. Derive (4.52) and then show that if $\boldsymbol{\phi}$ is a solution of (4.52) $\boldsymbol{\phi}'(t) = (A + B(t))\boldsymbol{\phi}(t)$. [*Hint:* Apply the variation of constants formula to (4.50) using $B(t)\mathbf{y}$ as the inhomogeneous term. Then rewrite similarly to (4.42). The fact that $|\mathbf{y}(t)|$ is bounded ensures the existence of all the integrals in (4.52). (Prove this fact.)]

2. Prove that if r is a continuous nonnegative integrable (scalar) function on $0 \le t < \infty$ and if $\sigma > 0$, then

(a) $$\lim_{t \to \infty} \int_t^\infty r(s)\, ds = 0$$

(b) $$\lim_{t \to \infty} \int_0^t e^{-\sigma(t-s)} r(s)\, ds = 0$$

[*Hint:* To prove (b) write $e^{-\sigma t} \int_0^t e^{\sigma s} r(s)\, ds = e^{-\sigma t} \int_0^T e^{\sigma s} r(s)\, ds + \int_T^t e^{-\sigma(t-s)} r(s)\, ds$. The first of these integrals clearly can be made as small as described for each fixed $T > 0$ by choosing t sufficiently large. For the second integral, choose T large enough so that $\int_T^\infty r(s)\, ds$ is as small as desired, and then show that $\int_T^t e^{-\sigma(t-s)} r(s)\, ds \le \int_T^\infty r(s)\, ds$ for all $t \ge T$.]

Now, given a solution $\mathbf{x}(t)$ of (4.48) we can construct a solution $\mathbf{y}(t)$ of (4.50) by solving the integral equation

$$\mathbf{y}(t) = \mathbf{x}(t) + \int_0^t U_1(t-s)B(s)\mathbf{y}(s)\, ds - \int_t^\infty U_2(t-s)B(s)\mathbf{y}(s)\, ds \qquad (4.53)$$

where $\mathbf{x}(t) = e^{tA}\mathbf{c}$ for some constant vector $\mathbf{c}$.

• EXERCISE

3. Use the method of successive approximations to show that (4.53) has a bounded solution, which is also a bounded solution of (4.50). [*Hint:* See the proof of Theorem 4.6, which is more difficult than this.]

The result of Exercise 3, together with the above calculations, shows that $\lim_{t\to\infty} |\mathbf{y}(t) - \mathbf{x}(t)| = 0$. Given a solution $\hat{\mathbf{y}}(t)$ of (4.50), it is easy to construct a solution $\hat{\mathbf{x}}(t)$ of (4.48) such that $\lim_{t\to\infty} |\hat{\mathbf{y}}(t) - \hat{\mathbf{x}}(t)| = 0$, as the analogue of (4.53) now gives an explicit representation, rather than an integral equation, for $\hat{\mathbf{x}}(t)$, and this completes the proof of asymptotic equivalence. ▮

The reader should observe that Theorem 4.7 is applicable to the problem studied in Example 3. Since in Example 3 all the eigenvalues of the matrix A have real part zero, the proof is simplified by the fact that there is no need to split the matrix $\exp(tA)$ into two parts. The argument used in Theorem 4.7 can be used to obtain a result for the nonlinear system

$$\mathbf{y}' = A\mathbf{y} + \mathbf{g}(t, \mathbf{y}) \tag{4.54}$$

Theorem 4.8. *Let A be a real constant matrix satisfying the hypotheses of Theorem 4.7. Let $\mathbf{g}$, $\partial\mathbf{g}/\partial y_j$ $(j = 1, \ldots, n)$ be continuous for $0 \le t < \infty$, $|\mathbf{y}| < k$, for some $k > 0$. Suppose*

$$|\mathbf{g}(t, \mathbf{y})| \le \lambda(t)\, |\mathbf{y}|$$

for $0 \le t < \infty$, $|\mathbf{y}| < k$, where $\int_0^\infty \lambda(t)\, dt < \infty$. Then (4.48) and (4.54) are asymptotically equivalent.

• **EXERCISES**

4. Prove Theorem 4.8. [*Hint:* Study the proofs of Theorems 4.6 and 4.7; see also [21, p. 514].]
5. Extend Theorems 4.7, 4.8 to the case where the matrix $A = A(t)$ is periodic in t with period ω.

Note that Theorem 4.8 does not apply to the important case of autonomous perturbations. (Why not?)

4.7 Stability of Periodic Solutions

Consider the real system with n components

$$\mathbf{y}' = \mathbf{F}(t, \mathbf{y}) \tag{4.55}$$

where **F** is periodic of period ω in t. Suppose that **F** is continuous in $(t, \mathbf{y})$ and has continuous second partial derivatives with respect to the components of $\mathbf{y}$ in a domain $D = \{(t, \mathbf{y}) \mid 0 \le t \le \omega, |\mathbf{y}| < k\}$ where $k > 0$ is some constant. Suppose that (4.55) has a periodic solution $\mathbf{y} = \mathbf{p}(t)$ of period ω in t (it is in general a very difficult problem to obtain or even verify the existence of such a solution, but that is not the issue here). We wish to test the stability of the periodic solution $\mathbf{p}(t)$; in particular, we would like to determine sufficient conditions that assure its asymptotic stability.

We proceed much as in Section 4.4, where we investigated the stability of a critical point. Consider the solution $\boldsymbol{\psi}$ of (4.55) given by

$$\boldsymbol{\psi}(t) = \mathbf{p}(t) + \mathbf{z}(t) \tag{4.56}$$

Then $\boldsymbol{\psi}'(t) = \mathbf{F}(t, \mathbf{p}(t) + \mathbf{z}(t)) = \mathbf{p}'(t) + \mathbf{z}'(t)$. But $\mathbf{p}'(t) = \mathbf{F}(t, \mathbf{p}(t))$ and hence $\mathbf{z}$ satisfies the equation

$$\mathbf{z}'(t) = \mathbf{F}(t, \mathbf{p}(t)) + \mathbf{z}(t) - \mathbf{F}(t, \mathbf{p}(t))$$

Applying the mean value theorem shows that $\mathbf{z}$ satisfies the equation

$$\mathbf{z}' = \mathbf{F}_{\mathbf{y}}(t, \mathbf{p}(t))\mathbf{z} + \mathbf{g}(t, \mathbf{z}) \tag{4.57}$$

where $\mathbf{g}$ is periodic of period ω in t continuous in $(t, \mathbf{z})$ for all t and for $|\mathbf{z}|$ "small" and

$$\lim_{|\mathbf{z}| \to 0} \frac{|\mathbf{g}(t, \mathbf{z})|}{|\mathbf{z}|} = 0 \tag{4.58}$$

moreover the matrix $\mathbf{F}_{\mathbf{y}}(t, \mathbf{p}(t))$ is continuous and periodic in t of period ω. Clearly $\mathbf{z} \equiv \mathbf{0}$ is a solution of (4.57) and from the definitions of stability and asymptotic stability (see Definitions 3 and 4, Section 4.2) it follows that the periodic solution $\mathbf{p}(t)$ of the system (4.55) is stable or asymptotically stable if and only if $\mathbf{z} \equiv \mathbf{0}$ is respectively a stable or asymptotically stable solution of the system (4.57). This suggests that we study almost linear systems of the form

$$\mathbf{z}' = A(t)\mathbf{z} + \mathbf{g}(t, \mathbf{z}) \tag{4.59}$$

where $A(t)$ is a periodic matrix of period ω in t and where $\mathbf{g}(t, \mathbf{z})$ is a continuous function in $(t, \mathbf{z})$ periodic in t of period ω and which is small in the sense of (4.58). Clearly (4.57) is a special case of (4.59) with $A(t) = \mathbf{F}_{\mathbf{y}}(t, \mathbf{p}(t))$.

The linear part of (4.59) is the linear system $\mathbf{x}' = A(t)\mathbf{x}$ with periodic coefficients of period ω. As we have seen (see Theorem 2.12, p. 96 and Corollary 1, p. 98), such a system can be transformed to a linear system with constant coefficients by a suitable nonsingular linear transformation. Let $P(t)$ and R be the matrices in Floquet's theorem (p. 96) with respect to the linear system $\mathbf{x}' = A(t)\mathbf{x}$. Define the change of variable

$$\mathbf{z} = P(t)\mathbf{u} \tag{4.60}$$

This transforms the almost linear system (4.59) to the almost linear system

$$\mathbf{u}' = R\mathbf{u} + P^{-1}(t)\mathbf{g}(t, P(t)\mathbf{u}) \tag{4.61}$$

which has constant coefficients.

• EXERCISE

1. Establish (4.61). [*Hint:* Use (4.60) in (4.59) and follow the proof of Theorem 2.12, Corollary 1, p. 98.]

We may now apply the theory of Section 4.4 to the system (4.61) in order to obtain asymptotic stability criteria for the zero solution of the system (4.59) (and hence of the periodic solution $\mathbf{p}(t)$ of (4.55)). By looking at the linear system $\mathbf{x}' = A(t)\mathbf{x}$, we have the following result as a consequence of Floquet theory and Theorem 4.3.

Theorem 4.9. *Let* $\mathbf{g}$ *and* $\partial\mathbf{g}/\partial z_j$ $(j = 1, \ldots, n)$ *be periodic in* t *of period* ω *and continuous in* $(t, \mathbf{z})$ *for* $|\mathbf{z}| < k_1$ $(k_1 > 0$ *a constant*$)$. *Let* (4.58) *be satisfied. Let* $A(t)$ *be a continuous n-by-n periodic matrix of period* ω *in* t. *Let the multipliers* $\lambda_1, \lambda_2, \ldots, \lambda_n$ (counting multiplicities) *of the linear system* $\mathbf{x}' = A(t)\mathbf{x}$ *have magnitude* $|\lambda_k| < 1$ $(k = 1, \ldots, n)$. *Then the zero solution of* (4.59) *is asymptotically stable.*

• EXERCISES

2. Prove Theorem 4.9. [*Hints:* (a) Consider the transformed system (4.61) and show that the nonlinear term $P^{-1}(t)\mathbf{g}(t, P(t)\mathbf{u})$ satisfies condition (4.58) if $\mathbf{g}$ does. (b) Apply Theorem 4.3 and the relation (2.67), p. 98, between the eigenvalues of R and the multipliers $\lambda_1, \lambda_2, \ldots, \lambda_n$.]
3. Use Theorem 4.9 to state and prove an asymptotic stability theorem for the periodic solution $\mathbf{p}(t)$ of (4.55).
4. Let $\mathbf{p}(t)$ be a periodic solution of period ω of the autonomous system

$$\mathbf{y}' = \mathbf{F}(\mathbf{y})$$

Show that in this case the analysis that led to (4.57) now yields the system

$$\mathbf{z}' = F_y(\mathbf{p}(t))\mathbf{z} + \mathbf{g}(\mathbf{z})$$

in which t does not enter explicitly and where g satisfies (4.58).

5. In Exercise 4, show that $\mathbf{x} = \mathbf{p}'(t)$ is a solution of the linear system $\mathbf{x}' = F_y(\mathbf{p}(t))\mathbf{x}$. [*Hint:* Substitute.]

6. Use the result of Exercise 5 to show that 1 is a multiplier of the linear system $\mathbf{x}' = F_y(\mathbf{p}(t))\mathbf{x}$. [*Hint:* Use Exercise 3, Section 2.9, p. 99.]

Exercises 4–6 show that Theorem 4.9 is **never applicable** in the study of asymptotic **stability of a periodic solution of an autonomous system.** It is for this reason that for periodic solutions of autonomous systems, a different type of stability, called **orbital stability**, is introduced. The interested reader is referred to [4, Ch. 13].

• EXERCISES

7. State and prove the analogue of Theorem 4.4 when $A = A(t)$ is a periodic matrix of period ω.

8. State and prove the analogue of Theorem 4.5 when $A = A(t)$ is a periodic matrix of period ω.

9. State and prove the analogue of Theorem 4.6 when $A = A(t)$ is a periodic matrix of period ω.

Chapter 5 | LYAPUNOV'S SECOND METHOD

5.1 Introductory Remarks

Instead of trying to determine the stability of a nonlinear system by first examining the linear approximation (as was done in Chapter 4), we now explore an entirely different approach. This technique, discovered by Lyapunov at the end of the nineteenth century, was rediscovered and has been applied effectively to entirely new problems, especially during the past 20 years. The technique is also called the **direct method** because the technique can be applied directly to the differential equation, without any knowledge of the solutions, provided the person using the method is clever enough to construct the right auxiliary functions. As we will show by means of examples (for example, Example 4, Section 5.2), the right choice is not at all obvious. In addition to giving us criteria for stability, asymptotic stability, and instability of solutions (critical points), the method gives us a way of estimating the region of asymptotic stability, as we shall see in Section 5.5. This is something the linear approximation can never hope to do, because, as we have seen in Section 4.3, stability properties of linear systems are global but the addition of a nonlinear term may change the region of stability or asymptotic stability completely. (Recall the scalar equation $u' = -u + u^2$, where the origin is globally asymptotically stable for the linear approximation, but that is changed drastically when the nonlinear term u^2 is added.)

The idea behind the method can be traced to a result of Lagrange (stated by him about 1800 and proved later by Dirichlet): *If in a certain rest position a conservative mechanical system has minimum potential energy, then this position corresponds to a stable equilibrium; if the rest position does not correspond to minimum potential energy, then the equilibrium is unstable.* For simplicity we consider a particle of mass m moving in a straight line under the action of a force $f(y)$ that depends on the position y but not on the time t. Then the equation of motion is

$$my'' = f(y) \tag{5.1}$$

The statement that the system is conservative means that the force function f is determined by a potential function U by the relation $f(y) = -\text{grad } U = -dU/dy$; $U(y)$ is called the potential energy. In a particular case, the potential energy may have the graph shown in Figure 5.1, with rest positions corresponding to points A, 0, B, C, D. Then Lagrange's theorem says that the

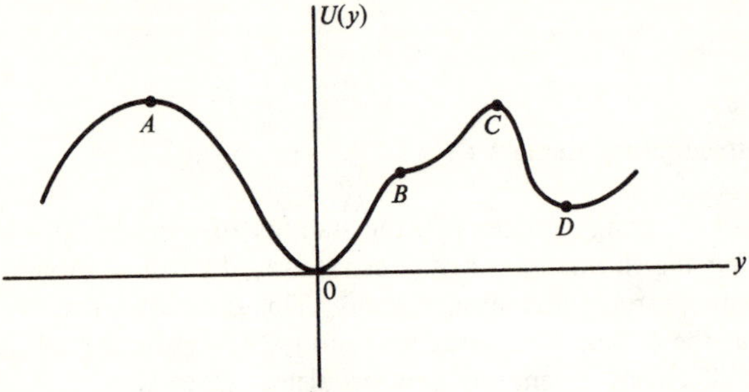

Figure 5.1

rest points corresponding to 0, D are stable, while those corresponding to A, B, C are unstable. For simplicity we assume, as indicated in the figure, that $U(0) = 0$ and $f(0) = -U'(0) = 0$ (note that U is determined only up to a constant); and that $U(y) > 0$ if $y \neq 0$ so that $y = 0$ is a stable equilibrium point corresponding to a minimum of $U(y)$. We can write (5.1) as a system of first-order autonomous equations by letting $y_1 = y$, $y_2 = y'$;

$$\begin{aligned} y_1' &= y_2 \\ my_2' &= f(y_1) = -\frac{dU(y)}{dy} \end{aligned} \tag{5.2}$$

The **kinetic energy** of the system is $T = (m/2)y'^2 = (m/2)y_2{}^2$ and the **total energy** V is

$$V(y_1, y_2) = U(y_1) + \frac{m}{2} y_2^2 \tag{5.3}$$

We remark that $\partial V/\partial y_2 = my_2$, $\partial V/\partial y_1 = U'(y_1) = -f(y_1)$, so that (5.2) can be written in the form (the so-called **Hamiltonian form**)

$$y_1' = \frac{1}{m}\frac{\partial V}{\partial y_2} \qquad y_2' = -\frac{1}{m}\frac{\partial V}{\partial y_1}$$

$$\text{or} \tag{5.4}$$

$$y' = \frac{\partial H}{\partial z} \qquad z' = -\frac{\partial H}{\partial y}$$

where we let $y_1 = y$, $y_2 = z$, and $H(y, z) = (1/m)V(y, z)$ is called the Hamiltonian for (5.2). Because of the importance of such systems in certain applications, we digress to make a few remarks about **Hamiltonian systems.** More generally, a system of $2n$ equations determined by a single scalar function $H(y_1, \ldots, y_n, z_1, \ldots, z_n)$ is called Hamiltonian if it is of the form

$$\begin{cases} y_i' = \dfrac{\partial H}{\partial z_i} \\[2ex] z_i' = -\dfrac{\partial H}{\partial y_i} \end{cases} \qquad (i = 1, \ldots, n) \tag{5.5}$$

- **EXERCISES**

1. If $\boldsymbol{\phi} = (\phi_1, \ldots, \phi_{2n})$ is any solution of the Hamiltonian system (5.5), prove that $H(\phi_1, \ldots, \phi_{2n})$ is a constant.

REMARK. A function constant along solution curves of a system is called an **integral** of the system.

2. With reference to (5.1), show that if $f(y)$ is continuous for $|y|$ near zero and if $yf(y) < 0$ for $y \neq 0$ and $|y|$ small, then the potential energy $U(y) = -\int_0^y f(\sigma)\, d\sigma$ has a minimum at $y = 0$.

3. In the conservative system described above, suppose that $m = 1$ and that the motion takes place in three dimensions. The force $\mathbf{f}(\mathbf{y})$ is now given by $\mathbf{f}(\mathbf{y}) = -\mathbf{grad}\, W(\mathbf{y})$; that is, $f_i(y_1, y_2, y_3) = -\partial W/\partial y_i\, (y_1, y_2, y_3)$, $i = 1, 2, 3$, and the motion is prescribed by the equations $y_i'' = f_i(y_1, y_2, y_3)$, $i = 1, 2, 3$.
   (a) Write the equivalent system of six first-order equations by letting $y_i' = z_i$ $(i = 1, 2, 3)$.
   (b) Show that if the Hamiltonian H is defined as the total energy: $H(y_i', y_2, y_3, z_1, z_2, z_3) = W(y_1, y_2, y_3) + \sum_{i=1}^{3} z_i^2$, the equations of motion form a Hamiltonian system in the sense of (5.5). (Exercise 1 above says that along any solution of a Hamiltonian system the total energy is constant. This is the principle of conservation of energy.)

4. Show that the equation of motion for an undamped simple pendulum, $y'' + (g/L)\sin y = 0$ (see Section 1.1, p. 6), can be written as a Hamiltonian system. What is the total energy? What does Lagrange's principle tell you about the critical points of the system?

Let us now return to an intuitive demonstration of Lagrange's theorem for the single system (5.1) or (5.2). This demonstration contains the intuitive ideas behind Lyapunov's second method. We restrict ourselves to the simple case of an equilibrium point at $y_1 = y_2 = 0$, when the potential energy has a minimum at $y_1 = 0$. We assume that $U(0) = 0$ (as in Figure 5.1); thus by the minimum property, $U(y_1) > 0$ for $y_1 \neq 0$ and $|y_1|$ small. We wish to show that the equilibrium point $y_1 = 0, y_2 = 0$ of (5.1) is stable. If V is defined by (5.3) we know (Exercise 1) that V is constant along a solution. Consider now the family of curves $V(y_1, y_2) = c$ (constant) in the (y_1, y_2) phase plane. If $c < 0$ there are no real curves. If $c = 0$ we get the single point $y_1 = y_2 = 0$. If $c > 0$ but c is sufficiently small, then the set $V(y_1, y_2) = c$ is a family of curves. There is a neighborhood of the origin that contains exactly one of these curves Γ_c, given by the equation $y_2 = \pm[(2/m)(c - U(y_1))]^{1/2}$. This curve Γ_c is closed, surrounds the origin, and is symmetric with respect to the y_1 axis, as shown in Figure 5.2. Clearly, by the property

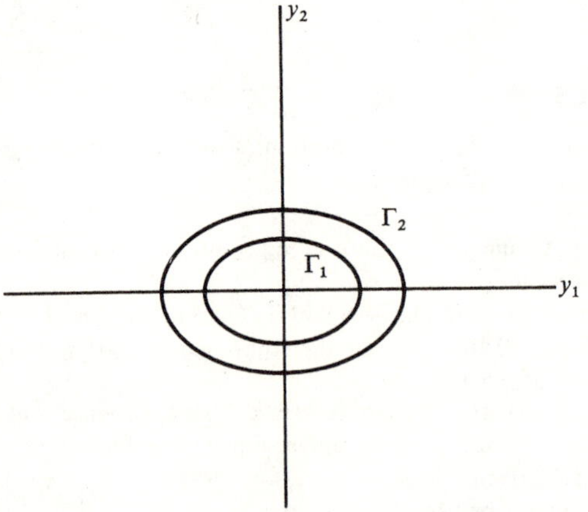

Figure 5.2

of a minimum if c_1 and c_2 are small with $c_1 < c_2 \leq c$, then the corresponding curves Γ_1 and Γ_2 are situated as shown, shrinking down to the origin as c decreases to zero. If a solution $(y_1(t), y_2(t))$ begins at time t_0 with $|y_1(t_0)|$ and $|y_2(t_0)|$ small, it stays close to the origin because it lies on the curve

Γ given by the equation

$$U(y_1) + \frac{m}{2} y_2^2 = U(y_1(t_0)) + \frac{m}{2} y_2^2(t_0)$$

for which $U(y_1(t_0)) + (m/2)y_2^2(t_0)$ can be made as small as desired by choosing $|y_1(t_0)|$ and $|y_2(t_0)|$ small enough. Thus the curve Γ can be made to remain arbitrarily close to the origin, which by definition of stability (Definition 1, Section 4.2, p. 146) says that the origin is stable, and completes a sketch of the proof. Notice, however, that the origin cannot be asymptotically stable in the case discussed above, because each nonzero solution lies on a curve Γ, and these curves certainly do not approach the origin. It is clear, intuitively, that in order to have the origin asymptotically stable, the total energy would at least have to decrease to zero as a function of time as $t \to \infty$ and that could happen only in the presence of a damping term, such as friction, which would cause dissipation of energy (for example, a damped pendulum). We will see, when we discuss the Liénard equation, that these statements are essentially correct. We will also see that it is the V function (5.3) that is basic to the method we are about to describe.

• **EXERCISE**

5. Show that if the potential energy has a maximum at the equilibrium point $y_1 = y_2 = 0$, then the origin is unstable.

We remark that it can be shown that if the potential energy has an inflection point at $y_1 = y_2 = 0$, then the origin is also unstable.

5.2 Lyapunov's Theorems

For clarity of exposition we will consider first autonomous systems of the form

$$\mathbf{y}' = \mathbf{f}(\mathbf{y}) \tag{5.6}$$

where $\mathbf{f}$ and $\partial\mathbf{f}/\partial y_j$ $(j = 1, \dots, n)$ are continuous in a region D of n-dimensional $\mathbf{y}$-space. (D may be the whole space.) We will assume that D contains the origin in its interior, that $\mathbf{f}(\mathbf{0}) = \mathbf{0}$ (that is, the origin is a critical point of (5.6) so that $\mathbf{y} \equiv \mathbf{0}$ is a solution of (5.6)), and that the origin is an isolated critical point of (5.6). This means that there is a neighborhood of the origin containing no other critical points. We will present criteria for the stability and instability of the zero solution. The consideration of the zero solution is no restriction since, as we have seen previously, the problem of investigating the stability of any critical point $\mathbf{y} = \mathbf{y}_0$ can always be transformed to an investigation of the zero solution (see, for example, Section 4.2, p. 150).

We have seen in Section 2.8 (p. 85) that solutions of autonomous systems such as (5.6) are conveniently represented as orbits in phase space. In the presentation of stability theory for autonomous systems, it is convenient to introduce certain additional terminology and some simple facts concerning orbits. If C is an orbit of (5.6) corresponding to the solution $\boldsymbol{\phi}(t)$ existing on $-\infty < t < \infty$, we denote by C^+ **(the positive semiorbit)** the set of points of C with coordinates $\boldsymbol{\phi}(t)$ where $t_0 \le t < \infty$ for any t_0 and by C^- **(the negative semiorbit)** the set of points of C with coordinates $\boldsymbol{\phi}(t)$, where $-\infty < t \le t_0$. Then $C = C^+ \cup C^-$ (the union of C^+ and C^-) is often referred to as the **full orbit**.

There is a close connection between uniqueness of solutions of the initial value problem for (5.6) (as guaranteed by Theorem 1.1, p. 26) and the following simple facts, which are of general interest. In what follows it will be convenient to denote by $\boldsymbol{\phi}(t, \boldsymbol{\eta})$ the solution of $\mathbf{y}' = \mathbf{f}(\mathbf{y})$ that satisfies the initial condition $\boldsymbol{\phi}(t_0, \boldsymbol{\eta}) = \boldsymbol{\eta}$.

Lemma 5.1. *If $\boldsymbol{\eta}$ is any point of D that is not a critical point of* (5.6), *then through the point $\boldsymbol{\eta}$ there passes at most one orbit of* (5.6).

Proof. Suppose there are two orbits C and $\tilde{C}$ through $\boldsymbol{\eta}$. Let C be generated by the solution $\boldsymbol{\phi}(t, \boldsymbol{\eta})$ with $\boldsymbol{\phi}(t_0, \boldsymbol{\eta}) = \boldsymbol{\eta}$; let $\tilde{C}$ be generated by the solution $\boldsymbol{\psi}(t, \boldsymbol{\eta})$ with $\boldsymbol{\psi}(t_1, \boldsymbol{\eta}) = \boldsymbol{\eta}$. We suppose that $t_1 \neq t_0$ (otherwise we finish by a trivial argument). Since (5.6) is invariant under translations of t, the function $\boldsymbol{\xi}(t) = \boldsymbol{\psi}(t + t_1 - t_0, \boldsymbol{\eta})$ is another solution of (5.6) representing $\tilde{C}$ and $\boldsymbol{\xi}(t_0) = \boldsymbol{\psi}(t_1, \boldsymbol{\eta}) = \boldsymbol{\eta}$. By uniqueness of the solution of the initial value problem (Theorem 1.1, p. 26), $\boldsymbol{\xi}(t) \equiv \boldsymbol{\phi}(t, \boldsymbol{\eta})$ and therefore the orbits C and $\tilde{C}$ coincide.

Lemma 5.2. *If an orbit C of* (5.6) *passes through some ordinary point of D, then C cannot reach any critical point $\boldsymbol{\alpha}$ in D in finite time.* (More precisely, if C is generated by a solution $\boldsymbol{\phi}$ and if $\lim_{t \to a} \boldsymbol{\phi}(t) = \boldsymbol{\alpha}$, $\boldsymbol{\alpha}$ in D, then $a = \pm\infty$.)

• **EXERCISE**

1. Prove Lemma 5.2. [*Hint:* Suppose it did; use a simple continuation argument and Lemma 5.1 to obtain a contradiction.]

A simple consequence of Lemmas 5.1 and 5.2 is the following result.

Lemma 5.3. *An orbit C of* (5.6) *that passes through at least one ordinary point of D cannot cross itself, unless it is a closed curve in D. In this case, C corresponds to a periodic solution of* (5.6).

• **EXERCISE**

2. Prove Lemma 5.3. [Recall that a solution $\boldsymbol{\phi}(t)$ of (5.6) is periodic with period ω if $\boldsymbol{\phi}(t+\omega) = \boldsymbol{\phi}(t)$ $(-\infty < t < \infty)$. It is obvious that if $\boldsymbol{\phi}$ is a periodic solution, its orbit is a closed curve. The point is to prove the converse!]

We will postpone to Section 5.4 certain generalizations of what follows for autonomous systems and to Section 5.6 the generalizations to nonautonomous systems. As has been our practice, our objective is to present the salient features of the method and not the most sophisticated results. The interested reader is referred to [8, 9, 15, or 25] for more advanced treatments.

We will be concerned with the construction of certain scalar functions and we require a number of definitions. Let $V(\mathbf{y})$ be a scalar continuous function (that is, a real-valued function of the variables $y_1, y_2, \ldots, y_n$) defined on some region Ω containing the origin; again, Ω could be the whole space.

Definition 1. *The scalar function* $V(\mathbf{y})$ *is said to be positive definite on the set* Ω *if and only if* $V(\mathbf{0}) = 0$ *and* $V(\mathbf{y}) > 0$ *for* $\mathbf{y} \neq 0$ *and* $\mathbf{y}$ *in* Ω.

Definition 2. *The scalar function* $V(\mathbf{y})$ *is negative definite on the set* Ω *if and only if* $-V(\mathbf{y})$ *is positive definite on* Ω.

Example 1. If $n = 3$ the function $V(\mathbf{y}) = y_1^2 + y_2^2 + y_3^2$ is obviously positive definite (on the whole space), but the function $V(\mathbf{y}) = y_1^2$, while obviously nonnegative, is not positive definite because $V(\mathbf{y}) = 0$ on the plane $y_1 = 0$ (that is, at every point of the (y_2, y_3) plane).

• **EXERCISE**

3. If $n = 2$ and $V(y_1, y_2) = \int_0^{y_1} g(\sigma)\, d\sigma + y_2^2/2$, where g is continuous and $\sigma g(\sigma) > 0$ for $\sigma \neq 0$, prove that V is positive definite on the whole (y_1, y_2) plane. What can you say if $\sigma g(\sigma) \geq 0$?

We will assume throughout that the scalar function $V(\mathbf{y})$ has continuous first-order partial derivatives at every point of the region Ω.

Definition 3. *The derivative of V with respect to the system* $\mathbf{y}' = \mathbf{f}(\mathbf{y})$ *is the scalar product*

$$V^*(\mathbf{y}) = \mathbf{grad}\, V(\mathbf{y}) \cdot \mathbf{f}(\mathbf{y}) = \frac{\partial V}{\partial y_1}(\mathbf{y}) f_1(\mathbf{y}) + \frac{\partial V}{\partial y_2}(\mathbf{y}) f_2(\mathbf{y}) + \cdots + \frac{\partial V}{\partial y_n}(\mathbf{y}) f_n(\mathbf{y}) \tag{5.7}$$

The reader is urged to note that $V^*(\mathbf{y})$ can be computed directly from the differential equation without any knowledge of the solutions. Herein lies the power of the Lyapunov method. We observe that if $\boldsymbol{\phi}(t)$ is any solution of (5.6), then by the chain rule, the definition of solution, and (5.7) we have

$$\frac{d}{dt} V(\boldsymbol{\phi}(t)) = \frac{\partial V}{\partial y_1}(\boldsymbol{\phi}(t))\phi_1'(t) + \cdots + \frac{\partial V}{\partial y_n}(\boldsymbol{\phi}(t))\phi_n'(t)$$

$$= \sum_{k=1}^{n} \frac{\partial V}{\partial y_k}(\boldsymbol{\phi}(t)) f_k(\boldsymbol{\phi}(t)) = V^*(\boldsymbol{\phi}(t)) \tag{5.8}$$

In words, along a solution $\boldsymbol{\phi}$ the total derivative of $V(\boldsymbol{\phi}(t))$ with respect to t coincides with the derivative of V with respect to the system evaluated at $\mathbf{y} = \boldsymbol{\phi}(t)$.

Example 2. For the system

$$y_1' = y_2$$

$$y_2' = -y_1 - 2y_2$$

and the given function

$$V(y_1, y_2) = \tfrac{1}{2}({y_1}^2 + {y_2}^2)$$

we obtain, using the definition (5.7) with $f_1(y) = y_2, f_2(y) = -y_1 - 2y_2$,

$$V^*(y_1, y_2) = y_1 y_2 + y_2(-y_1 - 2y_2) = -2{y_2}^2.$$

REMARK. The system of this example is equivalent to the scalar equation

$$u'' + 2u' + u = 0$$

which can be regarded as a model for a damped linearized oscillator. Thus V represents the total energy ($\frac{1}{2}{y_1}^2 = \frac{1}{2}u^2$ is the potential energy and $\frac{1}{2}{y_2}^2 = \frac{1}{2}(u')^2$ is the kinetic energy). From (5.8) and the above calculation, we have

$$\frac{d}{dt}[V(\phi_1(t), \phi_2(t))] = V^*[\phi_1(t), \phi_2(t)] = -2[\phi_2(t)]^2 \leq 0$$

This immediately tells us that the total energy is a nonincreasing function of t.

• **EXERCISE**

4. Consider the function $V(y_1, y_2)$ defined by (5.3). Compute the derivative of $V(y_1, y_2)$ with respect to the system (5.2).

We now state and illustrate Lyapunov's original theorems for autonomous systems; we postpone their proofs to Section 5.3.

Theorem 5.1. *If there exists a scalar function* $V(\mathbf{y})$ *that is positive definite and for which* $V^*(\mathbf{y}) \leq 0$ (that is, the derivative (5.7) with respect to $\mathbf{y}' = \mathbf{f}(\mathbf{y})$ is nonpositive) *on some region* Ω *containing the origin, then the zero solution of* $\mathbf{y}' = \mathbf{f}(\mathbf{y})$ *is stable.*

Theorem 5.2. *If there exists a scalar function* $V(\mathbf{y})$ *that is positive definite and for which* $V^*(\mathbf{y})$ *is negative definite on some region* Ω *containing the origin, then the zero solution of* $\mathbf{y}' = \mathbf{f}(\mathbf{y})$ *is asymptotically stable.*

We also have two instability results.

Theorem 5.3. *If there exists a scalar function* $V(\mathbf{y})$, $V(\mathbf{0}) = 0$, *such that* $V^*(\mathbf{y})$ *is either positive definite or negative definite on some region* Ω *containing the origin and if there exists in every neighborhood* N *of the origin,* $N \subset \Omega$, *at least one point* $\mathbf{a} \neq \mathbf{0}$ *such that* $V(\mathbf{a})$ *has the same sign as* $V^*(\mathbf{y})$, *then the zero solution of* $\mathbf{y}' = \mathbf{f}(\mathbf{y})$ *is unstable.*

Theorem 5.4. *If there exists a scalar function* V *such that in a region* Ω *containing the origin,*

$$V^* = \lambda V + W$$

where $\lambda > 0$ *is a constant and* W *is either identically zero or* W *is a nonnegative or a nonpositive function such that in every neighborhood* N *of the origin,* $N \subset \Omega$, *there is at least one point* $\mathbf{a}$ *such that* $V(\mathbf{a}) \cdot W(\mathbf{a}) > 0$, *then the zero solution of* $\mathbf{y}' = \mathbf{f}(\mathbf{y})$ *is unstable.*

The reader will notice immediately that each of these theorems depends on the existence of a scalar function V with certain properties. Four points must be emphasized. First, nothing is said about how the function V is to be constructed, and this is the principal limitation of this method—there are no general methods for the construction of such functions. Second, the theorems give sufficient conditions for stability and instability, but these conditions are by no means necessary, as we shall soon see. In Section 5.4 we shall present certain improvements. Third, in the case of Theorems 5.1, 5.2, we see that, in view of the definition of stability, solutions starting close to the equilibrium position exist and are bounded for all $t \geq 0$. Finally, in the case of Theorem 5.2, nothing is said about the size of the region of asymptotic stability (the set of initial values for which solutions tend to zero). We shall return to this problem in Section 5.5.

Next, let us look at these theorems geometrically. In particular, let us discuss the condition $V^*(\mathbf{y}) \leq 0$ in Theorem 5.1, where V is positive definite in a region Ω. Let c be a constant and consider the equation $V(\mathbf{y}) = c$; because V is positive definite in Ω we need consider only values $c \geq 0$, and for $c = 0$ the equation $V(\mathbf{y}) = 0$ gives only the origin in phase space. If $c > 0$, the equation $V(\mathbf{y}) = c$ represents a surface (see Figure 5.3 in the case

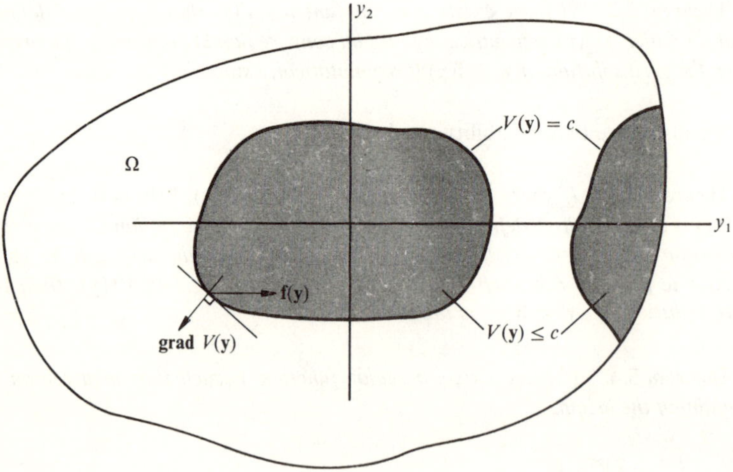

Figure 5.3

$n = 2$). Since V is positive definite, if $c > 0$ is sufficiently small, this surface has a component whose interior contains the origin, and since V is continuous this component shrinks to the origin as $c \to 0$. (In most simple examples, such as Example 2, the surface $V(\mathbf{y}) = c$ consists only of the component surrounding the origin.) Now, by definition $V^*(\mathbf{y}) = \mathbf{grad}\, V(\mathbf{y}) \cdot \mathbf{f}(\mathbf{y})$, where

grad $V(\mathbf{y})$ is a vector normal to the surface $V(\mathbf{y}) = c$ in the direction shown in Figure 5.3. The assumption $V^*(\mathbf{y}) \leq 0$ means that the vector $\mathbf{f}(\mathbf{y})$ cannot point into the "exterior" of the region bounded by the surface $V(\mathbf{y}) = c$ (at any rate it cannot point into the exterior of that component of the region bounded by the surface $V(\mathbf{y}) = c$ that contains the origin for sufficiently small $c > 0$). But the vector $\mathbf{f}(\mathbf{y})$ is the tangent vector to the orbit of the system $\mathbf{y}' = \mathbf{f}(\mathbf{y})$ at each point $\mathbf{y}$. Therefore, for sufficiently small c, the orbit of a solution starting close enough to the origin cannot leave the region bounded by the surface $V(\mathbf{y}) = c$; that is, for sufficiently small c the orbit must remain close to the origin, so that the origin is stable. This is almost a proof of Theorem 5.1. If $V^*(\mathbf{y})$ is negative definite, as in the hypothesis of Theorem 5.2, the orbits actually cross from the exterior to the interior of the region bounded by the surface $V(\mathbf{y}) = c$ for every $c > 0$, no matter how small, and this leads to asymptotic stability. The instability results can be discussed in a similar way.

Example 3. Consider the equation $u'' + g(u) = 0$, where g is continuously differentiable for $|u| < k$, with some constant $k > 0$, and $ug(u) > 0$ if $u \neq 0$. Thus, by continuity, $g(0) = 0$. (This condition is satisfied if $g(u) = \sin u$, the case of the undamped simple pendulum.) Writing the equation as a system of first-order equations, we have

$$\begin{aligned} y_1' &= y_2 \\ y_2' &= -g(y_1) \end{aligned} \tag{5.9}$$

and the origin $y_1 = y_2 = 0$ is an isolated critical point. To investigate the stability of this equilibrium point we wish to see if one of Lyapunov's theorems stated above can be applied. To do that we must try to select a suitable V function. If we think of $g(u)$ as the restoring force of a spring or pendulum acting on the particle of unit mass at a displacement u from equilibrium and of u' as the velocity of the particle, then the potential energy at a displacement u from equilibrium is

$$\int_0^u g(\sigma)\, d\sigma$$

On the other hand, the kinetic energy is $\frac{1}{2}(u')^2$ so that the total energy is

$$\tfrac{1}{2}(u')^2 + \int_0^u g(\sigma)\, d\sigma$$

This suggests that for the system (5.9) we might try this total energy as the V function

$$V(y_1, y_2) = \tfrac{1}{2}y_2^2 + \int_0^{y_1} g(\sigma)\, d\sigma \tag{5.10}$$

This function is defined on the region $\Omega = \{(y_1, y_2) \mid |y_1| < k, |y_2| < \infty\}$, $V(0, 0) = 0$ and since $\sigma g(\sigma) > 0$, the graph of g has the form suggested in Figure 5.4, so that

$$\int_0^{y_1} g(\sigma)\, d\sigma > 0 \qquad \text{for } 0 < |y_1| < k$$

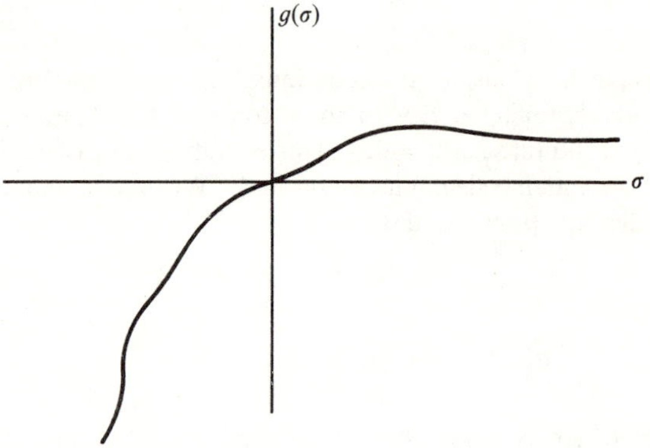

Figure 5.4

Therefore $V(y_1, y_2)$ is positive definite on Ω. Now the derivative of V with respect to the system (5.9) is (see formula (5.7))

$$V^*(y_1, y_2) = y_2 y_2' + g(y_1)y_1' = y_2(-g(y_1)) + g(y_1)y_2 = 0 \qquad [(y_1, y_2) \text{ in } \Omega] \tag{5.11}$$

Thus we have found a function V that satisfies the hypothesis of Theorem 5.1 and therefore the equilibrium solution $y_1 = y_2 = 0$ is stable. We have in this case also found much more. Namely, the V function (5.10) can be used to obtain a complete phase plane portrait of the system. Because of (5.8) and (5.11) we know that if (y_{10}, y_{20}) is any point in Ω and if $(\phi_1(t), \phi_2(t))$ is any solution of (5.9) through this point, then $d/dt\, V(\phi_1(t), \phi_2(t)) = V^*(\phi_1(t), \phi_2(t)) = 0$, and integration gives

$$V(\phi_1(t), \phi_2(t)) \equiv \text{constant} = V(y_{10}, y_{20})$$

Thus the orbit of every solution starting at (y_{10}, y_{20}) in Ω is a curve C whose equation is given by

$$\frac{y_2^2}{2} + \int_0^{y_1} g(\sigma)\, d\sigma = \frac{y_{20}^2}{2} + \int_0^{y_{10}} g(\sigma)\, d\sigma$$

Because of the hypothesis concerning g these orbits are, for $|y_{10}|, |y_{20}|$ sufficiently small, closed curves about the origin, symmetric with respect to the y_1 axis, such as those shown in Figure 5.5. Therefore (see Lemma 5.3), every such solution of (5.9) is periodic, and we cannot hope to prove more; for example, asymptotic stability of the zero solution.

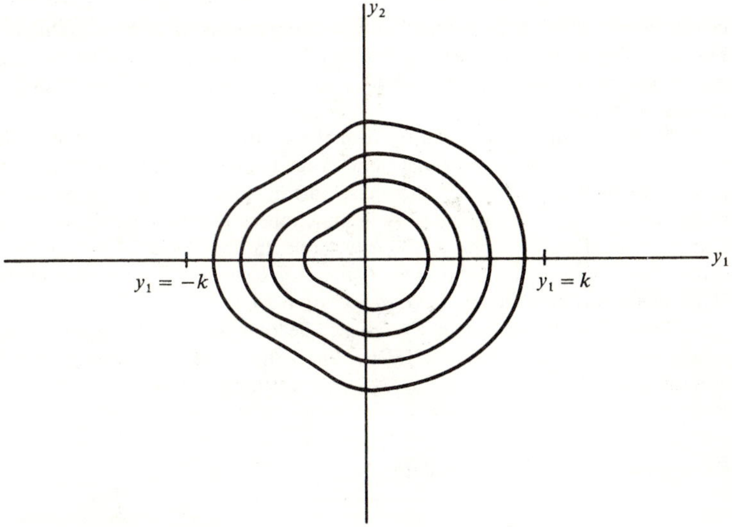

Figure 5.5

• **EXERCISES**

5. Carry out a similar analysis for the case of the simple pendulum equation $\theta'' + \sin\theta = 0$, written as the system

$$y_1' = y_2$$
$$y_2' = -\sin y_1$$

in the neighborhood of $y_1 = y_2 = 0$.
(a) Sketch the phase portrait for $|y_1| < \pi$. Are all the orbits closed?
(b) Sketch the phase portrait for $|y_1| < 2\pi$.

[REMARK. The reader should note that $\sin y_1 = y_1 + h(y_1)$ where $h(y_1) = O(y_1^2)$ as $|y_1| \to 0$, so that the system can be written in the almost linear form

$$\mathbf{y}' = A\mathbf{y} + \mathbf{f}(\mathbf{y})$$

where

$$A = \begin{pmatrix} 0 & 1 \\ -1 & 0 \end{pmatrix}, \qquad \mathbf{f}(\mathbf{y}) = \begin{pmatrix} 0 \\ h(y_1) \end{pmatrix}$$

however, since the eigenvalues of A are $\lambda = \pm i$ **none** of the theory of Section 4.4 applies to this system. Thus Lyapunov's second method can sometimes handle these most critical cases. The reader will benefit from a comparison of the orbits sketched above to those of the linearized system $y' = Ay$ near the origin. Notice that every orbit of the linear system is a circle. However, the character of the orbits of the nonlinear system changes drastically. This is an example of what we referred to in Section 4.1: a nonlinear system will exhibit phenomena that a linear system cannot hope to achieve. For a more complete discussion, see Sections 6.1, 6.2.]

6. In the system of Exercise 5 try to make a similar analysis for the critical point $y_1 = \pi$, $y_2 = 0$ and show that none of the theorems 5.1, 5.2, 5.3, 5.4 apply. [*Note:* We shall see later, from the phase portrait that the interested reader can make for himself by using the same V function as the one already employed, that the critical point $y_1 = \pi$, $y_2 = 0$ is unstable, in fact a saddle point.] What can you say about the critical point $y_1 = n\pi$, $y_2 = 0$ where n is a positive integer? [This problem is discussed in detail in Section 6.2.]

7. Show that the system in Exercise 3, Section 5.1, has the zero solution $y_i = z_i = 0$ $(i = 1, 2, 3)$ stable. [*Hint:* Show that the total energy can be used as a V function in Theorem 5.1, if one assumes that the potential energy $W(y_1, y_2, y_3)$ is positive definite in some neighborhood of $y_1 = y_2 = y_3 = 0$.]

Example 4. Consider the system

$$\begin{aligned} y_1' &= -y_1 - y_2 \\ y_2' &= y_1 - y_2^3 \end{aligned} \tag{5.12}$$

and investigate the stability of $y_1 = y_2 = 0$. Here we have no physical motivation for trying any particular V function. In such cases we might try a function such as $V = y_1^2 + y_2^2$, which is obviously positive definite. The question is, what is V^* with respect to this particular system? We have

$$V^*(y_1, y_2) = 2y_1(-y_1 - y_2) + 2y_2(y_1 - y_2^3) = -2y_1^2 - 2y_2^4$$

and this is obviously negative definite. Thus we were very lucky (as the next example will show, such is not always the case) and we can conclude from Theorem 5.2 that $y_1 = y_2 = 0$ is an asymptotically stable solution of our system. Notice that we have no information as yet about the region of asymptotic stability. We will return to this problem later.

• EXERCISES

8. Write the system (5.12) in Example 4 in the almost linear form $y' = Ay + f(y)$ and see if you can draw any conclusion about the solution $y_1 = y_2 = 0$ by using the appropriate theorem on almost linear systems.
9. Find the other critical points (if any) of the system (5.12). What conclusions can you draw about their stability? Does this tell you anything about the region of asymptotic stability of the zero solution of (5.12)?

Example 5. The Liénard Equation. Consider the scalar equation (Liénard's equation)

$$u'' + u' + g(u) = 0 \tag{5.13}$$

or, written as a system,

$$\begin{aligned} y_1' &= y_2 \\ y_2' &= -g(y_1) - y_2 \end{aligned} \tag{5.14}$$

where g satisfies the hypothesis of Example 3 (g is continuously differentiable for $|u| < k$, $k > 0$ some constant, $ug(u) > 0$, $u \neq 0$)—such a function is usually called a nonlinear spring. Physically, under this assumption (5.13) might represent the motion of a simple pendulum, with $g(u) = \sin u$, which encounters air resistance proportional to the velocity (Section 1.1, p. 7); or as another example, the motion of a mass-spring system (Section 1.1, p. 5) in which the restoring force of the spring is now $g(u)$ (rather than ku) and where the air resistance is proportional to the velocity. As in Example 3, we might quite naturally try "the total energy" as a V function. Thus

$$V(y_1, y_2) = \frac{{y_2}^2}{2} + \int_0^{y_1} g(\sigma)\, d\sigma \tag{5.15}$$

would seem like a good candidate. Indeed, it is positive definite in the region $\Omega = \{(y_1, y_2) \mid |y_1| < k, |y_2| < \infty\}$. Now the derivative of V with respect to (5.14) is

$$V^*(y_1, y_2) = y_2(-g(y_1) - y_2) + g(y_1)y_2 = -{y_2}^2 \tag{5.16}$$

Since $V^*(y_1, y_2) \leq 0$ in Ω we can certainly conclude, by an application of Theorem 5.1, that the zero solution of (5.14) is stable. But $V^*(y_1, y_2)$ is not negative definite in Ω (note that $V^*(y_1, y_2) = 0$ at all points $(y_1, 0)$, that is, on the y_1 axis) and therefore we cannot invoke Theorem 5.2 to conclude that the zero solution of (5.14) is asymptotically stable. Yet we certainly expect that this is the case; for, if $g(x)$ is linear ($g(x) = bx$, $b > 0$) or almost linear, we can establish this fact by an easy application of Theorems 4.1 (p. 151) and 4.3 (p. 161), respectively, but notice that we cannot infer this behavior here, even in simple cases, by using Lyapunov's theorems and V of (5.15).

• **EXERCISE**

10. If $g(x)$ is linear or almost linear ($g(x) = bx + o(|x|)$, as $|x| \to 0$, $b > 0$) show that the zero solution of (5.14) is asymptotically stable.

We recall that because we can apply Theorem 5.1 in both Example 3 and Example 5 to deduce the stability of the zero solution, we establish automatically the existence on $0 \leq t < \infty$ and the boundedness of those solutions (ϕ_1, ϕ_2) for which $|\phi_1(0)|$ and $|\phi_2(0)|$ are sufficiently small (see the definition of stability, Section 4.2, p. 146).

Example 6. We shall now show how the V function (5.15) constructed for the Liénard equation in Example 5 can be modified in order that asymptotic stability of the zero solution can be deduced directly from the Lyapunov method, even if $g(x)$ is not necessarily almost linear, but does satisfy the key hypothesis $ug(u) > 0$ $(u \neq 0)$. In Section 5.4 we will obtain this asymptotic stability by a different approach. Consider the function

$$U(y_1, y_2) = \frac{{y_2}^2}{2} + \beta g(y_1)y_2 + \int_0^{y_1} g(\sigma)\, d\sigma = V(y_1, y_2) + \beta g(y_1)y_2 \tag{5.17}$$

where (y_1, y_2) is in the region $\Omega = \{(y_1, y_2) \mid |y_1| < k, |y_2| < \infty\}$, and where β is a small positive number to be determined by several requirements. The first of these is that $U(y_1, y_2)$ should be positive definite. The motivation is very simple: if U is a quadratic form, say $ay_1^2 + by_1y_2 + cy_2^2$, then we can certainly choose real numbers a, b, c so that $ay_1^2 + by_1y_2 + cy_2^2$ is positive definite. Here we try to do the analogous thing and we can expect to be successful because $g(y_1)$ has the sign of y_1.

We recall the obvious inequality for real numbers A, B: $2|AB| \le A^2 + B^2$, and, more generally, for any $\gamma > 0$, letting $A = u/\sqrt{\gamma}$, $B = \sqrt{\gamma}v$, we have

$$2|uv| \le \frac{u^2}{\gamma} + \gamma v^2 \tag{5.18}$$

Thus from (5.17) and (5.18) with $\gamma = 1$ we have

$$\begin{aligned} U(y_1, y_2) &\ge V(y_1, y_2) - \frac{\beta}{2}\{g^2(y_1) + y_2^2\} \\ &= \tfrac{1}{2}(1-\beta)y_2^2 + \int_0^{y_1} g(\sigma)\, d\sigma - \frac{\beta}{2} g^2(y_1) \end{aligned} \tag{5.19}$$

for (y_1, y_2) in the region Ω. Now consider

$$\lim_{y_1 \to 0} \frac{g^2(y_1)}{\int_0^{y_1} g(\sigma)\, d\sigma} = \lim_{y_1 \to 0} \frac{2g(y_1)g'(y_1)}{g(y_1)} = 2g'(0)$$

(by l'Hospital's rule). Since g is continuously differentiable this limit exists. Since $\sigma g(\sigma) > 0$ for $\sigma \ne 0$ we have $\int_0^{y_1} g(\sigma)\, d\sigma > 0$ for $0 < |y_1| < k$. Therefore, there exists a constant $C > 0$ (where C depends on the constant k_1) such that

$$g^2(y_1) \le C \int_0^{y_1} g(\sigma)\, d\sigma \qquad (-k_1 \le y_1 \le k_1)$$

where k_1 is any positive constant less than k. Note that the ratio

$$\frac{g^2(y_1)}{\int_0^{y_1} g(\sigma)\, d\sigma}$$

is obviously positive for $0 < |y_1| < k$.

Substituting this in (5.19), we obtain the inequality

$$U(y_1, y_2) \geq \tfrac{1}{2}(1 - \beta)y_2^2 + \left(1 - \frac{C\beta}{2}\right)\int_0^{y_1} g(\sigma)\, d\sigma \tag{5.20}$$

Thus if we choose $0 < \beta < \min(1, 2/C)$, we have $U(y_1, y_2)$ positive definite for $-k_1 \leq y_1 \leq k_1$, $|y_2| < \infty$.

Let us now compute the derivative $U^*(y_1, y_2)$ with respect to the system (5.14) and see if we can also choose β to make $U^*(y_1, y_2)$ negative definite. Starting with (5.17) we have

$$\begin{aligned} U^*(y_1, y_2) &= y_2(-g(y_1) - y_2) + \beta g'(y_1)y_2^2 \\ &\quad + \beta g(y_1)[-g(y_1) - y_2] + g(y_1)y_2 \\ &= -y_2^2 + \beta g'(y_1)y_2^2 - \beta g(y_1)y_2 - \beta g^2(y_1) \end{aligned}$$

Therefore, using (5.18), we have

$$-U_1^*(y_1, y_2) \geq y_2^2 - \beta g'(y_1)y_2^2 - \frac{\beta}{2}\left(\frac{g^2(y_1)}{\gamma} + \gamma y_2^2\right) + \beta g^2(y_1)$$

We let $M = \max_{-k_1 \leq y_1 \leq k_1} |g'(y_1)|$, where $0 < k_1 < k$, and we obtain

$$-U^*(y_1, y_2) \geq y_2^2\left[1 - \beta\left(M + \frac{\gamma}{2}\right)\right] + \beta\left(1 - \frac{1}{2\gamma}\right)g^2(y_1) \tag{5.21}$$

Now choose γ sufficiently large, in particular $\gamma = 1$ will do, so that $(1 - 1/2\gamma) > 0$. Now choose β sufficiently small so that $0 < \beta < 1/(M + \gamma/2)$; then $U^*(y_1, y_2)$ is negative definite. Recall that we have already chosen β small enough so that $0 < \beta < \min(1, 2/C)$; this can certainly be done. Thus (5.20) and (5.21) show that the function $U(y_1, y_2)$ defined by (5.17) is positive definite and has $U^*(y_1, y_2)$ (with respect to the system (5.14)) negative definite for $-k_1 \leq y_1 \leq k_1$, $|y_2| < \infty$, where k_1 is any constant satisfying $0 < k_1 < k$. Therefore by Lyapunov's theorem (Theorem 5.2) the zero solution is asymptotically stable.

• EXERCISES

11. Discuss, by the above method, the stability of the zero solution of the equation $u'' + au' + bu + u^2 = 0$, where $a > 0, b > 0$ are given fixed constants. [*Hint:* Note that $g(u) = bu + u^2$, $b > 0$ and show that for $|u|$ sufficiently small, $ug(u) > 0$ $(u \neq 0)$.]

12. Discuss, by the above method, the stability of the zero solution of the Liénard equation

$$u'' + h(u, u')u' + g(u) = 0$$

where $g(u)$ is as in Example 5 and $h(u, u')$, which can be thought of as representing damping of a mechanical or electrical system that depends on the displacement and velocity, is continuously differentiable and satisfies

$$h(u, v) > 0 \qquad \text{for all } (u, v) \neq (0, 0) \text{ that satisfy } (|u|, |v| \leq k_1 < k)$$

that is, the damping is positive definite. [*Hint:* Use the same V function as in Example 5 and modify it in the same way.]

We shall return to Examples 5 and 6 in Sections 5.4, 5.5, and 6.3. For the moment the important point this example shows is that when a particular V function only permits us to conclude stability, it can sometimes be modified in such a way as to give asymptotic stability.

We close this section with a trivial example of instability.

Example 7. Consider the scalar equation $u' = u^3$, which has zero as a solution. By direct integration we can show that the zero solution is not stable. We can also see it by using the V function $V(u) = u^2$, which is certainly positive definite for $-\infty < u < \infty$ and for which $V^*(u) = 2u(u^3) = 2u^4$ is also positive definite. Therefore the conclusion follows from Theorem 5.3.

5.3 Proofs of Lyapunov's Theorems

We normally employ the notation $\boldsymbol{\phi}(t, t_0, \mathbf{y}_0)$ to denote the solution of (5.6) satisfying the initial condition $\boldsymbol{\phi}(t_0, t_0, \mathbf{y}_0) = \mathbf{y}_0$; however, (5.6) is autonomous and therefore it is no loss of generality to suppose that $t_0 = 0$ (the system is invariant under translation of time), and we shall simply write $\boldsymbol{\phi}(t)$ for the solution that satisfies the initial condition $\boldsymbol{\phi}(\mathbf{0}) = \mathbf{y}_0$.

Proof of Theorem 5.1. By positive definiteness of V on Ω, there exists a sphere of radius $r > 0$, contained in the region Ω center at the origin of the phase plane, such that

$$V(\mathbf{y}) > 0 \; (\mathbf{y} \neq \mathbf{0}, \|\mathbf{y}\| \leq r) \quad \text{and} \quad V^*(\mathbf{y}) \leq 0 \; (\|\mathbf{y}\| \leq r)$$

(Because of the easy geometric interpretation we employ the Euclidean norm $\| \;\|$; the proof is unchanged if we use the norm $| \;|$.) Let $\mathbf{y}_0 \neq 0$, $\|\mathbf{y}_0\| < r$, be given. Consider the solution $\boldsymbol{\phi}(t)$ of (5.6), with $\boldsymbol{\phi}(\mathbf{0}) = \mathbf{y}_0$. By local existence (Theorem 3.3, p. 123) this solution exists on $0 \le t < t_1$, for some $t_1 > 0$, and can be continued to the right (Theorem 3.6, p. 132) certainly for as long as $\|\boldsymbol{\phi}(t)\| \le r$. Suppose that $[0, t_1)$ is the largest interval of existence of the solution $\boldsymbol{\phi}(t)$ that can be achieved by continuation. Then either (i) $t_1 = +\infty$ or (ii) $0 < t_1 < +\infty$. We will show that for $\|\mathbf{y}_0\|$ chosen small enough, (ii) cannot arise.

By (5.8) we know that

$$\frac{d}{dt} V(\boldsymbol{\phi}(t)) = V^*(\boldsymbol{\phi}(t)) \le 0, \qquad (0 \le t < t_1)$$

and by integration

$$V(\boldsymbol{\phi}(t)) - V(\mathbf{y}_0) = \int_0^t V^*(\boldsymbol{\phi}(s))\, ds \le 0$$

Therefore,

$$0 < V(\boldsymbol{\phi}(t)) \le V(\mathbf{y}_0) \qquad (0 \le t < t_1) \tag{5.22}$$

where the inequality on the left follows from the assumption $\mathbf{y}_0 \ne 0$ (which implies, by uniqueness of solution of (5.6), that $\boldsymbol{\phi}(t) \ne 0$). Let $\varepsilon > 0$ be given with $0 < \varepsilon \le r$ and let S be the closed set between the spheres of radius ε and radius r; that is, $S = \{\mathbf{y} \mid \varepsilon \le \|\mathbf{y}\| \le r\}$. Then by continuity of V and the fact that S is closed, $\mu = \min_{\mathbf{y} \in S} V(\mathbf{y})$ exists and is strictly positive (it is in fact assumed for some point $\mathbf{y}$ in S). Since $\lim_{\mathbf{y} \to 0} V(\mathbf{y}) = 0$, we can choose a number δ, $0 < \delta < \mu$ such that for $\|\mathbf{y}_0\| \le \delta$, $V(\mathbf{y}_0) < \mu$. Then according to (5.22) the solution $\boldsymbol{\phi}(t)$, $\boldsymbol{\phi}(0) = \mathbf{y}_0$, $\|\mathbf{y}_0\| \le \delta$ satisfies

$$0 < V(\boldsymbol{\phi}(t)) \le V(\mathbf{y}_0) < \mu \qquad \text{for } 0 \le t < t_1 \tag{5.23}$$

By the definition of μ as the minimum value this implies $\|\boldsymbol{\phi}(t)\| < \varepsilon$ for $0 \le t < t_1$. But this then must mean that $t_1 = +\infty$. For, if at some first point $t_2 > t_0$ $\|\boldsymbol{\phi}(t_2)\| = \varepsilon$, then for $t = t_2$ we also have, from the definition of μ again and from (5.23),

$$\mu \le V(\boldsymbol{\phi}(t_2)) \le V(\mathbf{y}_0) < \mu$$

which is absurd. Thus $t_1 = +\infty$ and corresponding to the given $\varepsilon > 0$ we have found a $\delta > 0$ such that $\|\mathbf{y}_0\| < \delta$ implies $\|\boldsymbol{\phi}(t)\| < \varepsilon$ for $0 \le t < \infty$. This completes the proof. ∎

Proof of Theorem 5.2. By Theorem 5.1 the zero solution is stable. In particular, for $\varepsilon = r$ (see the above proof for notation), there exists a $\delta > 0$ such that all solutions $\boldsymbol{\phi}(t)$ of (5.6) with $\boldsymbol{\phi}(0) = \mathbf{y}_0$ and $\|\mathbf{y}_0\| < \delta$ exists on $0 \le t < \infty$ and satisfy

$$\|\boldsymbol{\phi}(t)\| < r \qquad (t \ge 0)$$

We already know from Theorem 5.1 that $V(\boldsymbol{\phi}(t))$ is a nonincreasing function of t which is bounded below. Therefore $\lim_{t \to +\infty} V(\boldsymbol{\phi}(t))$ exists. Suppose that for some $0 < \eta < r$ we could have

$$V(\boldsymbol{\phi}(t)) \ge \eta > 0 \qquad \text{for } t \ge 0 \tag{5.24}$$

We will show that (5.24) is impossible. By continuity, for the above η, there exists a $\delta > 0$, $0 < \delta < r$ such that

$$0 \le V(\mathbf{y}) < \eta \qquad \text{whenever } \|\mathbf{y}\| < \delta \tag{5.25}$$

Therefore, the solutions $\boldsymbol{\phi}(t)$ for which (5.24) holds must satisfy $\|\boldsymbol{\phi}(t)\| \ge \delta$ for $t \ge 0$. Let S be the set of $\mathbf{y}$ lying between the spheres of radius δ and r, that is, $S = \{\mathbf{y} \,|\, 0 < \delta \le \|\mathbf{y}\| \le r\}$. Consider the function $-V^*(\mathbf{y})$ on the closed bounded set S. By hypothesis on $\mathbf{f}$ and V, $-V^*(\mathbf{y})$ (defined by (5.7)) is continuous and positive definite. Let

$$\mu = \min_{\mathbf{y} \in S} (-V^*(\mathbf{y})) > 0$$

Since $\mathbf{0}$ is not a point of S we have (using also (5.8))

$$-\frac{d}{dt} V(\boldsymbol{\phi}(t)) = -V^*(\boldsymbol{\phi}(t)) \ge \mu, \qquad (t > 0)$$

Integrating, we obtain $V(\boldsymbol{\phi}(t)) \le V(\mathbf{y}_0) - \mu t$ for $t \ge 0$. But then clearly for t large enough $V(\boldsymbol{\phi}(t))$ is negative, which is an obvious contradiction. Thus (5.24) is impossible and we must have $\lim_{t \to \infty} V(\boldsymbol{\phi}(t)) = 0$, which implies $\lim_{t \to \infty} \boldsymbol{\phi}(t) = 0$. Since this holds for every solution $\boldsymbol{\phi}(t)$ with $\|\mathbf{y}_0\| < \delta$, this completes the proof. ∎

Proof of Theorem 5.3. Suppose that $V^*(\mathbf{y})$ is positive definite in Ω, and let $\|\mathbf{y}\| \le r$ be a sphere of radius r properly contained in Ω for some $r > 0$.

Then $V(\mathbf{y})$, being continuous in Ω, is bounded on every closed bounded subset of Ω and in particular there exists a constant $M > 0$ such that $|V(\mathbf{y})| < M$ for $\|\mathbf{y}\| \le r$. For any solution $\boldsymbol{\phi}(t)$ of (5.6) V^* positive definite implies $V(\boldsymbol{\phi}(t)) - V(\boldsymbol{\phi}(0)) = \int_0^t V^*(\boldsymbol{\phi}(s))\,ds > 0$, so that $V(\boldsymbol{\phi}(t)) > V(\boldsymbol{\phi}(0))$ for $t > 0$.

By hypothesis there exists a point $\mathbf{a} \neq 0$, $\|\mathbf{a}\| < r$ such that $V(\mathbf{a}) > 0$. Let $\boldsymbol{\phi}(t, \mathbf{a})$ be that solution of (5.6) which satisfies the initial condition $\boldsymbol{\phi}(0, \mathbf{a}) = \mathbf{a}$. Since $V(\mathbf{y})$ is continuous on Ω and $V(\mathbf{0}) = 0$, there exists a $\delta > 0$ such that $|V(\mathbf{y})| < V(\mathbf{a})$ for $\|\mathbf{y}\| < \delta$. By the existence theory of Chapter 3 the solution $\boldsymbol{\phi}(t, \mathbf{a})$ exists on some interval $[0, t_1)$, where we assume—if it exists—that t_1 is the first point at which $\|\boldsymbol{\phi}(t_1, \mathbf{a})\| = r$; if no such t_1 exists, let $t_1 = +\infty$. We show that $t_1 = +\infty$ is not possible. Consider the solution $\boldsymbol{\phi}(t, \mathbf{a})$ on the interval $[0, t_1)$. Since $V(\boldsymbol{\phi}(t, \mathbf{a})) \ge V(\mathbf{a}) > 0$, $V(\boldsymbol{\phi}(t, \mathbf{a}))$ is nondecreasing on $0 \le t < t_1$. But $\|\mathbf{y}\| < \delta$ implies $|V(\mathbf{y})| < V(\mathbf{a})$ and, therefore, we must also have $\|\boldsymbol{\phi}(t, \mathbf{a})\| \ge \delta$ for $0 \le t < t_1$. Define

$$\mu = \min_{0 < \delta \le \|\mathbf{y}\| \le r} V^*(\mathbf{y})$$

Since V^* is continuous on the closed region between the two concentric spheres, this minimum exists and is assumed at some point $\hat{\mathbf{y}}$, $0 < \delta \le \|\hat{\mathbf{y}}\| \le r$, and since V^* is positive definite, $\mu > 0$. We therefore have $V^*(\boldsymbol{\phi}(t, \mathbf{a})) \ge \mu > 0$ for $0 \le t < t_1$. Thus $V(\boldsymbol{\phi}(t, \mathbf{a})) \ge V(\mathbf{a}) + \mu t$ and therefore also $\lim_{t \to +\infty} V(\boldsymbol{\phi}(t, \mathbf{a})) = +\infty$. Since, however, $|V(\mathbf{y})| < M$ for $\|\mathbf{y}\| \le r$, there must exist a first t_1, $0 < t_1 < \infty$, such that $\|\boldsymbol{\phi}(t_1, \mathbf{a})\| = r$. Therefore, no matter how small $\|\mathbf{a}\| > 0$ is taken, the solution $\boldsymbol{\phi}(t, \mathbf{a})$ will reach the boundary $\|\mathbf{y}\| = r$ at some finite $t = t_1$ and the zero solution cannot be stable. ▌

We omit the proof of Theorem 5.4.

- **EXERCISE**

1. Prove Theorem 5.4. [*Hint:* Study the proof of Theorem 5.3.]

5.4 Invariant Sets and Stability

The Liénard equation, Example 5, Section 5.2 (p. 201) provides us with an example of a V function to which Lyapunov's original result (Theorem 5.2) cannot be applied to deduce asymptotic stability of the zero solution. Although it was possible to modify the V function for the Liénard equation

(Example 6, p. 202, and Exercises 11, 12, p. 204) in such a way that Theorem 5.2 could be applied to the modification, this modification, particularly in more complicated problems, can be technically quite tedious and may even be impossible. Our object now is to show that in many cases we can still conclude asymptotic stability **with the original Lyapunov V function** (certainly at least as far as Example 5, Section 5.2, is concerned).

The motivation for the present generalization is again found in the Liénard equation. Take the equation (5.13), written as the system

$$\begin{aligned} y_1' &= y_2 \\ y_2' &= -g(y_1) - y_2 \end{aligned} \tag{5.14}$$

and consider the original positive definite V function (5.15), which represents the "total energy" of a moving particle obeying the law (5.13), where $g(y_1)$ satisfies the hypothesis stated in Example 2, Section 5.2. Then the derivative of V with respect to (5.14) is

$$V^*(y_1, y_2) = -y_2^2 \tag{5.16}$$

and thus $V^*(y_1, 0) = 0$ even if $y_1 \neq 0$; $V^*(y_1, y_2)$ is negative everywhere except on the y_1 axis. It is clear intuitively that every solution, **except possibly those that start on the y_1 axis**, must approach the origin, provided, of course, they start close enough to the origin. This can be established rigorously by an argument paralleling the proof of Theorem 5.2. It is also clear by looking at (5.14) (see Figure 5.6) that if a solution starts on the y_1 axis, close to the origin, **but not at the origin**, then because at such a point $y_2' = -g(y_1) \neq 0$, this solution cannot stay on the y_1 axis and thus moves away (note also that $y_2 = 0$, $y_1 \neq 0$ is **not** a solution of (5.14)). But as soon as it moves away from the y_1 axis it is "captured" (because $V^*(y_1, y_2) = -y_2^2$) and is "pulled in" toward the origin; of course, in the process of approaching the origin it may cross the y_1 axis infinitely often. We will see that the asymptotic stability of the zero solution of (5.14) can also be deduced from a general principle that for the particular system (5.14) uses the V function (5.15) and the expression (5.16) and the concept of invariant set, which we will now introduce. We will make use of this concept to analyze the set of points $\mathbf{y}$ for which $V^*(\mathbf{y}) = 0$ (for the system (5.14) this was the set of all points on the y_1 axis).

We consider the autonomous system

$$\mathbf{y}' = \mathbf{f}(\mathbf{y}) \tag{5.6}$$

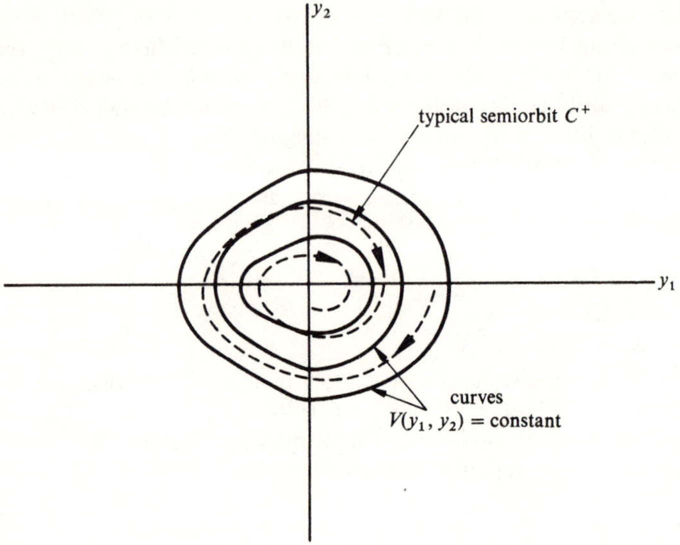

Figure 5.6

where $\mathbf{f}(\mathbf{y})$ and $\partial \mathbf{f}/\partial y_j$ $(j = 1, 2, \ldots, n)$ are continuous in the n-dimensional space E_n. We remind the reader that this means that every solution can be continued for as long as it remains bounded (see Corollary to Theorem 3.6, p. 133).

Definition 1. *A set Γ of points in E_n is (positively) invariant with respect to the system* (5.6) *if every solution of* (5.6) *starting in Γ remains in Γ for all future time.**

Geometrically (see Section 5.2) we denote by C^+ the positive semiorbit associated with a solution $\boldsymbol{\phi}$ of (5.6). If $\mathbf{p}_0 = \boldsymbol{\phi}(t_0)$ is a point of the set Γ and if Γ is invariant with respect to (5.6), then the semiorbit C^+ through $\mathbf{p}_0$ lies in Γ for all $t \geq t_0$ $(C^+ \subset \Gamma)$. For example, for the system (5.14) the origin $y_1 = y_2 = 0$ is clearly an invariant set. (Explain!) Similarly, each curve in Figure 5.5 (Section 5.2) is an invariant set of the system (5.9) if $g(y_1)$ satisfies the hypothesis of Example 2, Section 5.2. (Explain!) Naturally, much more complicated possibilities can arise. For example, let V be any continuously

* In what follows we will use the term invariant set in place of positively invariant set. We remark that the name invariant set is sometimes reserved for a set invariant both as $t \to +\infty$ and as $t \to -\infty$; see [16, p. 58].

differentiable function (V not necessarily positive definite) such that $V^*(\mathbf{y}) \leq 0$. Let S_k be the set of all points of R_n that satisfy the inequality $V(\mathbf{y}) \leq k$. (This can be a very complicated set with several different components even for $n = 2$; see, for example, Figure 5.3.) We use the notation $S_k = \{\mathbf{y} \mid V(\mathbf{y}) \leq k\}$. **However, for every k the set S_k, in fact, each of its components, is an invariant set with respect to the system** (5.6). For, if $\mathbf{y}_0 \in S_k$ and if $\boldsymbol{\phi}(t, \mathbf{y}_0)$ is the solution of (5.6) through the point $\mathbf{y}_0$, then $d/dt\,[V(\boldsymbol{\phi}(t, \mathbf{y}_0))] = V^*(\boldsymbol{\phi}(t, \mathbf{y}_0)$, and therefore $d/dt\,[V(\phi(t, \mathbf{y}_0))] \leq 0$, so that $V(\boldsymbol{\phi}(t, \mathbf{y}_0)) \leq V(\mathbf{y}_0)$. Since $\mathbf{y}_0 \in S_k$, $V(\boldsymbol{\phi}(t, y_0)) \leq k$ for all t, and thus $\boldsymbol{\phi}(t, \mathbf{y}_0) \in S_k$ for all $t \geq 0$, which shows that S_k is invariant. However, if $\mathbf{y}_0$ is in one component of S_k, $\boldsymbol{\phi}(t, \mathbf{y}_0)$ necessarily remains in the same component of S_k for all $t \geq 0$. Thus each component of S_k is an invariant set. We remark also that because no conditions are imposed on the set S_k, the solution $\boldsymbol{\phi}$ that lies in S_k can, of course, become infinite whenever any component of the set S_k is unbounded. We remark further that if, however, $V(\mathbf{y})$ is positive definite, the set S_k must have at least one component H_k **that for sufficiently small $k > 0$ contains the origin; moreover, this component H_k shrinks to the origin as $k \to 0$** (see Figure 5.3). But since H_k is an invariant set, every solution starting in H_k stays in H_k for k small enough, and this, incidentally, gives us another proof of Theorem 5.1.

Now when, for example, the hypothesis of this simple stability theorem (Theorem 5.1) is satisfied, solutions starting near the origin may, but do not necessarily, approach zero. To analyze what happens to these solutions we need another concept.

Let $\boldsymbol{\phi}$ be a solution of (5.6) and let C^+ be its positive semiorbit.

Definition 2. *A point* $\mathbf{p}$ *in* E_n *is said to lie in the positive limit set* $L(C^+)$ (or is said to be a limit point of the orbit C^+) *of the solution* $\boldsymbol{\phi}(t)$ *if and only if there exists a sequence* $\{t_n\} \to +\infty$ *as* $n \to \infty$ *such that* $\lim_{n\to\infty} \boldsymbol{\phi}(t_n) = \mathbf{p}$.

Thus $L(C^+)$ is the set of all limit points (accumulation points) of the orbit C^+. For example, the origin is the positive limit set (that is, $L(C^+) = (0, 0)$) of the semiorbit C^+ shown in Figure 5.6 spiraling toward the origin. Every critical point $\mathbf{p}$ of (5.6) is the limit set $L(C^+)$ of the semiorbit C^+ through $\mathbf{p}$. (Why?) Referring to Figure 5.5, Section 5.2, we see that every orbit C^+ of the system (5.9) has the property that $L(C^+) = C^+$. (Explain carefully!) Again, in general, the positive limit sets of positive semiorbits can be very complicated. We shall see how, using positive limit sets and invariant sets, we can make qualitative statements about solutions of general autonomous systems such as (5.6). Positive limit sets have a number of interesting geometric properties of which we state only those essential for our development. Let $\boldsymbol{\phi}(t, \mathbf{y}_0)$ be the solution of (5.6) satisfying the initial condition $\boldsymbol{\phi}(0, \mathbf{y}_0) = \mathbf{y}_0$, where $\mathbf{y}_0$ is a point of E_n, and let C^+ be the positive semiorbit of $\boldsymbol{\phi}(t, \mathbf{y}_0)$.

Lemma 5.4. *If the solution* $\boldsymbol{\phi}(t, \mathbf{y}_0)$ *is bounded for* $0 \le t < \infty$ (that is, if there exists a constant M such that $\|\boldsymbol{\phi}(t, \mathbf{y}_0)\| \le M$ for $0 \le t < \infty$), *then its positive limit set* $L(C^+)$ *is a nonempty invariant set* (with respect to (5.6)). *Moreover, the solution* $\boldsymbol{\phi}(t, \mathbf{y}_0)$ *approaches the set* $L(C^+)$ *as* $t \to +\infty$ (in the sense that for each $\varepsilon > 0$ there exists a $T > 0$ such that for every $t > T$ there exists a point $\mathbf{p}$ in $L(C^+)$ (possibly depending on t) such that $\|\boldsymbol{\phi}(t, \mathbf{y}_0) - \mathbf{p}\| < \varepsilon$; that is, for t sufficiently large the semiorbit of the solution $\boldsymbol{\phi}(t, \mathbf{y}_0)$ lies arbitrarily close to points of $L(C^+)$).

Proof. Since $\boldsymbol{\phi}(t, \mathbf{y}_0)$ is bounded, its orbit C^+ lies in the interior of some closed sphere S_M of sufficiently large radius. Consider the sequence of points $\{\boldsymbol{\phi}(n, \mathbf{y}_0)\}$ $(n = 1, 2, \ldots)$; this is an infinite sequence such that $\|\boldsymbol{\phi}(n, \mathbf{y}_0)\|$ is bounded. Hence there is a subsequence that converges to a point $\mathbf{p}$ of the closed sphere S_M. By definition $\mathbf{p}$ is in $L(C^+)$ and thus $L(C^+)$ is nonempty. Note that we have also shown that $L(C^+)$ is contained in S_M. (Why?)

To show that $L(C^+)$ is an invariant set, let $\mathbf{p} \in L(C^+)$. Then by definition there exists a sequence $\{t_n\} \to +\infty$ as $n \to \infty$ such that $\boldsymbol{\phi}(t_n, \mathbf{y}_0) \to \mathbf{p}$ as $n \to \infty$. Consider the solution $\boldsymbol{\phi}(t, \mathbf{p})$. By continuous dependence of solutions on their initial values (Theorem 3.7, p. 135) we have for any $t > 0$

$$\lim_{n \to \infty} \boldsymbol{\phi}(t, \boldsymbol{\phi}(t_n, \mathbf{y}_0)) = \boldsymbol{\phi}(t, \mathbf{p})$$

We must show that $\boldsymbol{\phi}(t, \mathbf{p})$ is a point of $L(C^+)$. But obviously

$$\boldsymbol{\phi}(t, \boldsymbol{\phi}(t_n, \mathbf{y}_0)) = \boldsymbol{\phi}(t + t_n, \mathbf{y}_0)$$

and therefore

$$\boldsymbol{\phi}(t, \mathbf{p}) = \lim_{n \to \infty} \boldsymbol{\phi}(t + t_n, \mathbf{y}_0)$$

is a point of $L(C^+)$.*

To show that $\boldsymbol{\phi}(t, \mathbf{y}_0)$ approaches $L(C^+)$ as $t \to \infty$, suppose this is false. Then there exists $\varepsilon > 0$ and a sequence $t_n \to \infty$ such that $\|\boldsymbol{\phi}(t_n, \mathbf{y}_0) - \mathbf{p}\| \ge \varepsilon$ for all $\mathbf{p}$ in $L(C^+)$; that is, there is a sequence of points $\boldsymbol{\phi}(t_n, \mathbf{y}_0)$ that remain

---

* We have actually proved more. Namely, the relation $\lim_{n \to \infty} \boldsymbol{\phi}(t, \boldsymbol{\phi}(t_n, y_0)) = \boldsymbol{\phi}(t, \mathbf{p})$ is true for every t. It follows that $\boldsymbol{\phi}(t, \mathbf{p})$ is a point of $L(C^+)$ for every t, and thus the full orbit through $\mathbf{p}$ is contained in $L(C^+)$. This shows that $L(C^+)$ is invariant in both directions.

at a positive distance from $L(C^+)$. But the solution $\phi(t, \mathbf{y}_0)$ is bounded; therefore the sequence $\{\phi(t_n, \mathbf{y}_0)\}_{n=1}^{\infty}$ has a subsequence $\{\phi(t_{n_k}, \mathbf{y}_0)\}_{k=1}^{\infty}$ that converges to some point $\mathbf{p}$ which must be in $L(C^+)$ by the definition of $L(C^+)$; this is a contradiction. ▌

Suppose now that $\phi(t, \mathbf{y}_0)$ is a bounded solution of (5.6) whose orbit C^+ can be shown by some method to lie in some set $\Omega \subset R^n$, **where Ω contains the origin.** Suppose that we have found some scalar function $V(\mathbf{y})$ that is continuously differentiable in Ω and for which the derivative with respect to (5.6) $V^* \leq 0$ (V may or may not be positive definite; it need not even be nonnegative). We now want to describe the behavior of the scalar function V on the limit set $L(C^+)$ of solutions $\phi(t, \mathbf{y}_0)$.

Lemma 5.5. *Let V be continuously differentiable in a set Ω containing the origin and let $V^*(\mathbf{y}) \leq 0$ at all points of Ω. Let $\mathbf{y}_0 \in \Omega$ and let $\phi(t, \mathbf{y}_0)$ be a bounded solution of* (5.6) *whose positive semiorbit C^+ lies in Ω for all $t \geq 0$ and let the positive limit set $L(C^+)$ of $\phi(t, \mathbf{y}_0)$ lie in Ω. Then $V^*(\mathbf{y}) = 0$ at all points of $L(C^+)$.*

Proof. Let $\mathbf{y}_1$, $\mathbf{y}_2$ be two points of $L(C^+)$. Then there exist sequences $\{t_n\}, \{s_n\} \to \infty$ as $n \to \infty$ such that

$$\lim_{n\to\infty} \phi(t_n, \mathbf{y}_0) = \mathbf{y}_1 \qquad \lim_{n\to\infty} \phi(s_n, \mathbf{y}_0) = \mathbf{y}_2 .$$

Since $d/dt\,[V(\phi(t, \mathbf{y}_0))] = V^*(\phi(t, \mathbf{y}_0)) \leq 0$, $V(\phi(t, \mathbf{y}_0))$ is a nonincreasing function of t and $V(\phi(t, \mathbf{y}_0))$ is bounded below (V is continuous and $\|\phi\|$ is bounded; note that if V is nonnegative, it is automatically bounded below, but for the lemma this is unnecessary). Thus $V(\phi(t, \mathbf{y}_0))$ has a limit as $t \to \infty$. Let this limit be A. Then

$$A = \lim_{n\to\infty} V[\phi(t_n, \mathbf{y}_0)] = \lim_{n\to\infty} V[\phi(s_n, \mathbf{y}_0)]$$

and so by continuity of V on Ω, $V(\mathbf{y}_1) = V(\mathbf{y}_2) = A$; that is, $V(\mathbf{y}) \equiv A$ on $L(C^+)$. But by Lemma 5.4, $L(C^+)$ is a positively invariant set: if $\mathbf{y} \in L(C^+)$, $\phi(t, \mathbf{y}) \in L(C^+)$ for all $t \geq 0$, where $\phi(0, \mathbf{y}) = \mathbf{y}$. But now for each $\mathbf{y} \in L(C^+)$

$$V^*(\mathbf{y}) = V^*(\phi(0, \mathbf{y})) = \frac{d}{dt}[V(\phi(t, \mathbf{y}))]_{t=0} = \frac{dA}{dt} = 0$$

and this completes the proof. ▌

We can now state a general principle on behavior of solutions of (5.6). As corollaries we obtain some useful stability results. As has been our practice throughout, we do not attempt to state the most general results known to date. In fact, any more general result than Theorem 5.5 is probably considerably more complicated.

Theorem 5.5. *Let $V(\mathbf{y})$ be a nonnegative scalar function defined on some set $\Omega \subset R_n$ containing the origin. Let V be continuously differentiable on Ω, let $V^*(\mathbf{y}) \le 0$ at all points of Ω, and let $V(\mathbf{0}) = 0$. For some real constant $\lambda \ge 0$ let C_λ be the component of the set $S_\lambda = \{\mathbf{y} \mid V(\mathbf{y}) \le \lambda\}$ which contains the origin. Suppose that C_λ is a closed bounded subset of Ω. Let E be the subset of Ω defined by $E = \{\mathbf{y} \mid V^*(\mathbf{y}) = 0\}$. Let M be the largest positively invariant subset of E (with respect to* (5.6)*). Then every solution of* (5.6) *starting in C_λ at $t = 0$ approaches the set M as $t \to +\infty$.*

Proof. Let $\mathbf{y}_0 \in C_\lambda$ and let $\phi(t, \mathbf{y}_0)$ be the solution of (5.6) satisfying the initial condition $\phi(0, \mathbf{y}_0) = \mathbf{y}_0$. Then by (5.8) and the hypothesis $d/dt\,[V(t, \phi(t, \mathbf{y}_0))] = V^*(\phi(t, \mathbf{y}_0)) \le 0$ and so $V(t, \phi(t, \mathbf{y}_0))$ is a decreasing function of t. Therefore $\phi(t, \mathbf{y}_0)$ remains in C_λ for all $t \ge 0$. Since C_λ is closed and bounded, the positive limit set $L(C^+)$ of the solution $\phi(t, \mathbf{y}_0)$ also lies in C_λ. By Lemma 5.5, $V^*(\mathbf{y}) = 0$ at all points $\mathbf{y}$ of $L(C^+)$ and so by definition $L(C^+)$ is contained in E. Since $L(C^+)$ is an invariant set, Lemma 5.4 insures that $L(C^+)$ is contained in M and also that $\phi(t, \mathbf{y}_0)$ tends to $L(C^+)$ (and hence to M) as $t \to +\infty$. This completes the proof. ∎

The problem that motivated the above considerations becomes a very special case of the following result, which in turn follows immediately from Theorem 5.5.

Corollary 1. For the system $\mathbf{y}' = \mathbf{f}(\mathbf{y})$, let there exist a positive definite, continuously differentiable scalar function V on some set Ω in E_n (containing the origin) *and let $V^*(\mathbf{y}) \le 0$ at all points of Ω. Let the origin be the only invariant subset* (with respect to (5.6)) *of the set $E = \{\mathbf{y} \mid V^*(\mathbf{y}) = 0\}$. Then the zero solution of* (5.6) *is asymptotically stable.*

• EXERCISE

1. Prove Corollary 1. [*Hint:* First apply Theorem 5.1, then Theorem 5.5.]

Example 1. Returning to the Liénard equation written as the system (5.14), with g satisfying the hypothesis in Example 2, Section 5.2, we use the V function (5.15) having $V^*(y_1, y_2) = -y_2{}^2$, so that $V^*(y_1, 0) = 0$, even though $y_1 \neq 0$. The set E is therefore the y_1 axis. However, as already shown at the beginning of this section, the origin is the only invariant subset of E. Thus by Corollary 1 the zero solution is asymptotically stable. The reader should compare this easy proof to the tedious one in Examples 5 and 6, Section 5.2, which proves the same result.

• **EXERCISE**

2. Prove the above result for the equation $u'' + f(u)u' + g(u) = 0$ where $f(u) > 0$ for $u \neq 0$ and $ug(u) > 0$ for $u \neq 0$.

If by some method (one such method will be given in the next section) we can show that all solutions of (5.6) remain bounded as $t \to \infty$, then the following result concerning **bounded solutions** of (5.6) is useful.

Corollary 2. Let $V(\mathbf{y})$ *be a nonnegative continuously differentiable function such that* $V^*(\mathbf{y}) \leq 0$ *for all* $\mathbf{y}$ *in* E_n *and let* $V(\mathbf{0}) = 0$. *Let* E *be the set in* E_n *defined by* $E = \{\mathbf{y} \mid V^*(\mathbf{y}) = 0\}$. *Let* M *be the largest invariant subset of* E. *Then all bounded solutions of* (5.6) *approach the set* M *as* $t \to +\infty$.

• **EXERCISE**

3. Prove Corollary 2.

5.5 The Extent of Asymptotic Stability—Global Asymptotic Stability

None of the results up to now, with the exception of Theorem 5.5, has given any indication of the size of the region of asymptotic stability. However, as previously indicated, this is perhaps the most important practical question. Let Ω be an open set in E_n containing the origin. Let there exist a positive definite scalar function $V(\mathbf{y})$ which, with respect to

$$\mathbf{y}' = \mathbf{f}(\mathbf{y}) \tag{5.6}$$

has $V^*(\mathbf{y}) \leq 0$ in Ω. Let the origin be the only invariant subset of the set $E = \{\mathbf{y} \mid V^*(\mathbf{y}) = 0\}$. Looking at Corollary 1 to Theorem 5.5, we might be led to the conjecture that the set Ω is contained in the region of asymptotic stability of the zero solution of (5.6). However, this set is in general too large and the conjecture is false, even if Ω is a bounded set, for the following reason. We consider again the set C_λ, the component of $S_\lambda = \{\mathbf{y} \mid V(\mathbf{y}) \leq \lambda\}$ containing the origin, for $\lambda \geq 0$. For $\lambda = 0$ we get the origin. When $n = 2$ (see Figure 5.7), for small $\lambda > 0$ we get closed bounded regions containing the origin and contained in Ω. But this can fail to be true when λ becomes too large. In this case (see Figure 5.7) the sets C_λ can extend outside Ω (and they may even

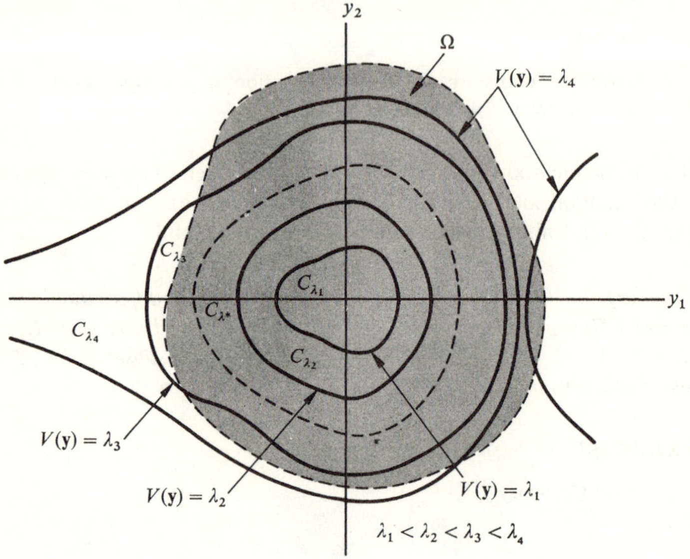

Figure 5.7

be unbounded); of course, the difficulty is that for such λ, C_λ contain points at which $V^*(\mathbf{y}) \leq 0$ need not hold. However, from Theorem 5.5 we can at least say that **every closed bounded region C_λ contained in Ω lies in the region of asymptotic stability**, and this is considerably better than anything we have been able to say up to now. We can compute the largest $\lambda = \hat{\lambda}$, so that $C_{\hat{\lambda}}$ has this property, as the largest value of λ for which the component of $S_{\hat{\lambda}} = \{\mathbf{y} \mid V(\mathbf{y}) = \hat{\lambda}\}$ containing the origin actually meets the boundary of Ω. We can also say, even when Ω is unbounded, that all those sets C_λ that are completely contained in Ω are positively invariant sets with respect to (5.6). Therefore a solution of (5.6) starting in C_λ at $t = 0$ is bounded, and thus tends to the origin by Corollary 2 of Theorem 5.5. This shows that such a set C_λ must be contained in the region of asymptotic stability. In fact we can certainly say: **The region of asymptotic stability is at least as large as the largest invariant set contained in Ω. In particular, the interior of $C_{\hat{\lambda}}$ is contained in the region of asymptotic stability.** We shall now illustrate these ideas by actual examples.

Example 1. Consider the scalar equation

$$u'' + u' + u + u^2 = 0 \tag{5.26}$$

which, written as a system, is

$$\begin{aligned} y_1' &= y_2 \\ y_2' &= -y_1 - y_1^2 - y_2 \end{aligned} \tag{5.27}$$

There are two critical points $(-1, 0)$ and $(0, 0)$, and by using the linear approximation and Theorems 4.3 (p. 161) and 4.6 (p. 171), we see that the first is unstable and the second is asymptotically stable. However, we know nothing about the **region of asymptotic stability**, except for the obvious remark that $(-1, 0)$ cannot be in it.

• **EXERCISE**

1. Show that the zero solution is asymptotically stable by using Corollary 1, Theorem 5.5, exactly as in Example 1, Section 5.4.

Our problem is to estimate the region of asymptotic stability. We note that (5.26) is a Liénard equation with $g(u) = u + u^2$; we try (see (5.15)) the V function

$$V(y_1, y_2) = \tfrac{1}{2}y_1^2 + \tfrac{1}{3}y_1^3 + \tfrac{1}{2}y_2^2 \tag{5.28}$$

This function is positive definite on the set $\Omega = \{(y_1, y_2) \mid y_2^2 > -y_1^2 - \tfrac{2}{3}y_1^3\}$ together with the origin and sketched in Figure 5.8.

We have $V^*(y_1, y_2) = y_1y_2 + y_1^2y_2 - y_2^2 - y_1y_2 - y_1^2y_2 = -y_2^2$, so that $V^*(y_1, y_2) \le 0$ on the whole plane, in particular, on Ω. To apply Theorem 5.5 and our discussion above on extent of asymptotic stability, we look at the subset E of Ω given by $E = \{(y_1, y_2) \in \Omega \mid V^*(y_1, y_2) = 0\}$. This is clearly the portion of the y_1 axis with $y_1 > -\tfrac{3}{2}$. Notice that E contains both of the critical points $(-1, 0)$ and $(0, 0)$ and these are both invariant subsets of E. By examining the system (5.27) on the y_1 axis, that is, the system (5.27) with $y_2 = 0$, namely

$$\begin{aligned} y_1' &= 0 \\ y_2' &= -y_1 - y_1^2 \end{aligned} \tag{5.29}$$

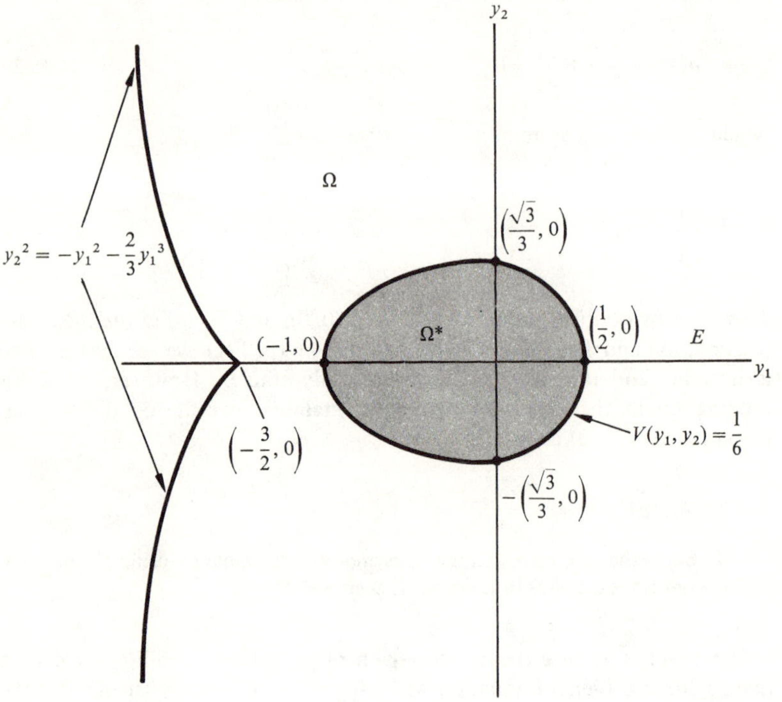

Figure 5.8

we see that the points $(-1, 0)$, $(0, 0)$ are the only invariant subsets of E, because at all other points of E, $y_2' \neq 0$. Clearly Ω cannot be the region of asymptotic stability. Define $\Omega_1 = \Omega - (-1, 0)$, that is, Ω with the point $(-1, 0)$ deleted. Then, at least, the origin is the only invariant subset of that part of E which lies in Ω_1. The boundary of Ω_1 consists of the curve $y_2{}^2 = -y_1{}^2 - \frac{2}{3}y_1{}^3$ and the point $(-1, 0)$. We now look at the regions $C_\lambda = \{(y_1, y_2) \mid V(y_1, y_2) \leq \lambda\}$ that lie in Ω_1. A little consideration shows that the curve $V(y_1, y_2) = V(-1, 0)$ passes through the boundary point $(-1, 0)$ of Ω_1 and is the closed curve shown in Figure 5.8. Since $V(-1, 0) = \frac{1}{2} - \frac{1}{3} = \frac{1}{6}$, the value $\hat{\lambda}$ in our discussion on extent of asymptotic stability is $\hat{\lambda} = \frac{1}{6}$ and the boundary curve is given by $V(y_1, y_2) = \frac{1}{6}$. **Thus the region of asymptotic stability certainly includes the bounded set**

$$\Omega^* = \{(y_1, y_2) \mid V(y_1, y_2) < \tfrac{1}{6}\} = \{(y_1, y_2) \mid y_1 > -1, 3y_1{}^2 + 3y_2{}^2 + 2y_1{}^3 < 1\}$$

However, it is almost certainly a larger set than Ω^*. Notice that we cannot hope to enlarge our estimate by enlarging Ω^* in such a way that the point $(-1, 0)$ would be included. But consider the region Ω^{**}, which is Ω^* together with the interior of the rectangle whose vertices are $(0, 0)$ $(-1, 0)$, $(-1, \sqrt{3}/3)$, $(0, \sqrt{3}/3)$, as shown in Figure 5.9. If we can show that no solution can leave Ω^{**} across the left and top edge of the rectangle, we will have

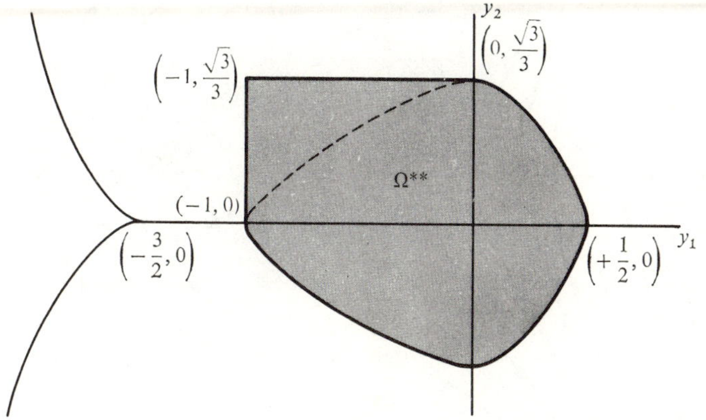

Figure 5.9

that Ω^{**} is contained in the region of asymptotic stability. On the left edge, $y_1 = -1$ and $0 \leq y_2 \leq \sqrt{3}/3$. Then from (5.27) $y_1' = y_2 \geq 0$ and $y_2' = -y_2 < 0$. On the top edge, $y_2 = \sqrt{3}/3$ while $-1 \leq y_1 \leq 0$, and here the function $-y_1 - y_1^2$ assumes its maximum value at the point $y_1 = -\frac{1}{2}$ and from (5.27) we have $y_1' = \sqrt{3}/3 > 0$, $y_2' = -y_2 - y_1 - y_1^2 \leq -\sqrt{3}/3 - \frac{1}{2} + \frac{1}{4} < 0$. Thus no solution starting in Ω^{**} can leave through the edge $y_1 = -1$ or the edge $y_2 = \sqrt{3}/3$ and this shows the desired property of Ω^{**}. Other refinements are possible. The actual region of asymptotic stability is shown in Figure 5.10. Here the origin is a spiral point and $(-1, 0)$ a saddle point.

• EXERCISE

2. Discuss the scalar equation $u'' + au' + bu + u^2 = 0$; $a > 0$, $b > 0$ are given constants. Determine an estimate for the region of asymptotic stability of the zero solution.

Example 2. Consider the Liénard equation

$$u'' + f(u)u' + g(u) = 0 \tag{5.30}$$

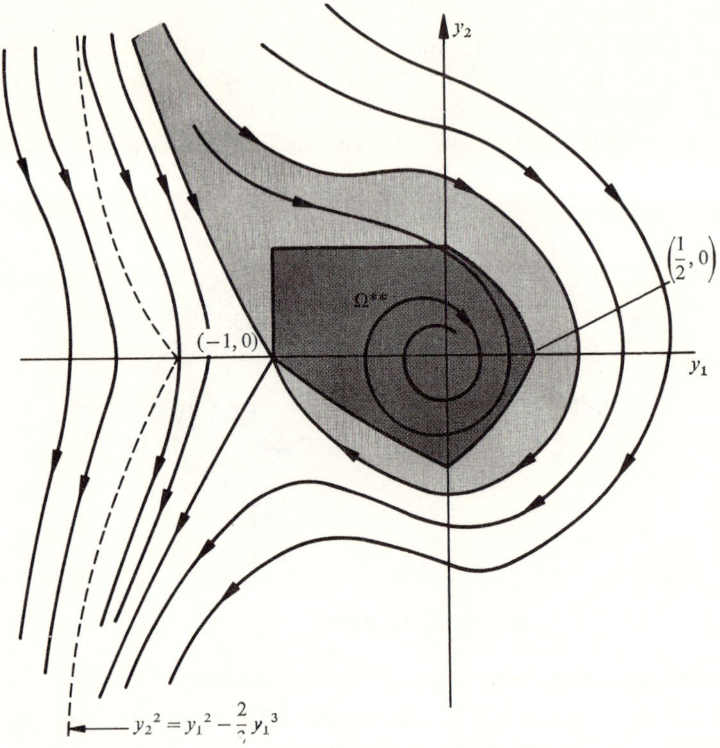

Figure 5.10

where $g(0) = 0$, $ug(u) > 0$ $(u \neq 0)$ for $-\infty < u < \infty$ and for some $a > 0$, $uF(u) > 0$ for $0 < |u| < a$, where $F(u) = \int_0^u f(\sigma)\, d\sigma$ with f and g continuously differentiable. It is easily shown by the method already employed (Corollary 1, Theorem 5.5) that the zero solution is asymptotically stable. Again, the problem is to determine the region of asymptotic stability. Here we employ a different equivalent system, namely

$$\begin{aligned} y_1' &= y_2 - F(y_1) \\ y_2' &= -g(y_1) \end{aligned} \tag{5.31}$$

where $F(y_1) = \int_0^{y_1} f(\sigma)\, d\sigma$.

• **EXERCISE**

3. Show that (5.30) and (5.31) are equivalent. [*Hint:* Put $y_1 = u$, $y_2 = u' + F(u)$.]

Define $G(y_1) = \int_0^{y_1} g(\sigma)\, d\sigma$ and try the V function

$$V(y_1, y_2) = \tfrac{1}{2}y_2{}^2 + G(y_1) \tag{5.32}$$

which is positive definite in the whole plane. Now $V^*(y_1, y_2) = y_2(-g(y_1)) + g(y_1)(y_2 - F(y_1)) = -g(y_1)F(y_1)$, and since by hypothesis $g(y_1)$ has the sign of y_1, we have immediately that $V^*(y_1, y_2) \leq 0$ on the strip

$$\Omega = \{(y_1, y_2) \mid -a < y_1 < a,\ -\infty < y_2 < \infty\}$$

(see Figure 5.11).

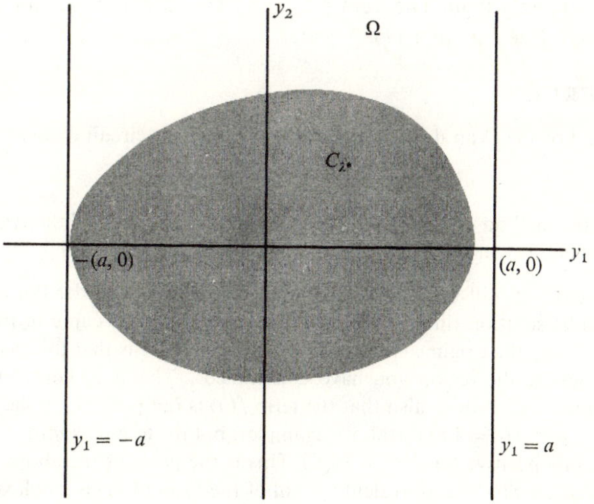

Figure 5.11

• **EXERCISE**

4. Show $V^*(y_1, y_2) \leq 0$ on Ω, considering the cases $0 \leq y_1 < a$ and $-a < y_1 \leq 0$ separately.

Therefore, by Exercise 4, the origin is stable, and from $V^*(y_1, y_2) = -g(y_1)F(y_1)$ we see that the set E of Theorem 5.5 is the y_2 axis:

$$E = \{(y_1, y_2) \mid y_1 = 0\}$$

But, since for (y_1, y_2) in E

$$y_1' = y_2$$
$$y_2' = 0$$

we see that the origin is the only invariant subset of E and thus the origin is (Corollary 1, Theorem 5.5) asymptotically stable.

We now wish to consider the curves $V(y_1, y_2) = \lambda$ for $-a < y_1 < a$ with increasing values of λ beginning with $\lambda = 0$. These are closed curves symmetric about the y_1 axis. For $-a < y_1 < 0$, $G(y_1)$ decreases and for $0 < y_1 < a$, $G(y_1)$ increases. Thus the curve $V(y_1, y_2) = \lambda$ first makes contact with the boundary of Ω at one of the points $(-a, 0)$ or $(a, 0)$. The best value of $\hat{\lambda}$ is $\hat{\lambda} = \min(G(a), G(-a))$ and $C_{\hat{\lambda}} = \{(y_1, y_2) \mid \frac{1}{2}y_2^2 + G(y_1) < \hat{\lambda}\}$. Hence, directly from Theorem 5.5, every solution starting in $C_{\hat{\lambda}^*}$ approaches the origin (see Figure 5.11).

• **EXERCISE**

5. For the (Van der Pol) equation of nonlinear circuit theory

$$u'' + \varepsilon(1 - u^2)u' + u = 0 \qquad \varepsilon > 0 \text{ a constant}$$

use the method of Example 2 to determine an estimate of the region of asymptotic stability in the phase plane. [*Answer:* $y_1^2 + y_2^2 < 3$.]

[NOTE. It will be shown in Chapter 6 that the Van der Pol equation has a periodic solution (limit cycle) in the form of a closed curve in the phase plane containing the origin. What you have shown here is that this periodic solution lies **outside** the region you have determined. This is of considerable practical importance. Notice also that the term $f(u)$ is not positive on the entire interval $-a < u < a$ ($a = 1$ in the above example); but the term $F(u)g(u) = \varepsilon(u - u^3/3)u$ is certainly positive for $0 < |u| < \sqrt{3}$. This is the point of the above technique and the reason why the equivalent system of the form (5.31) is employed!]

6. Another and more usual form of the Van der Pol equation is

$$(*) \quad v'' + \varepsilon(v^2 - 1)v' + v = 0 \qquad \varepsilon > 0 \text{ a constant}$$

Show that the identically zero solution is unstable. [*Hint:* (a) Show that the change of variable $\tau = -t$ transforms (*) to the equation

$$(**) \quad \frac{d^2u}{d\tau} + \varepsilon(1 - u^2)\frac{du}{d\tau} + u = 0$$

where $u(\tau) = v(-\tau)$, which is the equation of Exercise 5. Thus to study the behavior of solutions of (*) as $t \to +\infty$ is the same as studying that of (**) as $\tau \to -\infty$.

(b) Alternatively, use the equivalent system (5.31) for (*) and the energy function (5.32) to find

$$V^*(y_1, y_2) = \varepsilon y_1^2\left(1 - \frac{y_1^2}{3}\right)$$

from which the result follows directly.]

In many problems, in control theory or nuclear reactor dynamics, for example, in which the zero solution of (5.6) is known to be asymptotically stable, it is important to determine whether all solutions, no matter what their initial values may be, approach the origin. In other words, we wish to determine whether **the region of asymptotic stability is the whole space** E_n. If this is the case, we say that **the zero solution of** (5.6) **is globally asymptotically stable**. As we will see, everything depends on finding a "good enough" V function. For, with enough hypothesis on the V function, it is very easy to give a criterion for global asymptotic stability. We have already shown, as a consequence of Corollary 1 and Corollary 2, Theorem 5.5, that if $V(\mathbf{y})$ is positive definite, if, with respect to (5.6), $V^*(\mathbf{y}) \le 0$, and if in the set $E = \{\mathbf{y} \mid V^*(\mathbf{y}) = 0\}$ the origin is the only invariant subset, then **the zero solution of** (5.6) **is asymptotically stable** (Corollary 1) **and all bounded solutions of** (5.6) **approach zero as** $t \to +\infty$ (Corollary 2). Thus we only need to prove a result that insures that all solutions of (5.6) are bounded. (**The boundedness of all solutions of a system is often called Lagrange stability.**) Let the above V function have the additional property

$$V(\mathbf{y}) \to \infty \quad \text{as} \quad \|\mathbf{y}\| \to \infty \tag{5.33}$$

let $\mathbf{y}_0$ be any point in R_n and let $\boldsymbol{\phi}(t, \mathbf{y}_0)$ be the local solution of (5.6) through $\mathbf{y}_0$ existing for $0 \le t < t_1$. From (5.8) and the hypothesis we have, for as long as $\boldsymbol{\phi}(t, \mathbf{y}_0)$ exists,

$$\frac{d}{dt}[V(\boldsymbol{\phi}(t, \mathbf{y}_0))] = V^*(\boldsymbol{\phi}(t, \mathbf{y}_0)) \le 0$$

Therefore $V(\boldsymbol{\phi}(t, \mathbf{y}_0)) \le V(\mathbf{y}_0)$ (a constant) and the hypothesis (5.33) implies that $\|\boldsymbol{\phi}(t, \mathbf{y}_0)\|$ is bounded by a constant that depends only on $\mathbf{y}_0$ and not on t_1. Therefore, by the Corollary to Theorem 3.6 (p. 133), the solution $\boldsymbol{\phi}$ can be continued for all t, $0 \le t < \infty$, and for all such t the solution $\boldsymbol{\phi}$ remains bounded by this same constant. Since $\mathbf{y}_0$ is arbitrary this shows that all solutions of (5.6) are bounded. To summarize, we have proved the following result.

Theorem 5.6. *Let there exist a scalar function $V(\mathbf{y})$ such that:*

(i) *$V(\mathbf{y})$ is positive definite on E_n and $V(\mathbf{y}) \to \infty$ as $\|\mathbf{y}\| \to \infty$;*
(ii) *with respect to* (5.6), *$V^*(\mathbf{y}) \leq 0$ on R_n;*
(iii) *the origin is the only invariant subset of the set $E = \{\mathbf{y} \mid V^*(\mathbf{y}) = 0\}$.*

Then the zero solution of $\mathbf{y}' = \mathbf{f}(\mathbf{y})$ is globally asymptotically stable.

As an immediate consequence of Theorem 5.6, we have the following results.

Corollary 1. Let there exist a scalar function $V(\mathbf{y})$ that satisfies (i) *above and that has $V^*(\mathbf{y})$ negative definite. Then the zero solution of $\mathbf{y}' = \mathbf{f}(\mathbf{y})$ is globally asymptotically stable.*

Corollary 2. If only (i) *and* (ii) *of Theorem* 5.6 *are satisfied, then all solutions of $\mathbf{y}' = \mathbf{f}(\mathbf{y})$ are bounded for $t \geq 0$* (that is, (5.6) is Lagrange stable).

Example 3. Consider the Liénard equation $u'' + u' + g(u) = 0$ where we assume $g(u)$ continuously differentiable for all u and

$$ug(u) > 0 \qquad (u \neq 0) \tag{5.34}$$

$$G(x) = \int_0^x g(\sigma)\, d\sigma \to \infty \qquad \text{as } |x| \to \infty \tag{5.35}$$

We assert that the zero solution is globally asymptotically stable.

Proof. The equivalent system is

$$y_1' = y_2$$
$$y_2' = -g(y_1) - y_2$$

If we choose the familiar V function

$$V(y_1, y_2) = \tfrac{1}{2}y_2^2 + G(y_1)$$

then we have already shown, using it and Corollary 1, Theorem 5.5, that the origin is asymptotically stable (see Example 1, Section 5.5). Since hypothesis (i) of Theorem 5.6 is clearly fulfilled because of the requirement (5.35), the result follows from Theorem 5.6. ∎

• **EXERCISES**

7. What can you say about the solutions of

$$u'' + f(u)u' + g(u) = 0$$

where $g(u)$ is as in Example 3 and $f(u)$ is a continuously differentiable function that satisfies the condition $f(u) > 0$ for all $u \neq 0$?

8. Use the method of Exercise 7 to obtain conditions under which the results of Exercise 7 hold for the more general equation

$$u'' + h(u, u')u' + g(u) = 0$$

9. Replace condition (5.34) in Example 3 by the condition

$$G(x) = \int_0^x g(\sigma)\, d\sigma > 0 \qquad \text{for all } x \neq 0$$

and let condition (5.35) hold. Show that the conclusion of Example 3 still holds. Generalize and obtain similar results for the equations considered in Exercises 7 and 8.

It occasionally happens that it is somewhat difficult to find a V function that satisfies all three conditions of Theorem 5.6. In that case it can sometimes be useful to prove the boundedness of all solutions first, and separately establish the proposition that all bounded solutions approach zero. To illustrate this, consider the Liénard equation again, but under different assumptions. We now remove the requirement that $G(u) = \int_0^u g(\sigma)\, d\sigma \to \infty$, and replace it by a stronger requirement on the damping.

Example 4. Consider the equation

$$u'' + f(u)u' + g(u) = 0$$

where f, g are continuously differentiable functions for $-\infty < u < \infty$, and assume that

$$ug(u) > 0 \qquad \text{if } (u \neq 0) \tag{5.36}$$

$$f(u) > 0 \qquad \text{if } (u \neq 0) \tag{5.37}$$

$$|F(u)| \to \infty \quad \text{as} \quad |u| \to \infty \qquad \text{where } F(u) = \int_0^u f(\sigma)\, d\sigma \tag{5.38}$$

(Note that this is satisfied if $f(u) \equiv 1$.) **We claim that the zero solution is again globally asymptotically stable.**

Writing the equation as the system

$$\begin{aligned} y_1' &= y_2 \\ y_2' &= -g(y_1) - f(y_1)y_2 \end{aligned} \tag{5.39}$$

we define the V function $V(y_1, y_2) = G(y_1) + \frac{1}{2}y_2{}^2$ for which $V^*(y, y_2) = -f(y_1)y_2{}^2$. By the familiar argument we can deduce, from Corollary 2 to Theorem 5.5, only that **every bounded solution approaches zero**; we cannot use Theorem 5.6, because we cannot satisfy the hypothesis $V(\mathbf{y}) \to \infty$ as $\|y\| \to \infty$ unless we know that

$$G(u) = \int_0^u g(\sigma)\, d\sigma \to \infty \quad \text{as} \quad |u| \to \infty$$

which we are no longer assuming. However, we can use another method to prove that all solutions are bounded. Let $\lambda > 0$, $a > 0$ be arbitrary given numbers, and consider the region Ω defined by the inequalities

$$\begin{aligned} &V(y_1, y_2) < \lambda \\ &(y_2 + F(y_1))^2 < a^2 \end{aligned}$$

For each pair (λ, a), Ω is a bounded region as shown in Figure 5.12. Let $\mathbf{y}_0 = (y_{10}, y_{20})$ be any given point. Then by proper choice of (λ, a) we can assure that (y_{10}, y_{20}) lies strictly inside Ω. Let $\boldsymbol{\phi}(t, \mathbf{y}_0)$ be a solution of (5.39) such that $\boldsymbol{\phi}(0, \mathbf{y}_0) = \mathbf{y}_0$. It will be shown that $\boldsymbol{\phi}(t,\mathbf{y}_0)$ cannot leave Ω,

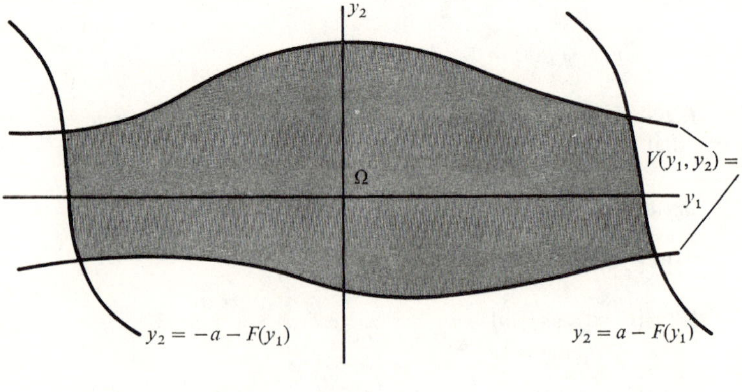

Figure 5.12

which will show that $\boldsymbol{\phi}(t, \mathbf{y}_0)$, which is an arbitrary solution, is bounded. To leave Ω, the solution must either cross $V(y_1, y_2) = \lambda$ or one of the curves $y_2 = \pm a - F(y_1)$, where we have chosen $a > 0$ so large that the part of the curve $y_2 = a - F(y_1)$ that is also the boundary of Ω corresponds to $y_1 > 0$, and similarly, the part of $y_2 = -a - F(y_1)$ corresponds to $y_1 < 0$. As $V^*(\boldsymbol{\phi}(t, \mathbf{y}_0)) \leq 0$, the solution $\boldsymbol{\phi}(t, \mathbf{y}_0)$ cannot cross $V(y_1, y_2) = \lambda$. To show that it does not cross either of the curves $y_2 = \pm a - F(y_1)$, we consider the function $W(t) = [\phi_2(t, \mathbf{y}_0) + F(\phi_1(t, \mathbf{y}_0))]^2$. Then we see easily that

$$W'(t) = 2(\phi_2(t, \mathbf{y}_0) + F(\phi_1(t, \mathbf{y}_0)))g(\phi_1(t, \mathbf{y}_0))$$

Suppose the solution $\boldsymbol{\phi}(t, y_0)$ reaches the "right" boundary curve $y_2 = a - F(y_1), y_1 > 0$; then along this part of the boundary $W'(t) = -2ag(\phi(t, \mathbf{y}_0)) < 0$, because here $y_1 > 0$ and $a > 0$; thus the solution $\boldsymbol{\phi}$ cannot cross outside Ω through the curve $y_2 = a - F(y_1)$. A similar argument applies to the "left" boundary curve, and completes the proof that every solution is bounded.

• EXERCISES

10. Show that at each point of the "left" boundary $y_2 = -a - F(y_1)$, $W'(t) < 0$, so that the solution $\boldsymbol{\phi}(t, \mathbf{y}_0)$ cannot leave Ω along the left boundary curve.

11. Consider the system*

$$\begin{aligned} x' &= -\sum_{i=1}^{n} a_i y_i \\ y_i' &= -h(x, \mathbf{y})y_i + b_i g(x) \qquad (i = 1, \ldots, n) \end{aligned} \tag{5.40}$$

where a_i, b_i are constants, x is a scalar, $\mathbf{y} = (y_1, \ldots, y_n)$, h, g are continuously differentiable for $-\infty < x < \infty$, $\|y\| < \infty$, $xg(x) > 0$ for all $x \neq 0$, $G(x) = \int_0^x g(\sigma)\, d\sigma \to \infty$ as $|x| \to \infty$, $h(x, \mathbf{y}) > 0$ for $x \neq 0$, $\mathbf{y} \neq 0$. Show that if either (a) there exists a constant c such that $a_i = cb_i$ $(i = 1, \ldots, n)$ with at least one $b_i \neq 0$, or (b) $a_i/b_i > 0$ $(i = 1, \ldots, n)$, then the solution $x = 0$, $\mathbf{y} = 0$ of (5.40) is globally asymptotically stable. [*Hint:* For (a) try the V function

$$V(x, \mathbf{y}) = \frac{1}{c} G(x) + \frac{1}{2} \sum_{i=1}^{n} y_i^2$$

for (b) modify V in a suitable way.]

12. Consider the scalar equation $u'' + u' + g(u) = 0$ where g is continuously differentiable for $-\infty < u < \infty$ and

$$G(u) = \int_0^x g(\sigma)\, d\sigma \to \infty \quad \text{as} \quad |u| \to \infty$$

* A certain problem in reactor dynamics can be transformed to this form.

[Note: **No assumption such as** $ug(u) > 0$ **for** $u \neq 0$ **is now made.**] Can you make any qualitative statement about the **global** behavior of solutions? [*Hint:* Start by trying to apply Corollary 2 to Theorem 5.6.]

13. Consider the scalar equation

$$u''' + f(u')u'' + au' + bu = 0$$

where $a > 0$, $b > 0$ are constants and where $f(z)$ is differentiable for $-\infty < z < \infty$. What condition must f satisfy in order that the zero solution be globally asymptotically stable? [*Hint:* For the equivalent system $y_1' = y_2, y_2' = y_3, y_3' = -by_1 - ay_2 - f(y_2)y_3$ try the V function

$$V(y_1, y_2, y_3) = \frac{a}{2} y_3{}^2 + by_2 y_3 + b \int_0^{y_2} \sigma f(\sigma)\, d\sigma + \tfrac{1}{2}(by_1 + ay_2)^2$$

Answer. There exists a constant $c > b/a$ such that $f(y_2) \geq c > b/a$ for $-\infty < y < \infty$. What can you say if $c \geq b/a$?]

5.6 Nonautonomous Systems

We consider the system

$$\mathbf{y}' = \mathbf{f}(t, \mathbf{y}) \tag{5.41}$$

in which $\mathbf{f}$ depends explicitly on t. The theory already developed in this chapter can be extended to nonautonomous systems. We will assume that $\mathbf{f}$ and $\partial\mathbf{f}/\partial y_j$ $(j = 1, \ldots, n)$ are continuous in a region D of $(n + 1)$-dimensional $(t, \mathbf{y})$ space. The region D may be the whole space. However, we will always require that D contains $H = \{(t, \mathbf{0}) \mid t \geq 0\}$ in its interior. We will also assume that $\mathbf{f}(t, \mathbf{0}) = \mathbf{0}$ so that $\mathbf{y} \equiv \mathbf{0}$ is a solution of (5.41). We remind the reader that the investigation of the stability of any solution can be reduced to this case (see Section 4.2, p. 150).

As in Section 5.2, we will study the stability by means of scalar V functions that may now depend on both t and $\mathbf{y}$. This naturally requires us to modify the earlier definitions as follows. Let $V(t, \mathbf{y})$ be a scalar continuous function having continuous first-order partial derivatives with respect to t and the components of $\mathbf{y}$ in a region Ω in $(t, \mathbf{y})$ space. We will also assume that Ω contains the set $H = \{(t, \mathbf{0}) \mid t \geq 0\}$.

Definition 1. *The scalar function $V(t, \mathbf{y})$ is said to be positive definite on the set Ω if and only if $V(t, \mathbf{0}) = 0$ and there exists a scalar function $W(\mathbf{y})$, independent of t, with $V(t, \mathbf{y}) \geq W(\mathbf{y})$ for $(t, \mathbf{y})$ in Ω and such that $W(\mathbf{y})$ is positive definite in the sense of Definition 1* (Section 5.2).

Definition 2. *The scalar function $V(t, \mathbf{y})$ is negative definite on Ω if and only if $-V(t, \mathbf{y})$ is positive definite on Ω.*

Example 1. If $n = 2$ let $V(t, \mathbf{y}) = y_1^2 + (1 + t)y_2^2 \geq y_1^2 + y_2^2 = W(\mathbf{y})$ for $0 \leq t < \infty$ and $W(\mathbf{y})$ is positive definite on the whole y-space. Thus $V(t, \mathbf{y})$ is positive definite on $\Omega = \{(t, \mathbf{y}) \mid t \geq 0\}$.

Example 2. If $n = 2$, let $V(t, \mathbf{y}) = y_1^2 + y_2^2/(1 + t)$. Consider $V(t, 0, a_2) = a_2^2/(1 + t)$. Since this approaches zero as $t \to \infty$, we cannot hope to find a suitable function W. This V function is not positive definite in the sense of Definition 1 even though $V(t, \mathbf{y}) > 0$ for $\mathbf{y} \neq \mathbf{0}$.

- **EXERCISE**

1. Let $n = 3$. Which of the following functions $V(t, \mathbf{y})$ are positive definite on a suitable set Ω?
   (a) $V(t, \mathbf{y}) = y_1^2 + y_3^2$.
   (b) $V(t, \mathbf{y}) = t(y_1^2 + y_2^2 + y_3^2)$.
   (c) $V(t, \mathbf{y}) = y_1^2 + y_2^2 - y_3^2$.
   (d) $V(t, \mathbf{y}) = y_1^2 + y_2^2 + y_3^2 + 2y_2 y_3 \cos t$.

Definition 3. *The derivative of $V(t, \mathbf{y})$ with respect to the system $\mathbf{y}' = \mathbf{f}(t, \mathbf{y})$ is*

$$V^*(t, \mathbf{y}) = \frac{\partial V}{\partial t}(t, \mathbf{y}) + \sum_{j=1}^{n} \frac{\partial V}{\partial y_j}(t, \mathbf{y}) f_j(t, \mathbf{y}) \tag{5.42}$$

If $\boldsymbol{\phi}$ is any solution of (5.41), we have

$$\frac{d}{dt} V(t, \boldsymbol{\phi}(t)) = V^*(t, \boldsymbol{\phi}(t)) \tag{5.43}$$

- **EXERCISE**

2. Prove formula (5.43).

The stability result for (5.41) is virtually identical to the result in the autonomous case:

Theorem 5.7. *If there exists a scalar function $V(t, \mathbf{y})$ that is positive definite and for which $V^*(t, \mathbf{y}) \leq 0$* (that is, the derivative (5.42) with respect to the system (5.41) is nonpositive) *on some region Ω that contains the set $H = \{(t, \mathbf{0}) \mid t \geq 0\}$, then the zero solution of $\mathbf{y}' = \mathbf{f}(t, \mathbf{y})$ is stable.*

• **EXERCISE**

3. Prove Theorem 5.7. [*Hint:* Follow the proof of Theorem 5.1.]

Example 3. Investigate the stability of the zero solution of the equation $y'' + a(t)y = 0$ assuming that $a(t) \geq \delta > 0$ and $a'(t) \leq 0$ $(0 \leq t < \infty)$.

Writing the equation as an equivalent system, we have

$$\begin{aligned} y_1' &= y_2 \\ y_2' &= -a(t)y_1 \end{aligned} \tag{5.44}$$

Consider the scalar function

$$V(t, y_1, y_2) = a(t){y_1}^2 + {y_2}^2$$

which is obviously positive definite on $\Omega = \{(t, y) \mid t \leq 0\}$ because

$$V(t, y_1, y_2) \geq \delta {y_1}^2 + {y_2}^2$$

Using (5.42) and (5.44), we have

$$V^*(t, y_1, y_2) = a'(t){y_1}^2 \leq 0$$

on Ω, because $a'(t) \leq 0$. Therefore, Theorem 5.7 implies that the solution $y_1 = y_2 = 0$ of (5.44) is stable.

We remark that if $\boldsymbol{\phi}(t) = (\phi_1(t), \phi_2(t))$ is any solution of (5.44), it follows from (5.43) and the above calculations that

$$0 \leq \delta {\phi_1}^2(t) + {\phi_2}^2(t) \leq V(t, \boldsymbol{\phi}(t)) \leq V(0, \boldsymbol{\phi}(0)) \leq a(0){\phi_1}^2(0) + {\phi_2}^2(0)$$

Thus every local solution of (5.44) can be continued (by Corollary to Theorem 3.6, p. 133) to the interval $0 \leq t < \infty$ and the solution remains bounded. Since (5.44) is linear, this result should not come as a surprise (see Exercise 12, Section 4.3, p. 154).

• **EXERCISES**

4. Let $g(t, y)$, $g_t(t, y)$, $g_y(t, y)$ be continuous in the region $\Omega = \{(t, y) \mid t \geq 0\}$. Let $g(t, y) \geq h(y) > 0$ if $y > 0$ and $g(t, y) \leq h(y) < 0$ if $y < 0$ on Ω (by continuity $g(t, 0) = 0$) and let $y g_t(t, y) \leq 0$ on Ω. Show that the zero of the equation $y'' + g(t, y) = 0$ is stable. [*Hint:* Write the equivalent system and try the function $V(t, y_1, y_2) = {y_2}^2/2 + \int_0^{y_1} g(t, \sigma)\, d\sigma$.]
5. Apply the result or technique of Exercise 4 to obtain a stability criterion (in terms of $a(t)$) for the zero solution of the equation $y'' + a(t)y^3 = 0$.
6. Generalize the result of Exercise 5 to the equation $y'' + a(t)h(y) = 0$.

We now turn to asymptotic stability of the zero solution of the system (5.41). One would be tempted to conjecture, in light of Theorem 5.2, that if the function $V(t, \mathbf{y})$ of Theorem 5.7 also satisfies the condition that $V^*(t, \mathbf{y})$ is negative definite on the region $\Omega = \{(t, \mathbf{y} \mid t \geq 0\}$, then the zero solution should be asymptotically stable. This turns out to be false unless an additional restriction is imposed. For this purpose we require another concept.

Definition 4. *A scalar function $U(t, \mathbf{y})$ is said to satisfy an infinitesimal upper bound if and only if for every $\varepsilon > 0$ there exists a $\delta > 0$ such that*

$$|U(t, \mathbf{y})| < \varepsilon \qquad \text{on } \{(t, \mathbf{y}) \mid t \geq 0, |\mathbf{y}| \leq \delta\}$$

Example 4. The function

$$U(t, y_1, y_2) = (1 + t)y_1{}^2 + y_2{}^2$$

is positive definite on the set $\Omega = \{(t, y_1, y_2) \mid t \geq 0\}$, but clearly does not satisfy an infinitesimal upper bound. On the other hand, the function

$$U(t, y_1, y_2) = \frac{1}{1+t} y_1{}^2 + y_2{}^2$$

also positive definite on Ω, does satisfy an infinitesimal upper bound.

• **EXERCISE**

7. Prove the statements made in Example 4.

We may now state Lyapunov's classical result.

Theorem 5.8. *If there exists a scalar function $V(t, \mathbf{y})$ that is positive definite, satisfies an infinitesimal upper bound, and for which $V^*(t, \mathbf{y})$ is negative definite, then the zero solution of $\mathbf{y}' = \mathbf{f}(t, \mathbf{y})$ is asymptotically stable.*

• **EXERCISE**

8. Prove Theorem 5.8. [*Hint:* See the proof of Theorem 5.2 and observe that the infinitesimal upper bound gives the additional information needed to take into account the dependence of $V(t, \mathbf{y})$ on t.]

Example 5. Consider the system

$$y_1' = - a(t)y_1 - by_2$$
$$y_2' = by_1 - c(t)y_2$$

where b is a real constant and where a, c are real continuous functions defined for $t \geq 0$ satisfying

$$a(t) \geq \delta > 0 \quad c(t) \geq \delta > 0 \qquad (0 \leq t < \infty)$$

We wish to show the solution $y_1 = y_2 \equiv 0$ is asymptotically stable. Consider the function

$$V(t, y_1, y_2) = y_1^2 + y_2^2$$

$$\begin{aligned} V^*(t, y_1, y_2) &= 2y_1(-a(t)y_1 - by_2) + 2y_2(by_1 - c(t)y_2) \\ &= -2a(t)y_1^2 - 2c(t)y_2^2 \leq -2\delta(y_1^2 + y_2^2) \end{aligned}$$

Clearly V is positive definite and satisfies an infinitesimal upper bound; moreover, $V^*(t, y_1, y_2)$ is negative definite. Thus Theorem 5.8 yields the result.

Example 6. Consider the system

$$\begin{aligned} y_1' &= -a(t)y_1 - by_2 + g_1(t, y_1, y_2) \\ y_2' &= by_1 - c(t)y_2 + g_2(t, y_1, y_2) \end{aligned} \tag{5.45}$$

where a, b, c are as in Example 5 above and where g_1, g_2 are real continuous functions defined on the region $\{(t, y_1, y_2) \mid 0 \leq t < \infty,\ 0 \leq y_1^2 + y_2^2 \leq r^2\}$ for some constant $r > 0$ that satisfy

$$\lim_{y_1^2 + y_2^2 \to 0} \frac{g_j(t, y_1, y_2)}{(y_1^2 + y_2^2)^{1/2}} = 0 \qquad (j = 1, 2) \tag{5.46}$$

This condition implies that $g_j(t, 0, 0) = 0$ $(j = 1, 2)$; thus $y_1 = y_2 \equiv 0$ is a critical point and we wish to establish its asymptotic stability. (We remark that the system (5.45) is a perturbation, in the sense of Theorem 4.3 (p. 161), of the linear system in Example 5. Note that Theorem 4.3 is not applicable because here the unperturbed system does not have constant coefficients.) To establish the asymptotic stability consider the same V function as before. However, with respect to the system (5.45) we now obtain

$$V^*(t, y_1, y_2) \leq -2[\delta(y_1^2 + y_2^2) + y_1 g_1(t, y_1, y_2) + y_2 g_2(t, y_1, y_2)]$$

Using (5.46), given any $\varepsilon > 0$ there exists a number $\eta > 0$ such that

$$|g_j(t, y_1, y_2)| \leq \varepsilon(y_1^2 + y_2^2)^{1/2} \quad \text{for} \quad y_1^2 + y_2^2 < \eta^2 \qquad (j = 1, 2)$$

Thus

$$|y_1 g_1(t, y_1, y_2) + y_2 g_2(t, y_1, y_2)| \le \varepsilon(y_1^2 + y_2^2)^{1/2}(|y_1| + |y_2|)$$

for $y_1^2 + y_2^2 < \eta^2$. Since $|y_j| \le (y_1^2 + y_2^2)^{1/2}$ $(j = 1, 2)$, we obtain

$$V^*(t, y_1, y_2) \le -2\delta(y_1^2 + y_2^2) + 4\varepsilon(y_1^2 + y_2^2)$$

provided $y_1^2 + y_2^2 < \eta^2$. Choosing $\varepsilon = \delta/4$ (any number less than $\delta/2$ will do) and determining the corresponding η we obtain

$$V^*(t, y_1, y_2) \le -\delta(y_1^2 + y_2^2) \qquad (y_1^2 + y_2^2 < \eta^2)$$

Thus V^* is negative definite on the set $\{(t, y_1, y_2) \,|\, t \ge 0, y_1^2 + y_2^2 < \eta^2\}$ and this shows that Theorem 5.8 yields the desired result.

• EXERCISES

9. Show that in the systems in Example 5 above all solutions tend to zero exponentially. [*Hint:* If $\boldsymbol{\phi} = (\phi_1, \phi_2)$ is any solution, show that

$$\frac{d}{dt} V(t, \phi_1(t), \phi_2(t)) \le -2\delta V(t, \phi_1(t), \phi_2(t))$$

Integrate this inequality, which gives

$$\phi_1^2(t) + \phi_2^2(t) \le (\phi_1^2(0) + \phi_2^2(0))e^{-2\delta t}.]$$

10. Formulate and prove an analogous result for the system (5.45) discussed in Example 6.

It is easy to give an example of a system to which Theorem 5.8 cannot be applied, and yet we expect that the zero solution is asymptotically stable.

Example 7. Consider the scalar equation

$$y'' + a(t)y' + y = 0 \qquad (5.47)$$

where $a(t)$ is a continuous function defined on $0 \le t < \infty$ satisfying $a(t) \ge \delta > 0$, where $\delta > 0$ is a constant. We may think of $a(t)y'$ as a damping term of a linear oscillator. It can be shown that if $a(t)$ is also bounded above, then every solution of the equation (5.47) together with its derivative approaches zero. This, however, does not follow from Theorem 5.8. For consider the equivalent system

$$y_1' = y_2$$
$$y_2' = -y_1 - a(t)y_2 \tag{5.48}$$

and the positive definite scalar function $V(t, y_1, y_2) = y_1^2 + y_2^2$ having an infinitesimal upper bound. However,

$$V^*(t, y_1, y_2) = -2a(t)y_2^2 \leq -2\delta y_2^2$$

Thus $V^*(t, y_1, y_2)$ is nonpositive, but not negative definite.

• EXERCISES

11. Use the method of Example 6, Section 5.2, to discuss the asymptotic stability of the zero solution of the system (5.48).
12. Generalize the result of Exercise 11 to the equation

$$y'' + a(t)y' + g(y) = 0$$

where g is a continuously differentiable function such that $yg(y) > 0$ for $y \neq 0$.

For problems such as the one considered in Example 7, it is natural to inquire whether the results and techniques discussed in Section 5.4 for autonomous systems can be carried over to the nonautonomous cases. One difficulty in doing this is that the notion of invariant set, natural for autonomous systems, cannot be defined directly for nonautonomous systems. However, for a class of problems that may be called **asymptotically autonomous** systems, analogous results do hold and we state one such theorem below. The proof, as well as other related theorems, may be found in [25, Ch. 3].

Let $\mathbf{f}(t, \mathbf{y})$ and $\mathbf{h}(\mathbf{y})$ be continuous together with their first derivatives with respect to the components of $\mathbf{y}$ in a set $\{(t, \mathbf{y}) \mid 0 \leq t < \infty, \mathbf{y} \in \Omega\}$ where Ω is some set in $\mathbf{y}$-space.

Definition. *The system*

$$\mathbf{y}' = \mathbf{f}(t, \mathbf{y}) \tag{5.41}$$

is said to be asymptotically autonomous on the set Ω *if and only if* (a) $\lim_{t\to\infty} \mathbf{f}(t, \mathbf{y}) = \mathbf{h}(\mathbf{y})$ *for* $\mathbf{y} \in \Omega$ *and this convergence is uniform for* $\mathbf{y}$ *in closed bounded subsets of* Ω.
(b) *For every* $\varepsilon > 0$ *and every* $\mathbf{y} \in \Omega$ *there exists a* $\delta(\varepsilon, \mathbf{y}) > 0$ *such that* $|\mathbf{f}(t, \mathbf{x}) - \mathbf{f}(t, \mathbf{y})| < \varepsilon$, *whenever* $|\mathbf{x} - \mathbf{y}| < \delta$ *for* $0 \leq t < \infty$.

Example 8. Consider the scalar equation

$$y' = -\left(1 + \frac{1}{1+t}\right)y^2 \qquad (0 \le t < \infty)$$

This equation is asymptotically autonomous on any closed bounded set $\Omega = \{y \,\big|\, |y| \le K\}$ where $K > 0$ is a constant. For:

$$\text{(a)}\ \lim_{t\to\infty} -\left(1 + \frac{1}{1+t}\right)y^2 = -y^2$$

uniformly with respect to y in Ω.

$$\text{(b)}\ |f(t, y) - f(t, x)| \le \left(1 + \frac{1}{1+t}\right)|y^2 - x^2|$$

$$\le 2|y^2 - x^2| \le 2|y + x|\,|y - x| < \varepsilon$$

whenever $|y - x| < \delta = \min(K, \varepsilon/6K)$.

• **EXERCISE**

13. Justify the above choice of δ. [*Hint:* $|y + x| \le |y| + |x| \le 3K$ whenever $|y| \le K$ and $|x - y| \le \delta$.]

Theorem 5.9. (See p. 214 and compare with Theorem 5.5.) *Suppose the system* (5.41) *is asymptotically autonomous on some set* Ω *in* **y**-*space. Suppose* $\mathbf{f}(t, \mathbf{y})$ *is bounded for* $0 \le t < \infty$ *whenever* **y** *lies in a closed bounded set* $Q = \{\mathbf{y} \,\big|\, |\mathbf{y}| \le K, K > 0\}$. *Suppose there exists a nonnegative scalar function* $V(t, \mathbf{y})$ *such that* $V^*(t, \mathbf{y}) \le -W(\mathbf{y})$ *where* $W(\mathbf{y}) \ge 0$ *with* $W(\mathbf{y}) = 0$ *only for* $\mathbf{y} \in \Omega$ (this defines Ω, that is, $\Omega = \{\mathbf{y} \mid W(\mathbf{y}) = 0\}$). *Let* M *be the largest positively invariant subset of* Ω *with respect to the limiting autonomous system*

$$\mathbf{y}' = \mathbf{h}(\mathbf{y}) \tag{5.49}$$

Then every bounded solution of (5.41) *approaches* M *as* $t \to \infty$. *In particular, if all solutions of* (5.41) *are bounded, then every solution of* (5.41) *approaches* M.

Example 9. We will now establish the asymptotic stability (global) of the zero solution of the system (5.48), considered in Example 7 above. In the notation of Theorem 5.9 we have

$$V(t, y_1, y_2) = y_1^2 + y_2^2$$
$$V^*(t, y_1, y_2) \leq -2\delta y_2^2$$

so that

$$W(y_1, y_2) = 2\delta y_2^2$$

and $\Omega = \{(y_1, y_2) \mid y_2 = 0\}$, that is, the y_1 axis. On this set Ω (5.48) is asymptotically autonomous and the corresponding limiting system is

$$\begin{aligned} y_1' &= 0 \\ y_2' &= -y_1 \end{aligned} \tag{5.50}$$

which is obtained from (5.48) by putting $y_2 = 0$. In order to apply Theorem 5.9 we must assume not only the condition $a(t) \geq \delta > 0$ for $0 \leq t < \infty$ (already made in Example 7) but also $|a(t)| \leq K$ for $0 \leq t < \infty$ where $K > 0$ is a constant. This insures that $|\mathbf{f}(t, \mathbf{y})|$ is bounded whenever $|\mathbf{y}|$ is bounded.

We observe next that every solution of (5.48) exists on $0 \leq t < \infty$ and is bounded. This may be seen as follows. Let $\boldsymbol{\phi} = (\phi_1, \phi_2)$ be any solution; then for as long as it exists we have from Example 7

$$\frac{d}{dt} V(t, \phi_1(t), \phi_2(t)) \leq -2\delta\phi_2^2(t) \leq 0$$

Therefore,

$$V(t, \phi_1(t), \phi_2(t)) \leq V(0, \phi_1(0), \phi_2(0)) = \phi_1^2(0) + \phi_2^2(0)$$

Thus by a familiar argument $\boldsymbol{\phi}$ exists on $0 \leq t < \infty$ and is bounded.

To apply Theorem 5.9 we need to obtain M, the largest invariant subset of Ω with respect to the system (5.50). Since every solution of (5.50) has the form $\phi_1(t) = c_1$, $\phi_2(t) = c_2 - c_1 t$, where c_1 and c_2 are arbitrary constants, M is clearly the origin and this, by Theorem 5.9, proves the result.

• EXERCISES

14. Generalize the result of Example 9 to the equation

$$y'' + a(t)y' + g(y) = 0$$

where g is a continuously differentiable function such that $yg(y) > 0$ $(y \neq 0)$.

15. For the equation considered in Exercise 14 show that all solutions, together with their first derivatives, tend to zero as $t \to +\infty$ if it is also assumed that $\lim_{|y| \to \infty} \int_0^y g(\sigma)\, d\sigma = \infty$; thus in this case the zero solution is globally asymptotically stable.

Chapter 6 | SOME APPLICATIONS

6.1 Introduction

In this chapter we consider several applications with extensions of the theory developed to this point. We will concentrate on two topics: (a) the existence of periodic solutions of unforced second-order equations; this will include both the damped and undamped pendulum, as well as the Van der Pol and Liénard equations, which have been referred to previously; (b) the regulator, also known as the Lur'e problem.

Our treatment is not at all exhaustive; our purpose is to indicate possible directions for further study. Thus, for example, in our discussion of the existence of periodic solutions, we do not study the general Poincaré–Bendixson theory and theory of the index (see, for example, [13]), methods involving fixed point theorems (see, for example, [20]), the methods of Poincaré and the method of averaging for perturbed almost linear systems (see [3, 4, 9 ,10, 19, 22, 23]), and singularly perturbed systems (see [24]). The latter topics are somewhat beyond the intended level of this book and require additional background material. The general Poincaré–Bendixson theory is readily accessible and beautifully treated in [13]. With respect to the regulator problem, the reader is referred to [1, 17, 18] for a more complete treatment, including also such questions as controllability and optimal control.

6.2 The Undamped Oscillator

We consider the equation

$$u'' + g(u) = 0 \tag{6.1}$$

where throughout we will assume that

$$ug(u) > 0 \qquad (u \neq 0) \tag{6.2}$$

and that g is continuously differentiable on $-\infty < u < \infty$. Let $y_1 = u$, $y_2 = u'$; then (6.1) with initial conditions $u(0) = \eta_1$, $u'(0) = \eta_2$ is equivalent to the system

$$\begin{aligned} y_1' &= y_2 & y_1(0) &= \eta_1 \\ y_2' &= -g(y_1) & y_2(0) &= \eta_2 \end{aligned} \tag{6.3}$$

Because of (6.2), $y_1 = 0$, $y_2 = 0$ is the only critical point of (6.3). We have already seen (see Example 3, Section 5.2, p. 197), using the positive definite scalar energy function,

$$V(y_1, y_2) = \frac{{y_2}^2}{2} + \int_0^{y_1} g(\sigma)\, d\sigma \tag{6.4}$$

having $V^*(y_1, y_2) \equiv 0$, that every solution of (6.3) with $|\eta_1| + |\eta_2|$ sufficiently small is periodic; the solution curves are simple closed curves in the (y_1, y_2) phase plane, symmetric about the y_1 axis. These solution curves are, of course, precisely those components of the curves

$$V(y_1, y_2) = V(\eta_1, \eta_2) \equiv \text{constant} \tag{6.5}$$

containing the origin in their interior; see Figure 6.1. We recall also that if, in addition to (6.2), we assume that $\lim_{|y_1| \to \infty} \int_0^{y_1} g(\sigma)\, d\sigma = \infty$, then every solution of (6.3) is periodic. We will now study the nature of these periodic solutions in more detail.

In order to study the dependence of the period of the periodic solutions on the amplitude, we will assume

$$g(u) = -g(-u) \tag{6.6}$$

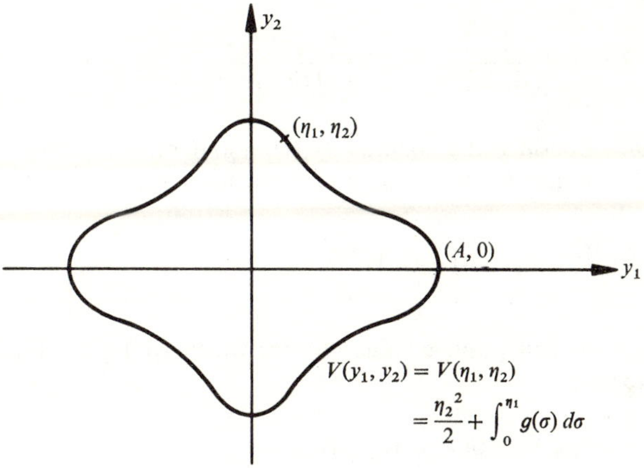

Figure 6.1

(that is, g is an odd function). This assumption used in (6.4) implies that

$$V(-y_1, y_2) = V(y_1, y_2)$$

so that the curves (6.5) are also symmetric about the y_2 axis. Thus it makes sense to speak of the amplitude of the periodic solution generated by a particular curve (6.5); this is simply a value of $y_1 = A > 0$, for which

$$V(A, 0) = V(\eta_1, \eta_2)$$

See Figure 6.1.

Notice that $\eta_2^2/2 + \int_0^{\eta_1} g(\sigma)\, d\sigma = \int_0^A g(\sigma)\, d\sigma$. We establish the following result.

Theorem 6.1. *Let* (6.2) *and* (6.6) *be satisfied. Then there exists a neighborhood N of the origin in the phase plane such that if* (η_1, η_2) *is in N, then the solution of* (6.3) *through* (η_1, η_2) *is periodic. Let* $T(A) > 0$ *be the least period of all periodic solutions that generate the solution curve through the point* $(A, 0)$ *in N, with* $A > 0$ *and sufficiently small. Then*

$$T(A) = 2\sqrt{2} \int_0^A \frac{d\sigma}{[G(A) - G(\sigma)]^{1/2}}$$

where $G(u) = \int_0^u g(\sigma)\, d\sigma$. *Moreover,*

$$T'(A) = \frac{2\sqrt{2}}{[G(A)]^{1/2}} - \sqrt{2}\int_0^A \frac{[g(A) - g(\sigma)]}{[G(A) - G(\sigma)]^{3/2}}\, d\sigma$$

Proof. Consider the solution $(y_1(t), y_2(t))$ of (6.3) through the point $(A, 0)$ (see Figure 6.1)). Because of the symmetry of the solution curve

$$\frac{y_2^2(t)}{2} + G(y_1(t)) = G(A) \tag{6.7}$$

$T(A)/4$ is the first positive value of t for which $y_1(t) = 0$. From (6.3) and (6.7) we have

$$y_1' = y_2 = [2(G(A) - G(y_1))]^{1/2} \tag{6.8}$$

Notice that $|y_1| \le A$ implies that $G(y_1) \le G(A)$. Separation of variables in (6.8) yields formally

$$\frac{T(A)}{4} = \frac{1}{\sqrt{2}}\int_0^A \frac{dy_1}{[G(A) - G(y_1)]^{1/2}} \tag{6.9}$$

which is the first result in Theorem 6.1, provided the improper integral converges; notice that the integrand becomes infinite as $y_1 \to A$. Therefore, consider

$$\lim_{\varepsilon \to 0+} \int_0^{A-\varepsilon} \frac{dy_1}{[G(A) - G(y_1)]^{1/2}} \tag{6.10}$$

We have for any $0 < B < A$ and small $\varepsilon > 0$

$$\int_0^{A-\varepsilon} \frac{dy_1}{[G(A) - G(y_1)]^{1/2}} = \int_0^B \frac{dy_1}{[G(A) - G(y_1)]^{1/2}} + \int_B^{A-\varepsilon} \frac{dy_1}{[G(A) - G(y_1)]^{1/2}}$$

The first of these is an ordinary Riemann integral of a continuous function on $0 \le y_1 \le B$, independent of ε. In the second, by the mean value theorem, there exists a ξ, $0 < B < \xi < A - \varepsilon$, such that

$$G(A) - G(y_1) = g(\xi)(A - y_1)$$

where we have used $g(\xi) = G'(\xi)$. Therefore

$$\int_B^{A-\varepsilon} \frac{dy_1}{[G(A) - G(y_1)]^{1/2}} \le \frac{1}{\min\limits_{B<\xi<A-\varepsilon} |g(\xi)|^{1/2}} \int_B^{A-\varepsilon} \frac{dy_1}{[A - y_1]^{1/2}}$$

Since $g(\xi) > 0$ for $\xi \neq 0$ and since $B > 0$, we can be sure that

$$\min_{B<\xi<A-\varepsilon} |g(\xi)|^{1/2} > 0.$$

Thus

$$\int_B^{A-\varepsilon} \frac{dy_1}{[G(A) - G(y_1)]^{1/2}} \le \frac{2}{\min\limits_{B<\xi<A-\varepsilon} |g(\xi)|^{1/2}} [\varepsilon^{1/2} - (A - B)^{1/2}]$$

which shows that the limit (6.10) exists and therefore the integral (6.9) converges.

To establish the second result in Theorem 6.1 we have to differentiate the improper integral (6.9). The fundamental theorem of calculus cannot be applied and using the definition of derivative directly without some preliminary trick is also not productive. We notice that the function

$$H(x, A) = \begin{cases} \dfrac{G(A) - G(A - x)}{x} & x \neq 0 \\ g(A) & x = 0 \end{cases} \tag{6.11}$$

is continuous at $x = 0$. Therefore, by the continuity of $H(x, A)$ in x, on $0 \le x \le A$,

$$\begin{aligned} T(A) &= 2\sqrt{2} \int_0^A \frac{dy_1}{[G(A) - G(y_1)]^{1/2}} = 2\sqrt{2} \int_0^A \frac{du}{[G(A) - G(A - u)]^{1/2}} \\ &= 2\sqrt{2} \int_0^A \frac{du}{[H(u, A)]^{1/2} u^{1/2}} \end{aligned} \tag{6.12}$$

We now remind the reader of the formula

$$\frac{d}{d\alpha}\left\{\int_{a(\alpha)}^{b(\alpha)} f(x, \alpha)\, dx\right\} = f(b(\alpha), \alpha)b'(\alpha) - f(a(\alpha), \alpha)a'(\alpha) + \int_{a(\alpha)}^{b(\alpha)} \frac{\partial f}{\partial \alpha}(x, \alpha)\, dx$$

which holds if f, $\partial f/\partial\alpha$ are continuous for $a(\alpha) \le x \le b(\alpha)$ and $a(\alpha)$, $b(\alpha)$ continuous for $c \le \alpha \le d$ (see, for example, [12, p. 360]). Applying this formula and proceeding purely formally, we obtain, using (6.11) and (6.12),

$$
\begin{aligned}
T'(A) &= \frac{2\sqrt{2}}{[H(A, A)A]^{1/2}} \\
&\quad + 2\sqrt{2}\int_0^A \frac{1}{\sqrt{u}}\left[\left(-\frac{1}{2}\right)(H(u, A))^{-3/2}\frac{\partial}{\partial A}H(u, A)\right] du \\
&= \frac{2\sqrt{2}}{[G(A)]^{1/2}} - \sqrt{2}\int_0^A \frac{1}{\sqrt{u}}[H(u, A)]^{-3/2}\left[\frac{g(A) - g(A-u)}{u}\right] du \\
&= \frac{2\sqrt{2}}{[G(A)]^{1/2}} - \sqrt{2}\int_0^A \frac{g(A) - g(A-u)}{[G(A) - G(A-u)]^{3/2}}\, du \\
&= \frac{2\sqrt{2}}{[G(A)]^{1/2}} - \sqrt{2}\int_0^A \frac{g(A) - g(\sigma)}{[G(A) - G(\sigma)]^{3/2}}\, d\sigma
\end{aligned}
$$

as asserted in the statement of Theorem 6.1. By a standard theorem (see, for example, [12, p. 361]), the formal differentiation is justified provided the differentiated integral converges uniformly with respect to A on some interval $\alpha \le A \le \beta$. Since $g(x)$ is continuously differentiable and $G'(x) = g(x)$, this is easily shown to be the case for the integral

$$\int_0^A \frac{g(A) - g(\sigma)}{[G(A) - G(\sigma)]^{3/2}}\, d\sigma \tag{6.13}$$

for $0 < \alpha \le A \le \beta$. This completes the proof of Theorem 6.1. ▌

• **EXERCISE**

1. Establish the convergence of the integral (6.13). [*Hint:* Proceed as in the proof of the existence of the limit (6.10), splitting the range of integration from 0 to $A - \varepsilon$ into two parts and using the mean value theorem on both the differences $g(A) - g(\sigma)$ and $G(A) - G(\sigma)$.]

We can deduce several consequences of Theorem 6.1.

Corollary 1. If the hypotheses of Theorem 6.1 *are satisfied and if in addition* $g(x)$ *is monotone increasing in some neighborhood of* $x = 0$ [for example, if $g'(0) > 0$], *then we also have*

$$T'(A) = -\frac{2\sqrt{2}g(A)}{G(A)} \int_0^A \left[\frac{G(\sigma)g'(\sigma)}{[g(\sigma)]^2} - \frac{1}{2}\right] \frac{d\sigma}{[G(A) - G(\sigma)]^{1/2}} \tag{6.14}$$

The proof of Corollary 1 begins with the formula for $T'(A)$ derived in Theorem 6.1 and involves a careful analysis of the integral

$$\int_\varepsilon^{A-\mu} \frac{g(A) - g(\sigma)}{[G(A) - G(\sigma)]^{3/2}} \, d\sigma \qquad (\varepsilon > 0, \mu > 0)$$

with the aid of integration by parts. For details, the reader is referred to [19, p. 5].

Corollary 2. Let the hypothesis of Corollary 1 be satisfied and assume in addition that $g''(x)$ is continuous. If $g''(x) \geq 0$ on $0 \leq x \leq A$ and if $g''(x)$ is not identically equal to zero, then $T'(A) < 0$. If $g''(x) \leq 0$ on $0 \leq x \leq A$ and if $g''(x)$ is not identically zero, then $T'(A) > 0$.

Proof. Consider the case $g''(x) \geq 0$. To show that $T'(A) < 0$ it suffices, from (6.14), to show that

$$W(\sigma) = \frac{G(\sigma)g'(\sigma)}{[g(\sigma)]^2} - \frac{1}{2} \geq 0 \tag{6.15}$$

and that $W(\sigma) \not\equiv 0$ for $0 \leq \sigma \leq A$. But a straightforward calculation shows that

$$\frac{d}{d\sigma}\{[g(\sigma)]^2 W(\sigma)\} = G(\sigma)g''(\sigma) \tag{6.16}$$

- **EXERCISE**

2. Prove (6.16).

Since $g^2 W$ is well defined at $\sigma = 0$ and is zero, then we have

$$[g(\sigma)]^2 W(\sigma) = \int_0^\sigma G(\tau)g''(\tau) \, d\tau$$

But $G(\tau) > 0$, $g''(\tau) \geq 0$, and g'' is not identically zero; therefore $[g(\sigma)]^2 W(\sigma) > 0$ for $\sigma > 0$. Since also $g(\sigma) \neq 0$ for $\sigma \neq 0$, this implies that $W(\sigma) > 0$ for $0 < \sigma \leq A$ and this completes the proof in the case $g''(x) \geq 0$. The other part is similar, and this completes the proof. ∎

• **EXERCISE**

3. Carry out the proof in the case $g''(x) \leq 0$.

A spring $g(u)$ is called **hard** if $g''(x) \geq 0$ and **soft** if $g''(x) \leq 0$. As we can see from Corollary 2, these definitions relate to the variation of the period with amplitude as follows: If $g''(x) \geq 0$ but is not identically zero on $0 \leq x \leq A$, then **the period $T(A)$ is a decreasing function of A as A increases,** and the opposite statement holds if $g''(x) \leq 0$, but is not identically zero as $0 \leq x \leq A$. For the simple undamped pendulum

$$g(x) = \frac{g}{L} \sin x$$

Thus $g''(x) = -g/L \sin x \leq 0$ for $0 \leq x \leq \pi$, so that the pendulum may be regarded as a soft spring and, as expected, its period increases as the amplitude increases. Notice, however, that a linear spring $g(x) = kx$, $k > 0$, is both hard and soft; here $g''(x) \equiv 0$, and as we know, the period $T = 2\pi/\sqrt{k}$ is a constant.

6.3 The Pendulum

In this section, we will discuss the simple undamped pendulum of length L (1.6), governed by the equation

$$u'' + \frac{g}{L} \sin u = 0 \tag{6.17}$$

In the notation of Theorem 6.1, $g(u) = (g/L) \sin u$. Thus $G(u) = (g/L)(1 - \cos u)$. The hypotheses of Theorem 6.1 are satisfied for $-\pi < u < \pi$ and the period of the pendulum is given by

$$T(A) = 2\left(\frac{2L}{g}\right)^{1/2} \int_0^A \frac{d\sigma}{[\cos \sigma - \cos A]^{1/2}}^{1/2} \qquad 0 \leq A < \pi$$

where, because $g(u)$ is odd, A is the amplitude.

• **EXERCISE**

1. Show that

$$T(A) = 4\left(\frac{L}{g}\right)^{1/2} \int_0^{\pi/2} \frac{d\phi}{(1 - k^2 \sin^2 \phi)^{1/2}}$$

where $k = \sin(A/2)$. [*Hint:* In the formula given by Theorem 6.1, let $\sin(\sigma/2) = k \sin \phi$ and use the identity $\cos \sigma = 1 - 2\sin^2(\sigma/2) = 1 - 2k^2 \sin^2 \phi$.]

2. The integral in Exercise 1 is called an elliptic integral of the first kind for $0 < k < 1$, and it cannot be evaluated explicitly. We have by Taylor's theorem for $0 \leq k < 1$

$$(1 - k^2 \sin^2 \phi)^{-1/2} = 1 + \frac{k^2}{2} \sin^2 \phi + \frac{3}{8} k^4 \sin^4 \phi + \cdots$$

Use this to estimate the period of the pendulum for any amplitude A for which $0 < k = \sin(A/2) \leq \frac{1}{2}$, that is, for $0 < A \leq \pi/3$, within an error of $1/10$.

We now wish to obtain a complete phase portrait of the pendulum. The above analysis and the theory of Section 6.1 (see also Example 3, Section 5.2, p. 197), are valid only for $-\pi < u < \pi$. The physical problem of a simple pendulum has a second equilibrium point at $u = \pi$. It is physically obvious and has also been shown in Exercise 4, Section 4.5, p. 177, that this equilibrium point is unstable. The complete phase portrait must reflect this fact.

The system

$$\begin{aligned} y_1' &= y_2 \\ y_2' &= -\frac{g}{L} \sin y_1 \end{aligned} \tag{6.18}$$

obtained from (6.17) by letting $y_1 = u$, $y_2 = u'$, is equivalent to (6.17). We shall obtain the complete phase portrait for (6.18) by a careful study of the energy function

$$V(y_1, y_2) = \frac{1}{2} y_2^2 + \frac{g}{L}(1 - \cos y_1) \tag{6.19}$$

which is (6.4) with $g(y_1) = g/L \sin y_1$. As we have seen in Example 3, Section 5.2, p. 197, $V^*(y_1, y_2) \equiv 0$ and all solutions of (6.18) lie on some curve

$$V(y_1, y_2) \equiv h$$

where h is a constant. If the solution passes through a point $(A, 0)$ (so that in case $0 < A < \pi$ the amplitude of the pendulum is A), $h = V(A, 0) = (g/L)(1 - \cos A)$. Thus to construct the complete phase portrait we simply construct the family of curves $V(y_1, y_2) = h$ for all values of $h \geq 0$. Before doing this let us recall that for $0 \leq A < \pi$ all solutions of (6.18) are periodic and their graphs in the phase are given by the equations

$$\frac{1}{2}y_2^2 + \frac{g}{L}(1 - \cos y_1) = \frac{g}{L}(1 - \cos A)$$

for $0 \leq A < \pi$. In fact this relation holds for the graphs of all solution curves, through the points $(A, 0)$ **for all values of** A. To determine these graphs we solve this relation for y_2 and we obtain

$$y_2 = \pm\left(\frac{2g}{L}\right)^{1/2}(\cos y_1 - \cos A)^{1/2} \tag{6.20}$$

For each $A \neq n\pi$, $n = 0, \pm 1, \pm 2, \ldots$, this gives a closed curve. If $-\pi \leq A \leq \pi$, relation (6.20) gives one of the closed curves shown in Figure 6.2. If $A = 0$, (6.20) gives only the origin in Figure 6.2.

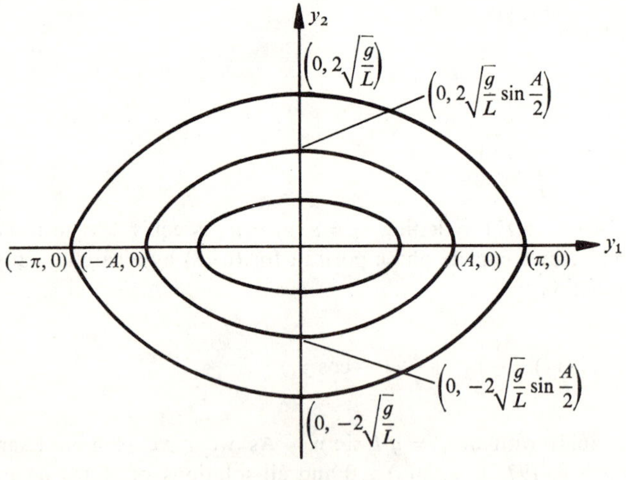

Figure 6.2

For solutions through $(A, 0)$ with $A > \pi$ (and similarly with $A < -\pi$) we take advantage of the periodicity of the function $(\cos y_1 - \cos A)$. Thus for each A we obtain one of the curves shown in Figure 6.3. If $A = 2n\pi$, the solution curve given by (6.20) reduces to the single point $(2n\pi, 0)$. The solution curves in Figure 6.3 through the point $(A, 0)$ correspond to the motion of an undamped pendulum with initial displacement A and initial velocity zero. Alternatively we may think of each closed curve in Figure 6.3

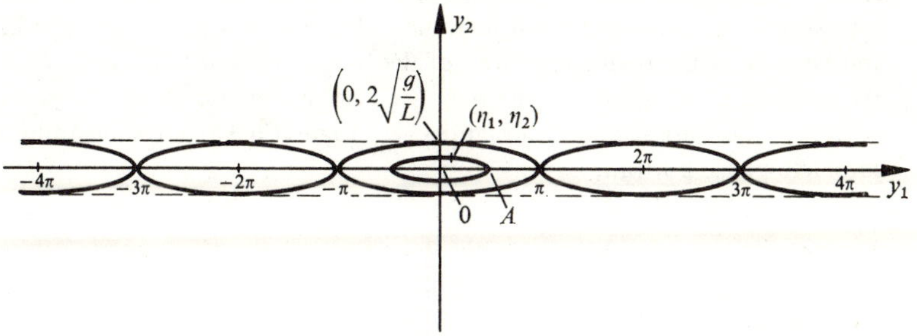

Figure 6.3

as corresponding to the motion of the pendulum with initial displacement η_1 and initial velocity η_2 where (η_1, η_2) is any point on the curve.

We may, however, also use the energy function (6.19) to obtain the graphs of all solutions of the system (6.18), not merely those through the points $(A, 0)$. To do this, consider the family of curves

$$V(y_1, y_2) = \frac{1}{2} y_2{}^2 + \frac{g}{L}(1 - \cos y_1) = h$$

or

$$y_2 = \pm\left\{2\left[h - \frac{g}{L}(1 - \cos y_1)\right]\right\}^{1/2} \tag{6.21}$$

for $0 \leq h < \infty$. If $0 \leq h \leq 2g/L$, we merely reproduce Figure 6.3 already obtained. For in each such case $y_2 = 0$ whenever $h - g/L\,(1 - \cos y_1) = 0$ or whenever $\cos y_1 = 1 - h\,(L/g)$; but if $0 \leq h \leq 2g/L$, then $-1 \leq 1 - h(L/g) \leq 1$ and therefore for each $0 \leq h \leq 2g/L$ there is a $y_1 = A$ for which $y_2 = 0$; in fact, by periodicity $y_2 = 0$ for $y_1 = A \pm 2n\pi$ $(n = 0, 1, \ldots)$. Therefore, the solution of (6.18) through any point $(A + 2n\pi, 0)$ has for its graph one of the closed curves in Figure 6.3, or possibly the single point $(2n\pi, 0)$. If $h > 2g/L$,

$$h - \frac{g}{L}(1 - \cos y_1) = h - \frac{g}{L} + \frac{g}{L}\cos y_1 \geq h - \frac{g}{L} - \frac{g}{L} > 0$$

Thus from (6.21), it is clear that there is no value of y_1 for which $y_2 = 0$. Therefore such solutions are not included in Figure 6.3. In fact, the interiors of all the closed curves correspond to the case $h < 2g/L$. Moreover, the curve $V(y_1, y_2) \equiv h$ with $h > 2g/L$ cannot be a closed curve, and hence these corresponding solutions are not periodic. The curves are still given by (6.21) and because of the periodicity in y_1 of the function $h - g/L(1 - \cos y_1)$, they are periodic in y_1 of period 2π; they are also symmetric about the y_1 axis. Several of these are sketched in Figure 6.4. Figures 6.3 and 6.4 constitute the complete phase portrait.

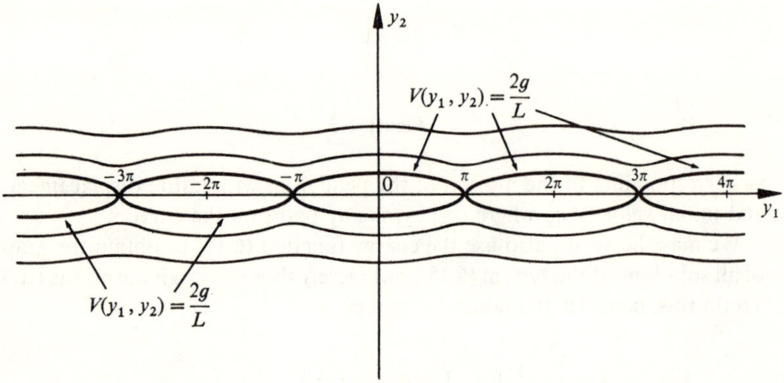

Figure 6.4

There is a simple way to construct the phase portrait associated with any scalar equation of the form

$$u'' + g(u) = 0 \tag{6.1}$$

where g and g' are real and continuous, having the equivalent system

$$\begin{aligned} y_1' &= y_2 \\ y_2' &= -g(y_1) \end{aligned} \tag{6.3}$$

where it is not necessarily assumed that $ug(u) > 0$, $(u \neq 0)$. Consider the energy function

$$V(y_1, y_2) = \frac{{y_2}^2}{2} + G(y_1) \tag{6.22}$$

where $G(y_1) = \int_0^{y_1} g(\sigma)\, d\sigma$. It is still the case that

$$V^*(y_1, y_2) = y_2(-g(y_1)) + g(y_1)(y_2) \equiv 0$$

Therefore the graph of any solution of (6.3) lies on some real curve $V(y_1, y_2) \equiv h$ in the (y_1, y_2) plane, where h is a constant; in particular, the solution through the point (η_1, η_2) is given by the curve $V(y_1, y_2) \equiv V(\eta_1, \eta_2)$. The real curves $V(y_1, y_2) \equiv h$ are given by equations

$$y_2 = \pm[2(h - G(y_1))]^{1/2}$$

for all y_1 for which $h - G(y_1) \geq 0$. To construct the phase portrait we construct the curve $y_2 = G(y_1)$ in the (y_1, y_2) plane, and proceed as in Figures 6.5, 6.6, where we have carried out the construction for the case of the pendulum with $G(y_1) = (g/L)(1 - \cos y_1)$. On the same graph in Figure 6.5,

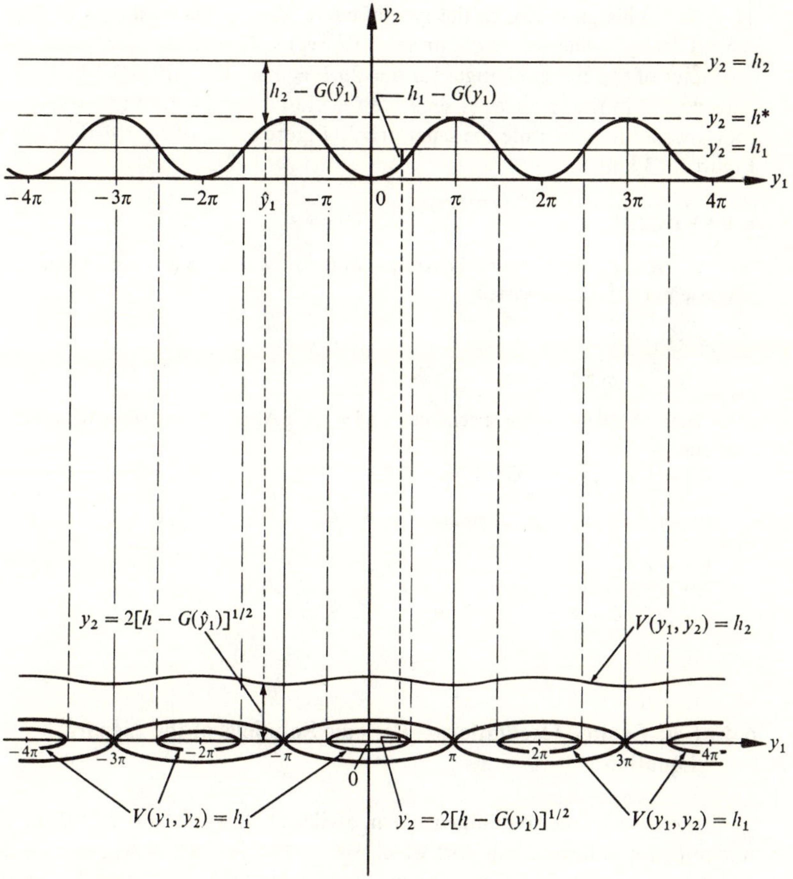

Figure 6.5 and 6.6

construct several lines $y_2 = h$. For a particular $h = h_1$, consider the differences $h_1 - G(y_1)$ for those admissible y_1 for which $h_1 - G(y_1)$ is positive. These differences can be measured in Figure 6.5. For the same y_1 in Figure 6.6, directly below Figure 6.5, construct the ordinates $y_2 = \pm[2(h_1 - G(y_1)]^{1/2}$. By doing this for all admissible y_1 we obtain all the curves $V(y_1, y_2) \equiv h_1$. By repeating this for all $0 \le h < \infty$, we obtain the entire phase portrait. In particular, for $h = 0$ we obtain the points $(\pm 2n\pi, 0)$, $n = 0, 1, 2, \ldots$, which are, of course, the equilibrium points of the system (6.18).

For $0 < h_1 < 2g/L$ we obtain closed curves such as those shown for $h = h_1$ in Figure 6.6. For $h_2 > 2g/L$, we consider the difference $h_2 - G(y_1) = h_2 - g/L(1 - \cos y_1)$, which, as we have seen, is strictly positive for $-\infty < y_1 < \infty$. This gives rise to the typical curve $V(y_1, y_2) \equiv h_2$ shown in Figure 6.6. Clearly none of the curves $V(y_1, y_2) = h_2 > 2g/L$ is closed. The character of the curves changes for $h = 2g/L = h^*$. The curves $V(y_1, y_2) \equiv h$, constructed in the same way, separate the closed curves from the others and they are said to constitute the **separatrix**. Figure 6.6 is, of course, the same as Figures 6.3 and 6.4.

• EXERCISE

3. Use the general method corresponding to Figures 6.5 and 6.6 to obtain the phase portrait of the system

$$y_1' = y_2$$

$$y_2' = -g(y_1)$$

for each of the following special cases of $g(y_1)$. Also, find the separatrix if there is one.

(a) $g(y_1) = y_1$ (compare Figure 6.6).

(b) $g(y_1) = y_1 - \dfrac{y_1^3}{6}$ (compare Figure 6.6).

(c) $g(y_1) = y_1^2$.

(d) $g(y_1) = y_1 - y_1^2$.

(e) $g(y_1) = y_1^3$.

6.4 Self-Excited Oscillations—Periodic Solutions of the Liénard Equation

In the case of the undamped linear oscillator $u'' + k^2 u = 0$, $k > 0$, or the undamped nonlinear oscillator, which we studied in Section 6.2, every solution (at least, every solution with sufficiently small initial position and initial velocity) is periodic; it is of constant period $2\pi/k$ in the linear case and of variable period depending on the amplitude in the nonlinear case. Since in most physical systems friction is present, the above phenomena are exceptional. For a damped linear oscillator of the form $u'' + au' + k^2u = 0$, every solution approaches zero as $t \to +\infty$ if $a > 0$, while every solution becomes unbounded as $t \to +\infty$ if $a < 0$, thus the only periodic solution is the identically zero one. A similar situation holds for the (autonomous) nonlinear equation $u'' + f(u, u')u' + g(u) = 0$, where $g(u)$ is a nonlinear spring and where the damping is of a fixed sign. We have already discussed several results for this equation in Chapter 5.

An entirely different situation arises when the damping changes sign. The equation

$$u'' + \varepsilon(u^2 - 1)u' + u = 0$$

where $\varepsilon > 0$ is a parameter, arises in the theory of feedback electronic circuits (see [3, pp. 154–155] or [22, pp. 119–128]). It was first studied extensively by Van der Pol, and is therefore known as the Van der Pol equation, but it has also received much attention from other authors. As we have seen in Exercise 6, Section 5.5, p. 222, the identically zero solution of the Van der Pol equation is unstable. However, much more than this can be learned from a careful analysis of the above-mentioned exercise. In the notation of (5.31), p. 220, the equivalent system to the Van der Pol equation is $y_1' = y_2 - \varepsilon((y_1^3/3) - y_1)$, $y_2' = -y_1$. We take $V(y_1, y_2) = \frac{1}{2}(y_1^2 + y_2^2)$, and from Definition 3, Section 5.2 (p. 193),

$$V^*(y_1, y_2) = -\varepsilon y_1^2((y_1^2/3) - 1) = \varepsilon(y_1^2 - (y_1^4/3))$$

Thus $V^*(y_1, y_2)$ is positive for $y_1^2 < 3$. From this we easily see that every solution starting at a point (y_{10}, y_{20}) inside the circle $y_1^2 + y_2^2 < 3$ in the phase plane moves, as t increases, in the direction of increasing V, that is, outward ($V^*(y_1, y_2) \geq 0$). We therefore say that the "flow" is outward. On the other hand, every solution starting at a point (y_{10}, y_{20}) in the region $y_1^2 > 3$ moves inward in order to decrease $V(y_1, y_2)$ ($V^*(y_1, y_2) < 0$), and we say that the flow is inward. When such a solution crosses one of the lines $|y_1| = \sqrt{3}$, it may move away from the origin. However, from the equations $y_2' = -y_1$, $|y_1| < \sqrt{3}$, we see that $|y_2'| \leq \sqrt{3}$. Thus such a solution remains bounded and therefore it can be continued across the entire strip $-\sqrt{3} \leq y_1 \leq \sqrt{3}$. Once it leaves this strip it again moves inward. Thus it seems plausible that the Van der Pol equation has a nonzero periodic solution. This is indeed the case, as the following more general development will show.

We consider the Liénard equation

$$u'' + f(u)u' + g(u) = 0 \tag{6.23}$$

where f and g are continuously differentiable for $-\infty < u < \infty$. We let $G(u) = \int_0^u g(\sigma)\, d\sigma$, $F(u) = \int_0^u f(\sigma)\, d\sigma$.

Theorem 6.2. *Suppose*

(i) $ug(u) > 0, \qquad (u \neq 0)$

(ii) $\lim_{|u|\to\infty} |F(u)| = +\infty$

and that for some $a, b > 0$

(iii) $F(u) < 0, \qquad (u < -a,\ 0 < u < b)$
$F(u) > 0, \qquad (-a < u < 0,\ u > b)$

Then (6.23) *has a nontrivial periodic solution.*

The reader will note that under the above hypotheses, the origin in the phase plane is the only critical point of (6.23) and of its equivalent system (6.24).

• **EXERCISE**

1. Show that the Van der Pol equation satisfies the hypotheses of Theorem 6.2. Determine a and b.

Proof of Theorem 6.2. The system

$$\begin{aligned} y_1' &= y_2 - F(y_1) \\ y_2' &= -g(y_1) \end{aligned} \tag{6.24}$$

is equivalent to (6.23) (see Exercise 3, Section 5.5, p. 221). Under our hypotheses the origin is the only critical point of (6.24). Consider the energy function

$$V(y_1, y_2) = \frac{y_2^{\,2}}{2} + G(y_1)$$

for which, with respect to the system (6.24), we have

$$V^*(y_1, y_2) = y_2(-g(y_1)) + g(y_1)(y_2 - F(y_1)) = -g(y_1)F(y_1)$$

(see Definition 3, Section 5.2, p. 193). For small values of V_0 the curves $V(y_1, y_2) = V_0$ in the (y_1, y_2) plane have components that are closed curves

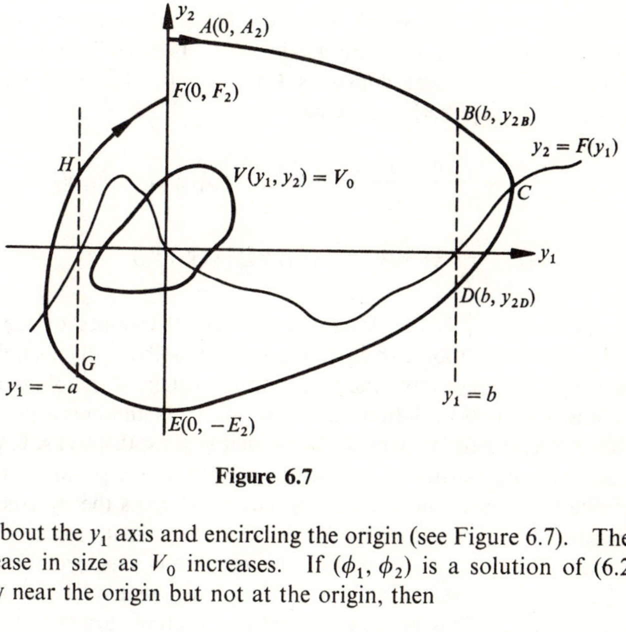

Figure 6.7

symmetric about the y_1 axis and encircling the origin (see Figure 6.7). These curves increase in size as V_0 increases. If (ϕ_1, ϕ_2) is a solution of (6.24) starting very near the origin but not at the origin, then

$$\frac{d}{dt} V(\phi_1(t), \phi_2(t)) = V^*(\phi_1(t), \phi_2(t)) = -g(\phi_1(t))F(\phi_1(t))$$

(see (5.8), p. 194). The hypotheses imply that $d/dt\, V(\phi_1(t), \phi_2(t)) \geq 0$ and thus such a solution curve penetrates the curve $V(y_1, y_2) = V_0$ as t increases for any sufficiently small $V_0 > 0$ and moves outward in the direction of increasing values of V_0. The origin is therefore an unstable critical point.

Consider now the curve $y_2 = F(y_1)$ in the phase plane (Figure 6.7). On this curve, we have, from (6.24), $y_1' = 0$, $y_2' = -g(y_1)$. Suppose that a solution $(\phi_1(t), \phi_2(t))$ of (6.24) starts at a point $A(0, A_2)$ at $t = 0$ for some $A_2 > 0$. Initially, $\phi_1'(0) = A_2 > 0$, $\phi_2'(0) = 0$, and the solution curve moves to the right into the first quadrant as t increases. For as long as $y_2 > F(y_1)$ and

$y_1 > 0$, we have $\phi_1'(t) > 0$, $\phi_2'(t) < 0$, so that the solution curve continues to move to the right in the first quadrant with a negative slope until it meets the curve $y_2 = F(y_1)$ at a point C. The solution curve may or may not intersect the line $y_1 = b$; this depends on the size of A_2. On the curve $y_2 = F(y)$, $\phi_1'(t) = 0$ and $\phi_2'(t) < 0$ if $y_1 > 0$, and the solution curve therefore crosses into the region $y_2 < F(y_1)$, where $\phi_1'(t) < 0$, $\phi_2'(t) < 0$. We note that once the solution curve has crossed into this region at C, it cannot cross the curve $y_2 = F(y_1)$ again before it reaches the y_2 axis. For suppose it does cross again at some point $(\bar{y}_1, \bar{y}_2)$ with $\bar{y}_1 > 0$, $\bar{y}_2 = F(\bar{y}_1)$, where $F(y_1) - y_2 > 0$ on some interval $\bar{y}_1 < y_1 < \bar{y}_1 + h$, with small $h > 0$. Then on this interval $|f(y_1)| < k$, $g(y_1) \geq \lambda > 0$ for some constants $k > 0$, $\lambda > 0$. But then on the solution curve in this neighborhood we have

$$\frac{d\phi_2}{d\phi_1} = -\frac{g(\phi_1)}{\phi_2 - F(\phi_1)} \to +\infty, \qquad (\phi_1, \phi_2) \to (\bar{y}_1, \bar{y}_2)$$

$$f(\phi_1) - \frac{d\phi_2}{d\phi_1} \to -\infty, \qquad (\phi_1, \phi_2) \to (\bar{y}_1, \bar{y}_2)$$

($f(\phi_1) - d\phi_2/d\phi_1$ is the difference between the slopes of the curves $y_2 = F(y_1)$ and the solution curve), and this is impossible. By a similar argument, the solution cannot cross the y_2 axis at the origin of the phase plane. Once we know that the solution curve cannot cross the curve $y_2 = F(y_1)$ again before reaching the y_2 axis, we know that $\phi_1'(t) < 0$, $\phi_2'(t) < 0$, but $|d\phi_2/d\phi_1|$ is finite on this portion of the curve. Therefore the solution cannot become infinite between C and the y_2 axis, and must cross the y_2 axis at a point E (Figure 6.7). Proceeding with the same argument, we see that we may continue the solution and that it intersects the y_2 axis again at some point F.

Starting at F, we can recover the entire segment AF by continuing backward in time. Thus we may say that the solution curve revolves around the origin in a clockwise direction as t increases, and this is true for every solution except the identically zero one. By the continuity of solutions with respect to initial conditions, the coordinates of the points B, C, D, E, F can be regarded as continuous functions of the point A. As $A_2 \to +\infty$, the points B, C, D, E, F, G, H also "move out to infinity" with B, D, G, H remaining on the lines $y_1 = b$ and $y_1 = -a$.

From the fact $d/dt\, V(\phi_1(t), \phi_2(t)) = -g(\phi_1(t))F(\phi_1(t))$, and from the hypothesis, we conclude that for sufficiently small $V_0 > 0$, $V^*(\phi_1(t), \phi_2(t)) \geq 0$ at all points of the closed curve $V(y_1, y_2) = V_0$, with the equality holding only on the y_2 axis. Thus for all sufficiently small $A_2 > 0$ the point F on the solution curve lies above the point A in Figure 6.7. Next, it will be shown that for sufficiently large $A_2 > 0$, the point F lies below the point A in Figure 6.7. For this purpose we consider the energy function V at the points

A, B, C, D, E, F. Along the solution (ϕ_1, ϕ_2) we have

$$V(\phi_1, \phi_2) = \frac{{\phi_2}^2}{2} + G(\phi_1)$$

Considering ϕ_2 as a function of ϕ_1 along the solution curve, we have

$$\frac{dV}{d\phi_1} = \phi_2 \frac{d\phi_2}{d\phi_1} + g(\phi_1) = \phi_2\left[\frac{-g(\phi_1)}{\phi_2 - F(\phi_1)}\right] + g(\phi_1) = \frac{-g(\phi_1)F(\phi_1)}{\phi_2 - F(\phi_1)}$$

On the portion of the solution curve from A to B we have $g(\phi_1) > 0$, $F(\phi_1) < 0$, $\phi_2 - F(\phi_1) > 0$, so that

$$V_B - V_A = \int_0^b \frac{dV}{d\phi_1}\, d\phi_1 = -\int_0^b \frac{g(\phi_1)F(\phi_1)}{\phi_2(\phi_1) - F(\phi_1)}\, d\phi_1 > 0$$

As A_2 increases, $\phi_2 \to \infty$ and b remains fixed. Therefore

$$\lim_{A_2 \to +\infty} (V_B - V_A) = 0$$

Similarly, $\lim_{A_2 \to +\infty} (V_E - V_D) = 0$. Now, regarding ϕ_1 as a function of ϕ_2 on the portion $\widehat{BD}$ of the solution curve, we have

$$\frac{dV}{d\phi_2} = \phi_2 + g(\phi_1)\frac{d\phi_1}{d\phi_2} = \phi_2 + g(\phi_1)\left[-\frac{(\phi_2 - F(\phi_1))}{g(\phi_1)}\right] = F(\phi_1)$$

since $g(\phi_1) \neq 0$ on $\widehat{BD}$. Since by hypothesis $F(y_1) > 0$ for $y_1 > b$,

$$V_B - V_D = \int_{y_{2D}}^{y_{2B}} \frac{dV}{d\phi_2}\, d\phi_2 = \int_{y_{2D}}^{y_{2B}} F(\phi_1)\, d\phi_2 > 0$$

But $\lim_{|u| \to \infty} |F(u)| = +\infty$, and as $A_2 \to +\infty$, $y_{2B} \to +\infty$ while $y_{2D} \to -\infty$. Therefore

$$\lim_{A_2 \to +\infty} (V_B - V_D) = +\infty$$

Combining these limiting statements, we conclude that for large $A_2 > 0$, $V_E < V_A$. Thus, if E has coordinates $(0, -E_2)$ with $E_2 > 0$, we have $V_E = {E_2}^2/2 < V_A = {A_2}^2/2$, so that $E_2 < A_2$ for sufficiently large A_2. We may repeat the argument on the portion $\widehat{EF}$ of the solution curve. Letting F have coordinates $(0, F_2)$ with $F_2 > 0$, we find, by considering the differences $V_G - V_E$, $V_H - V_G$, $V_F - V_H$ exactly as above, that $F_2 < E_2$. Thus $0 < F_2 < E_2 < A_2$, and F lies below A on the y_2 axis.

• **EXERCISE**

2. Show that $F_2 < E_2$.

Since the displacement $A_2 - F_2$ is a continuous function of A_2 that is positive for large values of A_2 and negative for small values of A_2, there is a value $\hat{A}_2$ such that $\hat{A}_2 = F_2(\hat{A}_2)$. Clearly, the solution through the point $(0, \hat{A}_2)$ is periodic, and this completes the proof. ▮

We next consider the question when the periodic solution constructed in Theorem 6.2 is unique. This requires some additional hypotheses. In particular, we obtain the following result with the aid of Theorem 6.2.

Theorem 6.3. *Suppose*

(i) $ug(u) > 0, \qquad (u \neq 0)$

(ii) $g(u) = -g(-u), \qquad f(u) = f(-u)$

and that for some $b > 0$,

(iii) $F(u) < 0, \qquad (0 < u < b)$

$F(u) > 0, \qquad (u > b)$

(iv) $F(u)$ is monotone increasing for $u > b$ and $\lim_{u \to \infty} F(u) = \infty$.

Then the equation

$$u'' + f(u)u' + g(u) = 0 \tag{6.23}$$

has a unique nontrivial periodic solution $p(t)$.

We note that, compared to Theorem 6.2, the additional hypotheses here are that $g(u)$ is an odd function, $f(u)$ is an even function, and $F(u)$ is monotone increasing for $u > b$. Since the system is autonomous, the uniqueness of the periodic solution means, of course, up to translations in time.

Proof of Theorem 6.3. We again consider the equivalent system

$$\begin{aligned} y_1' &= y_2 - F(y_1) \\ y_2' &= -g(y_1) \end{aligned} \tag{6.24}$$

and we construct in Figure 6.8 the auxiliary curve $y_2 = F(y_1)$ just as in

Figure 6.7 ($F(y_1)$ is odd). We construct again the curve Γ: $ABCDE$ exactly as in Figure 6.7. This curve is given by the equations $y_1 = \phi_1(t)$, $y_2 = \phi_2(t)$, where ϕ_1, ϕ_2 is that solution of (6.24) for which $\phi_1(0) = 0$, $\phi_2(0) = A_2$. By symmetry considerations, it suffices to draw the portion $y_1 \geq 0$ of the phase

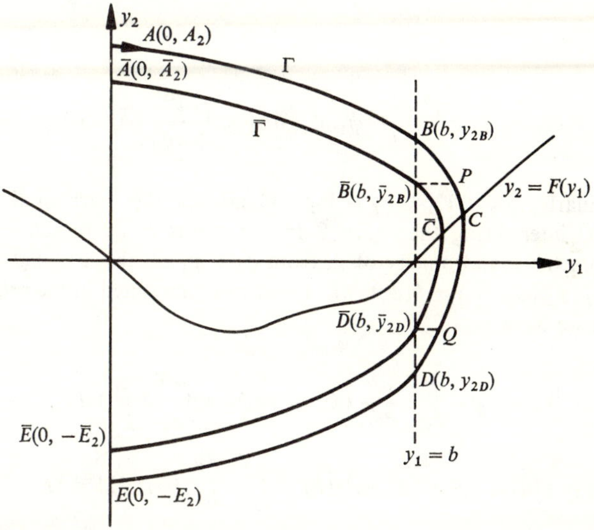

Figure 6.8

plane. Moreover, the solution is periodic if and only if $A_2 = E_2$. Recall that in proving Theorem 6.2 we showed that $E_2 < A_2$ for sufficiently large A_2 and $E_2 > A_2$ for sufficiently small A_2. It will now be shown that $E_2^2 - A_2^2$ is a strictly monotone decreasing function of A_2, and it therefore vanishes for only one value of A_2. This will imply that the periodic solution whose existence we already know from Theorem 6.2 is unique.

Consider another starting point $\bar{A}$ $(0, \bar{A}_2)$ in Figure 6.8 with $0 < \bar{A}_2 < A_2$. We construct the arc of the solution curve $\bar{\Gamma}$: $\overline{ABCDE}$ in the same way as $ABCDE$, taking $\bar{A}$ close enough to A so that this solution curve also crosses $y_1 = b$. The curve $\bar{\Gamma}$ is given by the equations $y_1 = \bar{\phi}_1(t)$, $y_2 = \bar{\phi}_2(t)$, where $\bar{\phi}_1(0) = 0$, $\bar{\phi}_2(0) = \bar{A}_2$ as shown. We shall compare the variation of the function $V(y_1, y_2) = y_2^2/2 + G(y_1)$ along the two solution curves Γ and $\bar{\Gamma}$. We first compare $V_B - V_A$ and $V_{\bar{B}} - V_{\bar{A}}$. From

$$\frac{dV}{d\phi_1} = \frac{-g(\phi_1)F(\phi_1)}{\phi_2 - F(\phi_1)}$$

which was computed in the proof of Theorem 6.2, we see that on the portion from A to B and $\bar{A}$ to $\bar{B}$, $\phi_2 > \bar{\phi}_2$ so that

$$\frac{dV}{d\phi_1}(\phi_1, \phi_2) < \frac{dV}{d\phi_1}(\bar{\phi}_1, \bar{\phi}_2)$$

Therefore

$$V_B - V_A = \int_0^b \frac{dV}{d\phi_1}(\phi_1, \phi_2)\, d\phi_1 < \int_0^b \frac{dV}{d\phi_1}(\bar{\phi}_1, \bar{\phi}_2)\, d\phi_1 = V_{\bar{B}} - V_{\bar{A}}$$

Similarly, $V_E - V_D < V_{\bar{E}} - V_{\bar{D}}$. Construct the horizontal lines through $\bar{B}$, $\bar{D}$ intersecting the curve Γ in the points P, Q, respectively, as shown. From the monotonicity of F, from $\phi_1 > \bar{\phi}_1$ on the arcs PQ, $\bar{B}\bar{D}$, and from $(dV/d\phi_2)(\bar{\phi}_1, \bar{\phi}_2) = F(\bar{\phi}_1)$, which was also computed in the proof of Theorem 6.2, we have

$$V_{\bar{B}} - V_{\bar{D}} = \int_{y_{2\bar{D}}}^{y_{2\bar{B}}} \frac{dV}{d\phi_2}(\bar{\phi}_1, \bar{\phi}_2)\, d\phi_2 = \int_{y_{2\bar{D}}}^{y_{2\bar{B}}} F(\bar{\phi}_1)\, d\phi_2$$

$$< \int_{y_{2\bar{D}}}^{y_{2\bar{B}}} F(\phi_1)\, d\phi_2 = \int_{y_{2\bar{D}}}^{y_{2\bar{B}}} \frac{dV}{d\phi_2}(\phi_1, \phi_2) = V_P - V_Q$$

But V decreases along Γ when $y_1 \geq b$, and therefore $V_B - V_P > 0$, $V_Q - V_D > 0$. Now we have $V_B - V_D = (V_B - V_P) + (V_P - V_Q) + (V_Q - V_D) > V_{\bar{B}} - V_{\bar{D}}$, or $V_D - V_B < V_{\bar{D}} - V_{\bar{B}}$.
Therefore, combining these inequalities, we obtain

$$V_E - V_A < V_{\bar{E}} - V_{\bar{A}}$$

From the definition of V, this last inequality means that

$$E_2^2 - A_2^2 < \bar{E}_2^2 - \bar{A}_2^2$$

which says that $E_2^2 - A_2^2$ is a strictly monotone decreasing function of A_2 (provided, of course, that the solution curves considered are such that the point C lies to the right of the line $y_1 = b$). As already remarked, this completes the proof of Theorem 6.3. ▌

Let Ω be the orbit of the periodic solution $p(t)$. If $(\phi_1(t), \phi_2(t))$ is any solution of the system (6.24), let C^+ be its positive semiorbit. Let us consider its positive limit set $L(C^+)$ (see Definition 2, Section 5.4, p. 211). We have the following consequence of the proof of Theorem 6.3.

Corollary to Theorem 6.3. $L(C^+) = \Omega$.

To see this we need only note that the flow is outward in the interior of Ω and inward in the exterior of Ω (by the monotonicity of $E_2^2 - A_2^2$).

• **EXERCISE**

3. Consider the equation

(*) $u'' + F(u') + u = 0$

where $F(w) = \int_0^w f(\sigma)\, d\sigma$ satisfies the hypotheses of Theorem 6.3. Show that (*) has a unique nontrivial periodic solution. [*Hint:* Let $u' = y$ and differentiate (*).] Note that the equation $u'' + \varepsilon(u'^2/3) - u') + u = 0$, which is called the Rayleigh equation, is of this form for $\varepsilon > 0$.

Consider again the equation

$$u'' + \varepsilon F(u') + u = 0 \qquad (\varepsilon > 0) \tag{6.25}$$

where F satisfies, for example, hypothesis (iii) of Theorem 6.3, but this is not at all necessary in the sequel. The system

$$\begin{aligned} y_1' &= y_2 \\ y_2' &= -\varepsilon F(y_2) - y_1 \end{aligned} \tag{6.26}$$

is equivalent to (6.25). Its orbits in the phase plane may be sketched conveniently by the following method, which is due to Liénard. Consider the curve Γ: $y_1 = -\varepsilon F(y_2)$ in the phase plane (see Figure 6.9). Since

$$\frac{dy_2}{dy_1} = -\frac{\varepsilon F(y_2) + y_1}{y_2} \tag{6.27}$$

it follows that the direction field has a horizontal tangent at each point of the curve Γ distinct from the origin (at the origin we have the only critical point of (6.26)). We may now construct the direction field at an arbitrary point P as follows: Construct the horizontal line through P intersecting the curve Γ at Q. Drop the perpendicular from Q intersecting the y_1 axis at R The direction of the field at P is then perpendicular to the segment $\overline{RP}$.

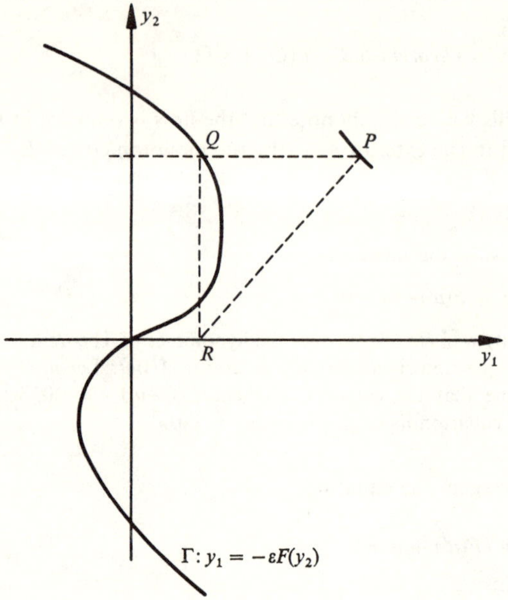

Figure 6.9

• EXERCISES

4. Justify the above construction. [*Hint:* Show that the slope of RP is $y_2/(y_1 + \varepsilon F(y_2))$ and compare with (6.27).]
5. Use this method to construct several orbits in the case that

$$F(y_2) = \frac{y_2^{\ 3}}{3} - y_2$$

and $\varepsilon = 1$, with some orbits starting near the origin and some "far" away from the origin.
6. Use the above construction to indicate that the equation

$$u'' - \varepsilon(\sin u') + u = 0$$

has infinitely many periodic solutions. Why does this not contradict Theorem 6.3?

The discussion of forced oscillations of equations of the types considered up to now in this chapter can be carried out by using topological methods, somewhat beyond the scope of this book; see, for example, [3, 9, 20].

Other approaches, particularly useful when the nonlinear system is almost linear, are the method of Poincaré and the method of averaging mentioned in the introduction to this chapter.

6.5 The Regulator Problem

We now study a different application. Consider a physical system governed by the real system of differential equations

$$\mathbf{y}' = \mathbf{f}(\mathbf{y}) \tag{6.28}$$

having an isolated stable equilibrium point that, without loss of generality, we can take as the origin. It is assumed that the real function $\mathbf{f}$ is defined in region D containing $\mathbf{0}$. It is desired to keep the system operating as closely to the origin as possible. The usual procedure is to linearize the system around the origin. This consists of writing (6.28) in the form

$$\mathbf{y}' = A\mathbf{y} + \mathbf{g}(\mathbf{y}) \tag{6.29}$$

where, with sufficient hypothesis on $\mathbf{f}$ (for example, $\mathbf{f} \in C^2(D)$), A is the real constant matrix that has $(\partial f_i/\partial y_j)(\mathbf{0})$ in the ith row and jth column, and the real vector function $\mathbf{g}$ satisfies

$$\lim_{|\mathbf{y}| \to 0} \frac{|\mathbf{g}(\mathbf{y})|}{|\mathbf{y}|} = 0$$

We now drop the nonlinear terms $\mathbf{g}(\mathbf{y})$. Notice that if all the eigenvalues of A have real parts negative, it follows from Theorem 4.3 (p. 161) that the zero solution of (6.29) is asymptotically stable. In other words, solutions of (6.29) that begin near the origin behave qualitatively like those of the linear system

$$\mathbf{z}' = A\mathbf{z} \tag{6.30}$$

Under this hypothesis, every solution of (6.29) behaves in future time like some solution of (6.30), if it comes sufficiently near the origin at some instant of time t_0.

We shall assume throughout that all the eigenvalues of A have negative real part. Such a matrix A is called a **stability matrix**. This means that every solution of the (uncontrolled) linear system (6.30) tends to zero as $t \to +\infty$.

To maintain the system near the origin and to make the solutions tend to zero more rapidly than in the uncontrolled case we add an external **control**. This is done mathematically in the following way. We replace (6.30) by the system

$$\mathbf{z}' = A\mathbf{z} - \mathbf{b}\xi \tag{6.31}$$

where $\mathbf{b}$ is a constant vector and where ξ is a scalar function that depends on the state of the system. The control of the system may be accomplished in different ways that correspond to various selections of the function ξ. One such way is to determine ξ from a scalar differential equation of the form

$$\xi' = \phi(\sigma) \tag{6.32}$$

where $\sigma = \mathbf{c}^T\mathbf{z} - \rho\xi$. Here ρ is a real constant, $\mathbf{c}$ is a constant column vector, and $\mathbf{c}^T$ denotes the transpose of $\mathbf{c}$. The function $\phi(\sigma)$ is called the **characteristic function of the control** mechanism, and σ is called the **feedback control signal**. The characteristic function, which depends on the nature of the control mechanism (for example, a servomechanism), is usually highly nonlinear.

We shall assume throughout that ϕ belongs to the class C of real admissible characteristic functions satisfying the following assumptions.

(i) $\phi(\sigma)$ is continuous for $-\infty < \sigma < \infty$.
(ii) $\sigma\phi(\sigma) > 0$ $(\sigma \neq 0)$.
(iii) $\lim_{|\sigma| \to \infty} \int_0^\sigma \phi(\xi)\, d\xi = +\infty$.

Assumption (ii) implies that $\phi(0) = 0$ and the graph of ϕ lies in the first and third quadrants. Assumption (iii) means that the graph of ϕ cannot approach the horizontal axis too rapidly as $|\sigma| \to \infty$.

The mathematical problem is to determine sufficient conditions that depend only on the known matrix A, the constant vectors $\mathbf{b}$ and $\mathbf{c}$, and the constant ρ such that **for all admissible characteristic functions ϕ in the class C every solution $\mathbf{z}(t)$, $\xi(t)$ of the system**

$$\begin{aligned} \mathbf{z}' &= A\mathbf{z} - \mathbf{b}\xi \\ \xi' &= \phi(\sigma), \qquad \sigma = \mathbf{c}^T\mathbf{z} - \rho\xi \end{aligned} \tag{6.33}$$

approaches zero as $t \to \infty$. A system exhibiting such behavior is said to be **absolutely stable**. We shall solve this problem by using Lyapunov's second method. Assuming that we have determined such conditions insures that errors in the measurement of initial conditions and a wide selection of admissible characteristic function do not change the qualitative behavior (performance) of the system in any significant way. We may, in fact, now wish to select an admissible function ϕ that insures that the system will satisfy some other criterion. For example, we may ask how to choose ϕ so that a solution $\mathbf{z}(t)$, $\xi(t)$ of (6.33) will enter a preassigned neighborhood of the origin in minimum time. We will not be able to investigate this important problem.

We remark that controlling the linear system (6.30) in the above manner does not, of course, insure the control of the nonlinear system (6.28) or (6.29). It is, however, reasonable to expect that it does insure the control of (6.28) in a small neighborhood of the origin.

Before carrying the discussion further, it is convenient to transform the system (6.33) under investigation to an equivalent form. Let

$$\mathbf{x} = A\mathbf{z} - \mathbf{b}\xi, \qquad \sigma = \mathbf{c}^T\mathbf{z} - \rho\xi \tag{6.34}$$

These equations define a linear transformation T from $(\mathbf{z}, \xi)$ space to $(\mathbf{x}, \sigma)$ space. The matrix of this transformation, which we call M, is

$$M = \begin{bmatrix} A & -\mathbf{b} \\ \mathbf{c}^T & -\rho \end{bmatrix} \tag{6.35}$$

We now make use of a result from matrix theory to the effect that if A is nonsingular, then

$$\det M = (-\rho + \mathbf{c}^T A^{-1}\mathbf{b}) \det A$$

This result is proved in Appendix 4. Since A is assumed to be a stability matrix, it is nonsingular, and thus the transformation T is nonsingular if and only if

$$\rho \neq \mathbf{c}^T A^{-1}\mathbf{b} \tag{6.36}$$

Comparing (6.33) and (6.34), we have $\mathbf{z}' = \mathbf{x}$. Differentiating the first equation of (6.34), we obtain, on using (6.33),

$$\mathbf{x}' = A\mathbf{z}' - \mathbf{b}\xi' = A\mathbf{x} - \mathbf{b}\xi' = A\mathbf{x} - \mathbf{b}\phi(\sigma)$$

In the same way, we obtain

$$\sigma' = \mathbf{c}^T\mathbf{x} - \rho\phi(\sigma)$$

Thus (6.33) and the system

$$\begin{aligned} \mathbf{x}' &= A\mathbf{x} - \mathbf{b}\phi(\sigma) \\ \sigma' &= \mathbf{c}^T\mathbf{x} - \rho\phi(\sigma) \end{aligned} \tag{6.37}$$

are equivalent, provided (6.36) is satisfied.

We note that $(\mathbf{x}_0, \sigma_0)$ is a critical point of (6.37) if and only if

$$\begin{aligned} A\mathbf{x}_0 &= \mathbf{b}\phi(\sigma_0) \\ \mathbf{c}^T\mathbf{x}_0 &= \rho\phi(\sigma_0) \end{aligned}$$

Since A is nonsingular, $\mathbf{x}_0 = A^{-1}\mathbf{b}\phi(\sigma_0)$, so that

$$(\mathbf{c}^T A^{-1}\mathbf{b} - \rho)\phi(\sigma_0) = 0$$

In view of (6.36) this implies $\phi(\sigma_0) = 0$; since ϕ is an admissible characteristic function, we conclude that $\mathbf{x}_0 = \mathbf{0}$ and $\sigma_0 = 0$. **Therefore, the origin in $(\mathbf{x}, \sigma)$ space is the only critical point of** (6.37).

In the sequel we shall employ **symmetric matrices.** An n-by-n matrix $B = (b_{ij})$ is said to be symmetric if and only if $b_{ij} = b_{ji}$ $(i, j = 1, \ldots, n)$ (that is, if and only if $B = B^T$).

• EXERCISE

1. If P and Q are matrices so that the product PQ is well defined, show that

$$(PQ)^T = Q^T P^T \tag{6.38}$$

With each n-by-n symmetric matrix $B = (b_{ij})$ we can associate the quadratic form

$$\mathbf{y}^T B\mathbf{y} = \sum_{j=1}^{n} \sum_{i=1}^{n} b_{ij} y_i y_j$$

Example 1. If $B = \begin{pmatrix} b_{11} & b_{12} \\ b_{12} & b_{22} \end{pmatrix}$, then

$$\mathbf{y}^T B\mathbf{y} = b_{11}{y_1}^2 + 2b_{12}y_1y_2 + b_{22}{y_2}^2$$

Definition. *The real symmetric n-by-n matrix B is said to be positive definite if and only if the quadratic form* $\mathbf{y}^T B\mathbf{y}$ *is positive definite.*

It can be shown that for a real symmetric matrix B the eigenvalues are all real [7, Vol. 1, p. 273], and that the matrix B is positive definite if and only if **all** the eigenvalues of B are positive [7, Vol. 1, p. 309].

• EXERCISES

2. (a) By applying the definition, derive a necessary and sufficient condition for the positive definiteness of the real matrix

$$B = \begin{pmatrix} b_{11} & 0 \\ 0 & b_{22} \end{pmatrix}$$

(b) Generalize to an arbitrary diagonal matrix.

3. Show that the real symmetric matrix

$$B = \begin{pmatrix} b_{11} & b_{12} \\ b_{12} & b_{22} \end{pmatrix}$$

is positive definite if and only if $b_{11} > 0$ and $b_{11}b_{22} - b_{12}^2 > 0$.

4. (a) Apply the criterion of Exercise 3 to the matrix

$$B = \begin{pmatrix} 2 & 2 \\ 2 & 3 \end{pmatrix}$$

(b) Compute the eigenvalues of B and verify the remarks immediately preceding Exercise 2.

Exercise 3 is a very special case of the following general result.

Sylvester's Theorem. *The real symmetric n-by-n matrix B is positive definite if and only if*

$$\det B_j = \det \begin{pmatrix} b_{11} & b_{12} & \cdots & b_{1j} \\ b_{12} & b_{22} & \cdots & b_{2j} \\ \vdots & & & \vdots \\ b_{1j} & & \cdots & b_{jj} \end{pmatrix} > 0, \qquad (j = 1, 2, \ldots, n)$$

For a proof we refer the reader to Appendix 4. The determinants $\det B_j$ ($j = 1, \ldots, n$) are called the **principal minors.**

We will also have occasion to use the following result, which is also of independent interest.

Lyapunov's Theorem on Matrices. *Let A be a given constant stable matrix and let C be a given symmetric positive definite matrix. Then there exists a symmetric positive definite matrix B such that*

$$A^T B + BA = -C$$

The proof of this theorem is also given in Appendix 4. The motivation of the theorem in the context of differential equations is the following.

Consider the linear system with constant coefficients

$$\mathbf{y}' = A\mathbf{y} \tag{6.39}$$

where A is a stable matrix (not necessarily symmetric). We look for a positive definite scalar function of the form

$$V(\mathbf{y}) = \mathbf{y}^T B\mathbf{y}$$

with the intention of applying Lyapunov's second method (Theorem 5.6, Corollary 1, p. 224), where B is some positive definite matrix (to be determined below). Now compute the derivative V^* of V with respect to the system (6.39) (see (5.8), p. 194). We have, for any solution $\mathbf{y}(t)$ of (6.39),

$$\begin{aligned}
V^*(\mathbf{y}(t)) &= \frac{d}{dt}[V(\mathbf{y}(t))] = [\mathbf{y}^T(t)B\mathbf{y}(t)]' \\
&= (\mathbf{y}^T(t))'B\mathbf{y}(t) + \mathbf{y}\ (t)B\mathbf{y}\ (t) \\
&= (\mathbf{y}'(t))^T B\mathbf{y}(t) + \mathbf{y}^T(t)BA\mathbf{y}(t) \\
&= (A\mathbf{y}(t))^T B\mathbf{y}(t) + \mathbf{y}^T(t)BA\mathbf{y}(t) \\
&= \mathbf{y}^T(t)A^T B\mathbf{y}(t) + \mathbf{y}^T(t)BA\mathbf{y}(t) \\
&= \mathbf{y}^T(t)(A^T B + BA)\mathbf{y}(t)
\end{aligned}$$

In this calculation we have used the easily established fact that $(\mathbf{y}^T)' = (\mathbf{y}')^T$ for any differentiable vector $\mathbf{y}$, as well as Exercise 1. Thus

$$V^*(\mathbf{y}) = \mathbf{y}^T(A^T B + BA)\mathbf{y}$$

Thus $V^*(\mathbf{y})$ will be negative definite if and only if the symmetric matrix

$$-C = A^T B + BA$$

is positive definite. Lyapunov's theorem, stated earlier, tells us that for any given positive definite matrix C and any given stable matrix A, we can satisfy this requirement with a positive definite matrix B.

As a matter of fact, it can also be shown (see [7, Vol 2, p. 189]) that the matrix A will be stable if and only if there exist real symmetric positive definite matrices B and C, such that $A^T B + BA = -C$.

• **EXERCISES**

5. Find a positive definite symmetric matrix B such that $A^T B + BA = -C$, where

$$A = \begin{pmatrix} -1 & 0 \\ 1 & -2 \end{pmatrix}, \qquad C = \begin{pmatrix} 1 & 0 \\ 0 & 1 \end{pmatrix}$$

[*Hint:* Let $B = \begin{pmatrix} a & b \\ b & c \end{pmatrix}$, substitute, and determine a, b, c; then check for positive definiteness of the result.]

6. Find a stable matrix A such that

$$A^T B + BA = -C$$

where

$$B = \begin{pmatrix} 7 & 1 \\ 1 & 3 \end{pmatrix}, \qquad C = \begin{pmatrix} 12 & 0 \\ 0 & 12 \end{pmatrix}$$

6.6 Absolute Stability of the Regulator System

In order to obtain conditions for the absolute stability of the original control system (6.33) we make the preliminary transformation (6.34) and obtain the equivalent system

$$\begin{aligned} \mathbf{x}' &= A\mathbf{x} - \mathbf{b}\phi(\sigma) \\ \sigma' &= \mathbf{c}^T\mathbf{x} - \rho\phi(\sigma) \end{aligned} \tag{6.37}$$

provided $\rho \neq \mathbf{c}^T A^{-1}\mathbf{b}$. We shall prove the following result.

Theorem 6.4. *Let A be a given stability matrix. Let C be any positive definite symmetric matrix and define B to be the positive definite symmetric matrix such that*

$$A^T B + BA = -C \tag{6.40}$$

Define $\mathbf{d} = B\mathbf{b} - \frac{1}{2}\mathbf{c}$. *Then if* $\rho \neq \mathbf{c}^T A^{-1}\mathbf{b}$ *and if* $\rho > \mathbf{d}^T C^{-1}\mathbf{d}$, *the system* (6.37), *and equivalently the original system* (6.33), *is absolutely stable for all admissible characteristic functions.* (If the system (6.37) is studied independent of the original system (6.33), the condition $\rho \neq \mathbf{c}^T A^{-1}\mathbf{b}$ is not needed.)

Proof. Consider the function

$$V(\mathbf{x}, \sigma) = \mathbf{x}^T B\mathbf{x} + \Phi(\sigma)$$

where

$$\Phi(\sigma) = \int_0^\sigma \phi(s)\, ds$$

The hypotheses on ϕ and B imply that V is positive definite on the entire $(\mathbf{x}, \sigma)$ space and that $V(\mathbf{x}, \sigma) \to \infty$ as $|\mathbf{x}|$ and $|\sigma|$ become infinite. Let $(\mathbf{x}(t), \sigma(t))$ be any solution of (6.37). Then by (5.8), p. 194, and (6.37), (6.40),

$$\begin{aligned} V^*(\mathbf{x}(t), \sigma(t)) &= \frac{d}{dt}[V(\mathbf{x}(t), \sigma(t))] \\ &= (\mathbf{x}'(t))^T B\mathbf{x}(t) + \mathbf{x}^T(t)B\mathbf{x}'(t) + \phi(\sigma(t))\sigma'(t) \\ &= [A\mathbf{x}(t) - \mathbf{b}\phi(\sigma(t))]^T B\mathbf{x}(t) + \mathbf{x}^T(t)B[A\mathbf{x}(t) - \mathbf{b}\phi(\sigma(t))] \\ &\qquad + \phi(\sigma(t))[\mathbf{c}^T\mathbf{x}(t) - \rho\phi(\sigma(t))] \\ &= \mathbf{x}^T(t)A^T B\mathbf{x}(t) - \mathbf{b}^T B\mathbf{x}(t)\phi(\sigma(t)) + \mathbf{x}^T(t)BA\mathbf{x}(t) \\ &\qquad - \mathbf{x}^T(t)B\mathbf{b}\phi(\sigma(t)) + \mathbf{c}^T\mathbf{x}(t)\phi(\sigma(t)) - \rho[\phi(\sigma(t))]^2 \\ &= -\mathbf{x}^T(t)C\mathbf{x}(t) - \phi(\sigma(t))[\mathbf{b}^T B\mathbf{x}(t) + \mathbf{x}^T(t)B\mathbf{b} - \mathbf{c}^T\mathbf{x}(t) \\ &\qquad + \rho\phi(\sigma(t))] \end{aligned}$$

Since B is symmetric and since $\mathbf{c}^T\mathbf{x}(t) = \mathbf{x}^T(t)\mathbf{c}$, we have, using the definition of $\mathbf{d}$,

$$\begin{aligned} \mathbf{b}^T B\mathbf{x}(t) + \mathbf{x}^T(t)B\mathbf{b} - \mathbf{c}^T\mathbf{x}(t) &= 2\mathbf{x}^T(t)B\mathbf{b} - \mathbf{x}^T(t)\mathbf{c} \\ &= \mathbf{x}^T(t)[2B\mathbf{b} - \mathbf{c}] = 2\mathbf{x}^T(t)\mathbf{d} \end{aligned}$$

Therefore,

$$V^*(\mathbf{x}, \sigma) = -\mathbf{x}^T C\mathbf{x} - 2\phi(\sigma)\mathbf{x}^T\mathbf{d} - \rho[\phi(\sigma)]^2 \tag{6.41}$$

We now define the real symmetric matrix

$$S = \begin{pmatrix} C & \mathbf{d} \\ \mathbf{d}^T & \rho \end{pmatrix}$$

Thus

$$-\dot{V}^*(\mathbf{x}, \sigma) = \begin{pmatrix} \mathbf{x} \\ \phi(\sigma) \end{pmatrix}^T S \begin{pmatrix} \mathbf{x} \\ \phi(\sigma) \end{pmatrix} \tag{6.42}$$

Since $\rho > \mathbf{d}^T C^{-1} \mathbf{d}$ and C is positive definite, it follows from Theorem 1, Appendix 4, and Sylvester's theorem that

$$\det S = (\rho - \mathbf{d}^T C^{-1} \mathbf{d}) \det C > 0$$

Using Sylvester's theorem again, we see that S is positive definite, hence $\dot{V}^*(\mathbf{x}, \sigma)$ is negative definite on the whole $(\mathbf{x}, \sigma)$ space. Since $\mathbf{x} = \mathbf{0}$, $\sigma = 0$ is the only critical point of the system (6.37), Corollary 1 of Theorem 5.6 (p. 224) implies that the solution $\mathbf{x} = 0$, $\sigma = 0$ of (6.37) is globally asymptotically stable for every admissible ϕ. Since $\rho \neq \mathbf{c}^T A^{-1} \mathbf{b}$, (6.37) and (6.33) are equivalent and the same result holds for (6.33). ▮

It will now be shown that the condition $\rho \neq \mathbf{c}^T A^{-1} \mathbf{b}$, which was used in the preliminary transformation, follows automatically from the main hypothesis $\rho > \mathbf{d}^T C^{-1} \mathbf{d}$ of Theorem 6.4.

Corollary. If $\rho > \mathbf{d}^T C^{-1} \mathbf{d}$, *then* $\rho \neq \mathbf{c}^T A^{-1} \mathbf{b}$.

Proof. Suppose $\rho > \mathbf{d}^T C^{-1} \mathbf{d}$, but $\rho = \mathbf{c}^T A^{-1} \mathbf{b}$. Consider the linear system

$$\begin{aligned} \mathbf{x}' &= A\mathbf{x} - \mathbf{b} \\ \sigma' &= \mathbf{c}^T \mathbf{x} - \rho\sigma \end{aligned} \tag{6.43}$$

By Theorem 6.4 and the remark following its statement, every solution of (6.43) tends to zero as $t \to \infty$. We use the condition $\rho = \mathbf{c}^T A^{-1} \mathbf{b}$, together with the result from matrix theory used earlier (see (6.35)), for the $(n + 1) \times (n + 1)$ matrix

$$M = \begin{pmatrix} A & -\mathbf{b} \\ \mathbf{c}^T & -\rho \end{pmatrix}$$

We recall that since A is nonsingular,

$$\det M = (-\rho + \mathbf{c}^T A^{-1} \mathbf{b}) \det A$$

Since $\rho = \mathbf{c}^T A^{-1}\mathbf{b}$, the matrix M of the system (6.43) is singular. Therefore the system of algebraic equations

$$A\mathbf{x} - \mathbf{b} = 0$$

$$\mathbf{c}^T\mathbf{x} - \rho\sigma = 0$$

has a nontrivial solution, say $\mathbf{x} = \mathbf{x}_0$, $\sigma = \sigma_0$. Thus $\mathbf{x}(t) \equiv \mathbf{x}_0$, $\sigma(t) \equiv \sigma_0$ is a solution of (6.43) that does not tend to zero, and this is a contradiction. ▮

We now make some additional remarks concerning the conditions of Theorem 6.4.

(i) The absolute stability condition $\rho > \mathbf{d}^T C^{-1}\mathbf{d}$ implies that $\rho > 0$. To see this we note that if C is positive definite, its (necessarily real) eigenvalues are all positive; hence the eigenvalues of C^{-1} (which are the reciprocals of the eigenvalues of C) are also all positive and C^{-1} is positive definite. The fact that ρ is positive is also to be expected from the expression (6.41), for $V^*(\mathbf{0}, \sigma) = -\rho[\phi(\sigma)]^2$.

(ii) The absolute stability condition $\rho > \mathbf{d}^T C^{-1}\mathbf{d}$ may be replaced by the conditions that $\rho > 0$ and $C - \mathbf{d}\mathbf{d}^T/\rho$ be positive definite. To see this, we note that the equation (6.41) can be written in the form

$$-V^*(\mathbf{x}, \sigma) = \mathbf{x}^T\left(C - \frac{\mathbf{d}\mathbf{d}^T}{\rho}\right)\mathbf{x} + \left(\sqrt{\rho}\,\phi(\sigma) + \frac{\mathbf{x}^T\mathbf{d}}{\sqrt{\rho}}\right)^2 \tag{6.44}$$

• EXERCISES

1. Show that (6.41) and (6.44) are equivalent.
2. Show that $-V^*(\mathbf{x}, \sigma)$, as given by (6.44), is positive definite if $\rho > 0$ and $C - \mathbf{d}\mathbf{d}^T/\rho$ is a positive definite matrix.

(iii) The class of admissible controls can be widened considerably without affecting the conclusion of Theorem 6.4. It can be shown that the system (6.37) is absolutely stable under the hypotheses of Theorem 6.4 even if the admissible controls are not required to satisfy the condition $\lim_{|\sigma| \to \infty} \int_0^\sigma \phi(\xi)\, d\xi = +\infty$. See [18, p. 28]. This is done by first showing that the condition $\rho > \mathbf{d}^T C^{-1}\mathbf{d}$ implies $\rho > \mathbf{c}^T A^{-1}\mathbf{b}$ (and not merely $\rho \neq \mathbf{c}^T A^{-1}\mathbf{b}$, as shown above). Subsequently we show, using this, that if $(\mathbf{x}(t), \sigma(t))$ is any solution of (6.37), then $|\mathbf{x}(t)| < \infty$, $|\sigma(t)| < \infty$. The result then follows from Theorem 5.6, p. 224.

We conclude with some remarks about the relation between Theorem 6.4 and the regulator problem in general.

(i) The absolute stability condition $\rho > (B\mathbf{b} - \frac{1}{2}\mathbf{c})^T C^{-1}(B\mathbf{b} - \frac{1}{2}\mathbf{c})$ of Theorem 6.4 depends on the choice of the positive matrix C. Once C is fixed, this determines the positive definite matrix B. This condition is least restrictive (that is, permits the largest range of values of ρ) if C is chosen such that $(B\mathbf{b} - \frac{1}{2}\mathbf{c})^T C^{-1}(B\mathbf{b} - \frac{1}{2}\mathbf{c})$ is a minimum. This question how to best choose C has not been resolved in general, except when A and C are diagonal, and we shall not go into it. See [9, p. 148].

(ii) If we permit only linear characteristic functions $\phi(\sigma) = \mu\sigma$ for all possible $\mu > 0$, we can obtain better conditions for global asymptotic stability. We illustrate this with the following example.

Example 1. Consider the two-dimensional system

$$\begin{aligned} x' &= -\alpha x - b\mu\sigma \\ \sigma' &= cx - \rho\mu\sigma \end{aligned} \tag{6.45}$$

corresponding to an uncontrolled one-dimensional system $x' = -\alpha x$. Here $x, \sigma, \alpha, \beta, c, \rho$ are all scalars. We first analyze the linear system (6.45) directly. The solution $x = 0$, $\sigma = 0$ of (6.45) is globally asymptotically stable for every $\mu > 0$ if and only if both eigenvalues of the matrix

$$\begin{pmatrix} -\alpha & -b\mu \\ c & -\rho\mu \end{pmatrix}$$

have negative real parts for every $\mu > 0$. The eigenvalues are the roots of the quadratic equation

$$\lambda^2 + (\alpha + \mu\rho)\lambda + \mu(bc + \alpha\rho) = 0$$

It is easy to verify that these eigenvalues both have negative real part if and only if

$$\alpha + \mu\rho > 0 \qquad \mu(bc + \alpha\rho) > 0$$

- **EXERCISE**

3. Verify the above condition.

If $\alpha < 0$, we cannot have $\alpha + \mu\rho > 0$ for small $\mu > 0$, and thus $\alpha \geq 0$. If $\rho < 0$, the condition $\alpha + \mu\rho > 0$ cannot be satisfied for large $\mu > 0$; thus $\rho \geq 0$. If $\alpha = 0$, the condition $\alpha + \mu\rho > 0$ reduces to $\rho > 0$. It follows that the solution $x = 0$, $\sigma = 0$ of (6.45) is globally asymptotically stable for every

$\mu > 0$ if and only if $\alpha \geq 0$, $\rho \geq 0$ (with α, ρ not both zero) and $bc + \alpha\rho > 0$. If $bc > 0$, then $bc + \alpha\rho > 0$ is satisfied whenever $\alpha \geq 0$, $\rho \geq 0$ and α, ρ are not both zero. Thus for $bc > 0$ we have global asymptotic stability of the solution $x = 0$, $\sigma = 0$ of (6.45) if and only if $\alpha \geq 0$, $\rho \geq 0$ (α, ρ not both zero). If $bc = 0$, we must have $\alpha > 0$, $\rho > 0$. If $bc < 0$, we let $z = -bc > 0$. In this case we have global asymptotic stability if and only if $\alpha\rho > z$, $\alpha \geq 0$, $\rho \geq 0$.

Now let us apply Theorem 6.4 to the problem (6.45). Here the matrices B and C are scalars. The equation (6.40) becomes $-2\alpha B = -C$, and thus B is determined by $B = c/2\alpha$. In the notation of Theorem 6.4, $d = Bb - \frac{1}{2}c = Cb/2\alpha - c/2$, and the condition $\rho > \mathbf{d}^T C^{-1}\mathbf{d}$ becomes

$$\rho > \frac{(Cb/2\alpha - c/2)^2}{C} = \frac{b^2 C}{4\alpha^2} + \frac{c^2}{4C} - \frac{bc}{2\alpha}$$

To obtain the best possible result, we must choose C so as to minimize $b^2C/4\alpha^2 + c^2/4C - bc/2\alpha$. It is easy to verify that this quantity is minimized by taking $C = \alpha|c/b|$, and the minimum value is $|bc|/2\alpha - bc/2\alpha$. Thus if $bc \geq 0$, the best possible global asymptotic stability condition given by Theorem 6.4 is $\rho > 0$. However, if $bc > 0$, the direct approach taken earlier gives global asymptotic stability under the less restrictive condition $\rho \geq 0$, provided $\alpha > 0$. The reader should note that if $\alpha = 0$, Theorem 6.4 is not applicable since the zero solution of the uncontrolled system is not globally asymptotically stable. The direct approach in the case $\alpha = 0$ gives a condition $\rho > 0$ under which the control can bring about global asymptotic stability.

If $bc < 0$, we again let $z = -bc > 0$. Then Theorem 6.4 gives the condition $\alpha\rho > |bc|/2 - bc/2 = z$. This is the same as the condition given by the direct approach.

• EXERCISE

4. Compare the results obtained by the direct approach and by application of Theorem 6.4 in case $bc = 0$.

(iii) A control system of the form (6.33) is not the only possibility. In (6.33) we are applying a feedback $-\mathbf{b}\xi$ to the uncontrolled system $\mathbf{z}' = A\mathbf{z}$, and this feedback is determined by an equation for its derivative. Such a system is called an **indirect control.** It would also be possible to consider a **direct control**, in which the feedback is determined directly. Such a control system would be governed by a system of differential equations

$$\begin{aligned} \mathbf{z}' &= A\mathbf{z} - \mathbf{b}\xi \\ \xi &= \phi(\sigma) \end{aligned} \qquad (6.46)$$

By the transformation $\sigma = \mathbf{c}^T\mathbf{z}$ we can show that (6.46) is equivalent to

$$\begin{aligned} \mathbf{z}' &= A\mathbf{z} - b\phi(\sigma) \\ \sigma &= \mathbf{c}^T\mathbf{z} \end{aligned} \tag{6.47}$$

Differentiation gives

$$\sigma' = \mathbf{c}^T\mathbf{z}' = \mathbf{c}^T(A\mathbf{z} - \mathbf{b}\phi(\sigma)) = \mathbf{c}^TA\mathbf{z} - \mathbf{c}^T\mathbf{b}\phi(\sigma)$$

and then (6.47) appears to have the form of (6.37), but with $\mathbf{c}^TA$ in place of $\mathbf{c}^T$ and $\mathbf{c}^T\mathbf{b}$ in place of ρ. It would appear that Theorem 6.4 can also be applied to the direct control problem (6.47). It turns out, however, that for the direct control problem, the derivative V^* of the relevant scalar function introduced in the proof of Theorem 6.4 is not negative definite and some modification of the procedure is necessary. However, we shall not discuss this problem further; see, for example, [1, 9, 17, or 18].

• EXERCISES

Apply the theory of this section and obtain criteria for the absolute stability of the following control systems

5. $x' = -kx - \phi(\sigma)$, $\sigma' = cx - \rho\phi(\sigma)$ where $k > 0$, c, ρ are real numbers, and $\phi(\sigma)$ is an admissible characteristic function.
6. $x' = -kx - \phi(\sigma)$, $\sigma = cx$, where $k > 0$, c are real numbers, and $\phi(\sigma)$ is an admissible characteristic function.

Appendix 1 GENERALIZED EIGENVECTORS, INVARIANT SUBSPACES, AND CANONICAL FORMS OF MATRICES

In the study of linear systems of differential equations with constant coefficients (Section 2.5, p. 55), we have made use of some results about matrices and linear transformations with which the reader may not be familiar. We therefore include the proofs of these results here for the sake of completeness. For a more detailed account and a discussion of related topics, we refer the reader to such sources as [7, 11, 14].

We will assume that the reader is familiar with the concepts of subspace, linear independence, basis, dimension, eigenvalue, and eigenvector. We recall that if T is a linear transformation of E_n into itself, then corresponding to any basis $\{\mathbf{v}_1, \mathbf{v}_2, \ldots, \mathbf{v}_n\}$ of E_n there is an $n \times n$ matrix A that represents the linear transformation T with respect to this basis. The elements a_{ij} $(i, j = 1, \ldots, n)$ of this matrix A are defined by

$$T\mathbf{v}_i = \sum_{j=1}^{n} a_{ji}\mathbf{v}_j \qquad (i = 1, \ldots, n) \tag{1}$$

Corresponding to a different basis $\{\mathbf{w}_1, \mathbf{w}_2, \ldots, \mathbf{w}_n\}$ of E_n there is a different matrix B that represents T. It can also be shown that the matrices A and B are similar; that is, there exists a nonsingular matrix P such that

$$B = P^{-1}AP$$

• EXERCISE

1. Establish the above relation. [*Hint:* In addition to (1), let $T\mathbf{w}_i = \Sigma_j b_{ji}\mathbf{w}_j$. Since $\{\mathbf{v}_i\}$ and $\{\mathbf{w}_i\}$ form bases, there exist nonsingular matrices P and Q such that $\mathbf{w}_i = \Sigma_j p_{ji}\mathbf{v}_j$ and $\mathbf{v}_i = \Sigma_j q_{ji}\mathbf{w}_j$. Compute $T\mathbf{v}_i = T(\Sigma_l q_{li}\mathbf{w}_l)$ in terms of the $\{\mathbf{v}_i\}$ in order to show $A = PBQ$. Then establish separately that $Q = P^{-1}$.]

The study of canonical forms of matrices involves the choice of a basis of E_n relative to which the matrix of a given linear transformation takes a particularly simple form. Equivalently, this study involves the determination of a matrix of a particularly simple form that is similar to a given matrix.

For example, suppose that a linear transformation T (or a matrix A) has n linearly independent eigenvectors $\mathbf{v}_1, \mathbf{v}_2, \ldots, \mathbf{v}_n$ corresponding to the eigenvalues $\lambda_1, \lambda_2, \ldots, \lambda_n$, respectively. These eigenvectors form a basis of E_n, and since $T\mathbf{v}_i = \lambda_i \mathbf{v}_i$ $(i = 1, \ldots, n)$, the matrix of T with respect to this basis is the diagonal matrix

$$B = \begin{bmatrix} \lambda_1 & & & \\ & \lambda_2 & & \\ & & \ddots & \\ & & & \lambda_n \end{bmatrix} \tag{2}$$

Thus a matrix with n linearly independent eigenvectors is always similar to a diagonal matrix, and we say that a matrix with n linearly independent eigenvectors has a diagonal canonical form. As the determination of the eigenvectors of a matrix and the verification that they are linearly independent is a tedious procedure, this criterion is not very practical. A more convenient condition that implies that a matrix can be diagonalized is given by the following result.

Theorem 1. *A set of k eigenvectors corresponding to any k distinct eigenvalues is linearly independent.*

Proof. We shall prove the theorem by induction on the number k of eigenvectors. For $k = 1$, the result is trivial. Now, assume that every set of $(p - 1)$ eigenvectors corresponding to $(p - 1)$ distinct eigenvalues of a given matrix A is linearly independent. Let $\mathbf{v}_1, \ldots, \mathbf{v}_p$ be eigenvectors of A corresponding to the eigenvalues $\lambda_1, \ldots, \lambda_p$, respectively, with $\lambda_i \neq \lambda_j$ for $i \neq j$. Suppose that $\mathbf{v}_1, \ldots, \mathbf{v}_p$ are linearly dependent, so that there exist constants $c_1, c_2, \ldots, c_p$, not all zero, such that

$$c_1\mathbf{v}_1 + c_2\mathbf{v}_2 + \cdots + c_p\mathbf{v}_p = \mathbf{0} \tag{3}$$

We may assume $c_1 \neq 0$. Applying $A - \lambda_1 E$ to both sides of this equation, we obtain

$$c_2(\lambda_2 - \lambda_1)\mathbf{v}_2 + c_3(\lambda_3 - \lambda_1)\mathbf{v}_3 + \cdots + c_p(\lambda_p - \lambda_1)\mathbf{v}_p = \mathbf{0}$$

But $\mathbf{v}_2, \mathbf{v}_3, \ldots, \mathbf{v}_p$ are linearly independent by the inductive hypothesis, and therefore $c_j(\lambda_j - \lambda_1) = 0$ $(j = 2, 3, \ldots, p)$. Since $\lambda_j \neq \lambda_1$ $(j = 2, 3, \ldots, p)$, we have $c_j = 0$ $(j = 2, 3, \ldots, p)$, and (3) becomes $c_1\mathbf{v}_1 = \mathbf{0}$. Since $c_1 \neq 0$, $\mathbf{v}_1 = \mathbf{0}$, which is a contradiction and shows that $\mathbf{v}_1, \ldots, \mathbf{v}_p$ are linearly independent. This proves the theorem by induction. ▮

Corollary. If a matrix A has n distinct eigenvalues $\lambda_1, \ldots, \lambda_n$, *then the corresponding eigenvectors form a basis of* E_n, *and A is similar to the diagonal matrix*

$$B = \begin{bmatrix} \lambda_1 & & & \\ & \lambda_2 & & \\ & & \ddots & \\ & & & \lambda_n \end{bmatrix}$$

If the eigenvalues of a matrix are not all distinct, there may still be n linearly independent eigenvectors, so that the matrix can be diagonalized. However, this is not always possible. For example, the matrix $A = \begin{bmatrix} 0 & 1 \\ 0 & 0 \end{bmatrix}$ has zero as its only eigenvalue (a double root of the characteristic equation) and every eigenvector is a scalar multiple of the vector $\begin{pmatrix} 1 \\ 0 \end{pmatrix}$.

To discuss the situation when there are fewer than n linearly independent eigenvectors, we introduce the concept of **generalized eigenvector.** If for some value λ and some integer $p \geq 1$, there is a vector $\mathbf{v}$ such that

$$(A - \lambda E)^p\mathbf{v} = \mathbf{0} \quad \text{but} \quad (A - \lambda E)^{p-1}\mathbf{v} \neq \mathbf{0} \tag{4}$$

then $\mathbf{v}$ is said to be a generalized eigenvector of index p corresponding to the generalized eigenvalue λ. When $p = 1$, λ is an eigenvalue and $\mathbf{v}$ a corresponding eigenvector. We note that since $\mathbf{u} = (A - \lambda E)^{p-1}\mathbf{v} \neq \mathbf{0}$, and since because of (4), $(A - \lambda E)\mathbf{u} = (A - \lambda E)^p\mathbf{v} = \mathbf{0}$, the "generalized eigenvalue" λ must be an eigenvalue of A with a corresponding eigenvector $\mathbf{u}$.

Lemma 1. *If* $\mathbf{v}$ *is a generalized eigenvector of index p, then the vectors* $\mathbf{v}$, $(A - \lambda E)\mathbf{v}, \ldots, (A - \lambda E)^{p-1}\mathbf{v}$ *are linearly independent.*

Proof. If the given vectors are linearly dependent, then one of them, say $(A - \lambda E)^k \mathbf{v}$ for some k $(0 \le k < p - 1)$, can be written as a linear combination of the later ones,

$$(A - \lambda E)^k \mathbf{v} = \sum_{j=k+1}^{p-1} c_j (A - \lambda E)^j \mathbf{v}$$

Since $(A - \lambda E)^q \mathbf{v} = 0$ for $q \ge p$, application of $(A - \lambda E)^{p-1-k}$ to both sides of this equation gives $(A - \lambda E)^{p-1} \mathbf{v} = 0$, which is a contradiction, proving the linear independence of the given vectors. ∎

Given an eigenvalue λ of A, we consider the subspace X of E_n consisting of all generalized eigenvectors corresponding to λ together with the zero vector. Let r be the largest index of any such generalized eigenvector, so that $(A - \lambda E)^r \mathbf{x} = \mathbf{0}$ for every $\mathbf{x} \in X$, but $(A - \lambda E)^{r-1} \mathbf{y} \neq \mathbf{0}$ for some $\mathbf{y} \in X$. Since, by Lemma 1, X contains at least r linearly independent vectors, r is finite (in fact, $r \le n$). The integer r is called the index of X, and dim $X \ge r$.

The subspace X may also be described as the null space of the linear transformation $(T - \lambda E)^r$ (or the null space of the matrix $(A - \lambda E)^r$). Let Y be the range of the linear transformation $(T - \lambda E)^r$. We recall that the range of the linear transformation $(T - \lambda E)^r$ is the set of vectors $\mathbf{y} \in E_n$ such that $(T - \lambda E)^r \mathbf{z} = \mathbf{y}$ for some $\mathbf{z} \in E_n$. By the theory of linear algebraic systems, dim X + dim $Y = n$. (The reader can find this in any text on linear algebra.) Next, we observe that X and Y are disjoint subspaces; for if $\mathbf{v}$ is in both X and Y, then $(T - \lambda E)^r \mathbf{v} = \mathbf{0}$ and $\mathbf{v} = (T - \lambda E)^r \mathbf{u}$ for some $\mathbf{u} \in E_n$. Thus $(T - \lambda E)^{2r} \mathbf{u} = (T - \lambda E)^r \mathbf{v} = \mathbf{0}$, and thus $\mathbf{u}$ is a generalized eigenvector; that is, $\mathbf{u} \in X$. But since r is the largest index of any generalized eigenvector corresponding to λ, $\mathbf{v} = (T - \lambda E)^r \mathbf{u} = \mathbf{0}$. This shows that X and Y are disjoint subspaces of E_n whose dimensions add up to n. It follows that E_n is the direct sum of X and Y (written $E_n = X \oplus Y$), that is, that every vector $\mathbf{v} \in E_n$ has a unique decomposition $\mathbf{v} = \mathbf{x} + \mathbf{y}$, with $\mathbf{x} \in X$, $\mathbf{y} \in Y$.

The subspaces X and Y are **invariant** under T, that is, $T\mathbf{x} \in X$ for every $\mathbf{x} \in X$ and $T\mathbf{y} \in Y$ for every $\mathbf{y} \in Y$. To see this we observe that X and Y are invariant under $(T - \lambda E)$, and it follows easily that they are invariant under T.

Next, suppose that the subspace X has dimension k, and let $\mathbf{v}_1, \ldots, \mathbf{v}_k$ be a basis of X. Then Y has dimension $(n - k)$, and if $\mathbf{v}_{k+1}, \ldots, \mathbf{v}_n$ is a basis of Y, the fact that $E_n = X \oplus Y$ implies that $\mathbf{v}_1, \ldots, \mathbf{v}_k, \mathbf{v}_{k+1}, \ldots, \mathbf{v}_n$ is a basis of E_n. Since X is invariant under T, $T\mathbf{v}_i \in X$ $(i = 1, \ldots, k)$, and since $\mathbf{v}_1, \ldots, \mathbf{v}_k$ is a basis of X,

$$T\mathbf{v}_i = \sum_{j=1}^{k} a_{ji} \mathbf{v}_j \qquad (i = 1, \ldots, k)$$

Note that the sum is from $j = 1$ to $j = k$, not to $j = n$. Similarly,

$$T\mathbf{v}_i = \sum_{j=k+1}^{n} a_{ji}\mathbf{v}_j \qquad (i = k+1, \ldots, n)$$

Comparing these formulas to (1), p. 274, we see that the matrix A of T with respect to the basis $\mathbf{v}_1, \ldots, \mathbf{v}_n$ of E_n is a "block diagonal" matrix,

$$A = \begin{bmatrix} A_1 & 0 \\ 0 & A_2 \end{bmatrix}$$

Here the $k \times k$ matrix A_1 represents the restriction T_1 of the transformation T to the subspace X and the $(n-k) \times (n-k)$ matrix A_2 represents the restriction T_2 of T to the subspace Y.

Since for the block diagonal matrix A

$$\det(A - \lambda E) = \det(A_1 - \lambda E)\det(A_2 - \lambda E_2)$$

where E_1 is the $k \times k$ identity matrix and E_2 is the $(n-k) \times (n-k)$ identity matrix, the characteristic polynomial of A (or of T) is the product of the characteristic polynomials of A_1 and of A_2 (or of T_1 and of T_2).

We have now developed the algebraic machinery needed to prove the following basic theorem, which is used in Section 2.5 (p. 64).

Theorem 2. *Let $\lambda_1, \lambda_2, \ldots, \lambda_k$ be the distinct eigenvalues of a matrix A, with multiplicities $n_1, n_2, \ldots, n_k$, respectively. Then E_n is the direct sum of the subspaces $X_1, \ldots, X_k$ of generalized eigenvectors corresponding to the eigenvalues $\lambda_1, \ldots, \lambda_k$, respectively. The subspace X_j is invariant under A, has dimension n_j, and*

$$(A - \lambda_j E)^{n_j}\mathbf{x} = \mathbf{0} \quad \textit{for every} \quad \mathbf{x} \in X_j \qquad (j = 1, \ldots, n)$$

Proof. On X_j, the transformation T corresponding to the matrix A has only the eigenvalue λ_j. For if λ is an eigenvalue and $\mathbf{v}$ a corresponding eigenvector in X_j, and if r_j is the index of X_j, then

$$\begin{aligned} \mathbf{0} &= (A - \lambda_j E)^{r_j}\mathbf{v} = (A - \lambda_j E)^{r_j - 1}(A - \lambda_j E)\mathbf{v} \\ &= (A - \lambda_j E)^{r_j - 1}(A\mathbf{v} - \lambda_j \mathbf{v}) = (A - \lambda_j E)^{r_j - 1}(\lambda - \lambda_j)\mathbf{v} \\ &= (\lambda - \lambda_j)(A - \lambda_j E)^{r_j - 1}\mathbf{v} = \cdots = (\lambda - \lambda_j)^{r_j}\mathbf{v} \end{aligned}$$

and $\lambda = \lambda_j$. Thus X_j contains all the eigenvectors of A corresponding to the eigenvalue λ_j, but no other eigenvector.

The characteristic polynomial of A is $p(\lambda) = \Pi_{i=1}^{k} (\lambda - \lambda_i)^{n_i}$. As we have seen immediately preceding the statement of Theorem 2, the characteristic polynomial of A is the product of the characteristic polynomial of A on the null space X_j of $(A - \lambda_j E)^{r_j}$ and of the characteristic polynomial of A on the range Y_j of $(A - \lambda_j E)^{r_j}$. The characteristic polynomial of A on X_j contains all the factors $(\lambda - \lambda_j)$ in $p(\lambda)$ because X_j contains all the eigenvectors of A corresponding to λ_j, and it contains no factor $(\lambda - \lambda_i)$ with $i \neq j$ because X_j contains no eigenvector of A corresponding to an eigenvalue $\lambda_i \neq \lambda_j$. Therefore the characteristic polynomial of A on X_j is exactly $\pm(\lambda - \lambda_j)^{n_j}$. Since the degree of the characteristic polynomial of a linear transformation is equal to the dimension of the space, this shows that X_j has dimension n_j.

To show that E_n is the direct sum of the subspaces $X_1, \ldots, X_k$, suppose that two of them, say X_i and X_j, have a nonzero intersection W. Then W is an invariant subspace under the transformation T corresponding to the matrix A. Thus T can be considered as a transformation on W and must have an eigenvector in W. However, this eigenvector must belong to both λ_i and λ_j, which is impossible. Thus no two of the subspaces $X_1, \ldots, X_k$ have a nonzero intersection and

$$\dim E_n = n = n_1 + \cdots + n_k = \dim X_1 + \cdots + \dim X_k$$

Thus E_n is the direct sum of $X_1, X_2, \ldots, X_k$, and this completes the proof of the theorem. ▮

By the argument preceding the statement of Theorem 2, we now obtain the following interpretation of Theorem 2.

Corollary to Theorem 2. The matrix A of the linear transformation T relative to a basis of E_n made up of bases of the subspaces $X_1, \ldots, X_k$ is a block diagonal matrix

$$A = \begin{bmatrix} A_1 & & & \\ & A_2 & & \\ & & \ddots & \\ & & & A_k \end{bmatrix}$$

where A_j is an $n_j \times n_j$ matrix that represents T on X_j $(j = 1, \ldots, k)$.

The **Jordan canonical** form of a matrix, also referred to in Section 2.6, p. 77, is obtained from the above representation by choosing bases of the subspaces $X_1, \ldots, X_k$ in a suitable manner. This requires a careful study of nilpotent transformations. A linear transformation L such that $L^r = 0$ but $L^{r-1} \neq 0$ is said to be **nilpotent** of index r. We recall that the subspace X_j is the null space of the transformation represented by the matrix $(A - \lambda_j E)^{r_j}$. Since X_j is invariant under $(A - \lambda_j E)$, we may regard $(A - \lambda_j E)$ as a linear transformation on X_j that is nilpotent of index r_j.

Let L be a nilpotent linear transformation of index r on a vector space X of dimension n. Then there is a vector $\mathbf{u}$ such that $L^r\mathbf{u} = \mathbf{0}$ but $L^{r-1}\mathbf{u} \neq \mathbf{0}$. By Lemma 1, the chain of vectors $\mathbf{u}, L\mathbf{u}, \ldots, L^{r-1}\mathbf{u}$ is linearly independent. We will form a basis for X consisting of several chains of this type. If $r = n$, then we have a basis of X consisting of the vectors $\mathbf{u}, L\mathbf{u}, \ldots, L^{r-1}\mathbf{u}$. If $r < n$, let U_1 be the subspace of X spanned by these vectors. For every $\mathbf{v} \notin U_1$ consider the chain $\mathbf{v}, L\mathbf{v}, \ldots, L^{s-1}\mathbf{v}$ where each vector $L^p\mathbf{v}$, $0 \leq p \leq s-1$, lies outside U_1 but $L^s\mathbf{v} \in U_1$. We choose a $\mathbf{v}$ that maximizes the length s of this chain, and we let $r_2 \leq r$ be the length of this maximal chain. Then $\mathbf{v}, L\mathbf{v}, \ldots, L^{r_2-1}\mathbf{v}$ are outside U_1 but $L^{r_2}\mathbf{v} \in U_1$. Since $\mathbf{u}, L\mathbf{u}, \ldots, L^{r-1}\mathbf{u}$ is a basis of U_1, we may write

$$L^{r_2}\mathbf{v} = \sum_{j=0}^{r-1} c_j L^j\mathbf{u} \tag{5}$$

We apply L^{r-r_2} to both sides of this equation and use $L^r\mathbf{u} = L^r\mathbf{v} = \mathbf{0}$ to see that

$$\mathbf{0} = L^r\mathbf{v} = \sum_{j=0}^{r-1} c_j L^{j+r-r_2}\mathbf{u} = \sum_{j=0}^{r_2-1} c_j L^{j+r-r_2}\mathbf{u}$$

Since $L^{r-r_2}\mathbf{u}, L^{r-r_2+1}\mathbf{u}, \ldots, L^{r-1}\mathbf{u}$ are linearly independent, $c_j = 0$ for $j = 0, 1, \ldots, r_2 - 1$. Thus (5) becomes

$$L^{r_2}\mathbf{v} = \sum_{j=r_2}^{r-1} c_j L^j\mathbf{u} \tag{6}$$

We define

$$\mathbf{u}_2 = \mathbf{v} - \sum_{j=r_2}^{r-1} c_j L^{j-r_2}\mathbf{u}$$

Then it is clear from (6) that $L^{r_2}\mathbf{u}_2 = \mathbf{0}$. On the other hand

$$L^k\mathbf{u}_2 = L^k\mathbf{v} - \sum_{j=r_2}^{r-1} c_j L^{k+j-r_2}\mathbf{u} \qquad (k = 0, 1, \ldots, r_2 - 1)$$

Since $L^{k+j-r_2}\mathbf{u}$ is in U_1 but $L^k\mathbf{v}$ is outside U_1, $L^k\mathbf{u}_2$ is outside U_1 ($k = 0, 1, \ldots, r_2 - 1$). Thus every nonzero linear combination of $\mathbf{u}_2$, $L\mathbf{u}_2, \ldots, L^{r_2-1}\mathbf{u}_2$ is outside U_1. Let U_2 be the subspace of X spanned by $\mathbf{u}_2$, $L\mathbf{u}_2, \ldots, L^{r_2-1}\mathbf{u}_2$; then U_1 and U_2 are disjoint. The direct sum $U_1 \oplus U_2$ is invariant under L.

• **EXERCISE**

2. Prove the last statement.

If this direct sum is not all of X, we construct a maximal chain outside $U_1 \oplus U_2$ by the same method. Continuing in this manner, we can write X as a direct sum of a finite number of subspaces $U_1, U_2, \ldots, U_t$, each of which is spanned by a chain of the type given above. Thus we have proved the following result.

Theorem 3. *Let L be a nilpotent linear transformation of index r_1 on a vector space X of dimension n. Then X has a basis of the form*

$$L^{r-1}\mathbf{u}, L^{r-2}\mathbf{u}, \ldots, L\mathbf{u}, \mathbf{u}, L^{r_2-1}\mathbf{u}_2, L^{r_2-2}\mathbf{u}_2, \ldots, \mathbf{u}_2, \ldots, L^{r_t-1}\mathbf{u}_t, \ldots, \mathbf{u}_t$$

with

$$r_1 \geq r_2 \geq r_3 \geq \cdots \geq r_t \geq 1 \quad \textit{and} \quad L^{r_k}\mathbf{u}_t = \mathbf{0} \qquad (k = 1, 2, \ldots, t)$$

To construct the matrix B that represents L with respect to the basis given by Theorem 3, we denote the basis elements respectively by $\mathbf{v}_1, \mathbf{v}_2, \ldots, \mathbf{v}_n$. Then we have

$$L\mathbf{v}_1 = \mathbf{0}, \quad L\mathbf{v}_2 = \mathbf{v}_1, \ldots, L\mathbf{v}_r = \mathbf{v}_{r-1}$$

$$L\mathbf{v}_{r+1} = \mathbf{0}, \quad L\mathbf{v}_{r+2} = \mathbf{v}_{r+1}, \ldots, L\mathbf{v}_n = \mathbf{v}_{n-1}.$$

From the definition (1) of the matrix of a linear transformation with respect to a given basis, we see that B has the form

$$B = \begin{bmatrix} B_1 & & & \\ & B_2 & & \\ & & \ddots & \\ & & & B_t \end{bmatrix} \tag{7}$$

where B_k is the $r_k \times r_k$ matrix given by

$$B_k = \begin{bmatrix} 0 & 1 & 0 & \cdots & 0 \\ 0 & & \ddots & \ddots & 0 \\ \vdots & \ddots & & & 1 \\ 0 & \cdots & 0 & & 0 \end{bmatrix}$$

The **Jordan canonical form** of a linear transformation T is now obtained by combining Theorems 2 and 3. According to Theorem 2, we can decompose E_n into a direct sum of subspaces $X_1, \ldots, X_k$. On the subspace X_j, the transformation $T - \lambda_j E$ is nilpotent of index r_j. We now use Theorem 3 to construct a basis for the subspace X_j. The matrix of the transformation $T - \lambda_j E$ restricted to X_j has the form (7). Thus with respect to this basis, the matrix of the restriction of T to X_j has the form

$$C = \begin{bmatrix} C_1 & & & \\ & C_2 & & \\ & & \ddots & \\ & & & C_t \end{bmatrix}$$

where C_j is the $r_j \times r_j$ matrix C_j.

$$C_j = \begin{bmatrix} \lambda_j & 1 & 0 & \cdots & 0 \\ 0 & & \ddots & \ddots & 0 \\ \vdots & \ddots & & & 1 \\ 0 & \cdots & 0 & & \lambda^j \end{bmatrix}$$

This gives the following important result.

Theorem 4. (Jordan Canonical Form.) *Let T be a linear transformation of E_n with eigenvalues $\lambda_1, \ldots, \lambda_k$ of multiplicities $n_1, \ldots, n_k$, respectively. Then there exists a basis of E_n relative to which T is given by a Jordan canonical matrix*

$$A = \begin{bmatrix} A_1 & & & \\ & A_2 & & \\ & & \ddots & \\ & & & A_k \end{bmatrix}$$

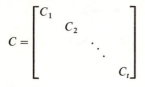

Here A_j is an $n_j \times n_j$ matrix that has all diagonal elements equal to λ_j, and that has chains of 1's *separated by single* 0's *immediately above the main diagonal, and all other elements zero.*

Corollary. Every matrix is similar to a Jordan canonical matrix.

It can also be shown that, except for the order of the blocks A_j, the Jordan canonical form is unique.

Appendix 2 | CANONICAL FORMS OF 2×2 MATRICES

We consider a 2×2 matrix A, which we regard as the coefficient matrix of a two-dimensional linear system with constant coefficients $\mathbf{y}' = A\mathbf{y}$. The change of variable $\mathbf{y} = T\mathbf{z}$, where T is a nonsingular matrix, transforms this to the system $\mathbf{z}' = T^{-1}AT\mathbf{z}$. Our object is to choose T in such a way as to make the coefficient matrix $T^{-1}AT$ as simple as possible. A simplification is always possible, as shown in the following result, which incidentally is entirely independent of differential equations and is purely algebraic.

Theorem 1. *Let A be a 2×2 matrix. Then there exists a nonsingular 2×2 matrix T such that $T^{-1}AT$ is one of the following:*

(i) $\begin{pmatrix} \lambda & 0 \\ 0 & \mu \end{pmatrix}$ $\quad (\lambda \neq \mu)$

(ii) $\begin{pmatrix} \lambda & 0 \\ 0 & \lambda \end{pmatrix}$

(iii) $\begin{pmatrix} \lambda & 1 \\ 0 & \lambda \end{pmatrix}$

Proof. *Case* (*i*). Suppose A has distinct eigenvalues λ, μ. Let $\mathbf{x} \neq \mathbf{0}$ be an eigenvector corresponding to λ and let $\mathbf{y} \neq \mathbf{0}$ be an eigenvector corresponding to μ; that is, $\mathbf{x}$ is a nonzero vector such that $A\mathbf{x} = \lambda\mathbf{x}$ and $\mathbf{y}$ is a nonzero vector such that $A\mathbf{y} = \mu\mathbf{y}$.

We define the 2×2 matrix T whose columns are the (column) vectors $\mathbf{x}$ and $\mathbf{y}$, which we write

$$T = (\mathbf{x}, \mathbf{y}) = \begin{pmatrix} x_1 & y_1 \\ x_2 & y_2 \end{pmatrix}$$

In order to show that T is nonsingular, we must prove that $\mathbf{x}$ and $\mathbf{y}$ are linearly independent. Suppose not; then there exist constants c_1, c_2 such that

$$c_1\mathbf{x} + c_2\mathbf{y} = \mathbf{0}$$

Since $\mathbf{x}$ and $\mathbf{y}$ are both different from zero, both c_1 and c_2 are different from zero. Thus, we may write

$$\mathbf{x} = -\frac{c_2}{c_1}\mathbf{y}$$

multiplying both sides on the left by A and using $A\mathbf{x} = \lambda\mathbf{x}$, $A\mathbf{y} = \mu\mathbf{y}$, we obtain

$$\lambda\mathbf{x} = A\mathbf{x} = -\frac{c_2}{c_1}A\mathbf{y} = -\frac{c_2}{c_1}\mu\mathbf{y}$$

Since $\lambda\mathbf{x} = -(c_2/c_1)\lambda\mathbf{y}$, this becomes

$$-\frac{c_2}{c_1}\lambda\mathbf{y} = -\frac{c_2}{c_1}\mu\mathbf{y}$$

This is impossible unless either $c_2 = 0$ or $\lambda = \mu$, both of which are false. Therefore, $\mathbf{x}$ and $\mathbf{y}$ are linearly independent, and T is nonsingular.

It is easy to verify that

$$T^{-1} = \frac{1}{x_1y_2 - x_2y_1}\begin{pmatrix} y_2 & -y_1 \\ -x_2 & x_1 \end{pmatrix}$$

Thus,

$$\begin{aligned} T^{-1}AT &= T^{-1}(A\mathbf{x}, A\mathbf{y}) = T^{-1}(\lambda\mathbf{x}, \mu\mathbf{y}) \\ &= \frac{1}{x_1y_2 - x_2y_1}\begin{pmatrix} y_2 & -y_1 \\ -x_2 & x_1 \end{pmatrix}\begin{pmatrix} \lambda x_1 & \mu y_1 \\ \lambda x_2 & \mu y_2 \end{pmatrix} = \begin{pmatrix} \lambda & 0 \\ 0 & \mu \end{pmatrix} \end{aligned}$$

This completes the proof in Case (i).

Case (*ii*). Suppose A has a double eigenvalue λ for which there are two linearly independent eigenvectors $\mathbf{x}$ and $\mathbf{y}$. Then defining T exactly as in Case (i), we see that T is nonsingular because its columns are given linearly independent. The same calculation as in Case (i) (but with μ replaced by λ) shows that

$$T^{-1}AT = \begin{pmatrix} \lambda & 0 \\ 0 & \lambda \end{pmatrix}$$

Case (*iii*). Suppose A has a double eigenvalue λ but any two eigenvectors are linearly dependent (that is, the space of eigenvectors has dimension 1). Let $\mathbf{v}$ be a nonzero vector which is not an eigenvector of A and let

$$\mathbf{u} = (A - \lambda E)\mathbf{v}$$

Since $\mathbf{v}$ is not an eigenvector, and $A - \lambda E$ is not the zero matrix, $\mathbf{u} \neq \mathbf{0}$; we will show that $\mathbf{u}$ is an eigenvector.

We assert that the vectors $\mathbf{u}$ and $\mathbf{v}$ are linearly independent. Suppose not; then there exist constants c_1, c_2 such that

$$c_1\mathbf{u} + c_2\mathbf{v} = \mathbf{0}$$

Since $\mathbf{u}$ and $\mathbf{v}$ are both different from zero, both c_1 and c_2 are different from zero. Using the definition of $\mathbf{u}$ and the fact that $c_1 \neq 0$, we may rewrite this as

$$c_1(A - \lambda E)\mathbf{v} + c_2\mathbf{v} = \mathbf{0}$$

or

$$c_1\left[\left(A - \lambda E + \frac{c_2}{c_1}E\right)\mathbf{v}\right] = \mathbf{0}$$

This says that $\lambda - c_2/c_1$ is an eigenvalue of A. Since λ is the only eigenvalue, $c_2 = 0$, which is a contradiction. Thus, $\mathbf{u}$ and $\mathbf{v}$ are linearly independent and span the space E_2.

Let $\mathbf{x}$ be any eigenvector of A; we may therefore write

$$\mathbf{x} = a\mathbf{u} + b\mathbf{v}$$

If $a = 0$, then $\mathbf{x}$ is a multiple of $\mathbf{v}$, and therefore $\mathbf{v}$ is an eigenvector (but by hypothesis it is not); thus $a \neq 0$. Now,

$$\mathbf{0} = (A - \lambda E)\mathbf{x} = a(A - \lambda E)\mathbf{u} + b(A - \lambda E)\mathbf{v}$$

$$= a(A - \lambda E)\mathbf{u} + b\mathbf{u} = a\left[\left(A - \lambda E + \frac{b}{a}E\right)\mathbf{u}\right]$$

This says that $\lambda - b/a$ is an eigenvalue, and since λ is the only eigenvalue $b = 0$. Therefore, $\mathbf{x}$ is a nonzero multiple of $\mathbf{u}$, and $\mathbf{u}$ must be an eigenvector.

Now, we define

$$T = (\mathbf{u}, \mathbf{v}) = \begin{pmatrix} u_1 & v_1 \\ u_2 & v_2 \end{pmatrix}$$

As in Case (i),

$$T^{-1} = \frac{1}{u_1 v_2 - u_2 v_1}\begin{pmatrix} v_2 & -v_1 \\ -u_2 & u_1 \end{pmatrix}$$

We have, using $(A - \lambda E)\mathbf{v} = A\mathbf{v} - \lambda\mathbf{v}$,

$$\begin{aligned} T^{-1}AT &= T^{-1}(A\mathbf{u}, A\mathbf{v}) \\ &= T^{-1}(\lambda\mathbf{u}, \mathbf{u} + \lambda\mathbf{v}) \\ &= T^{-1}(\lambda\mathbf{u}, \lambda\mathbf{v}) + T^{-1}(\mathbf{0}, \mathbf{u}) \\ &= \lambda T^{-1}T + \frac{1}{u_1 v_2 - u_2 v_1}\begin{pmatrix} v_2 & -v_1 \\ -u_2 & u_1 \end{pmatrix}\begin{pmatrix} 0 & u_1 \\ 0 & u_2 \end{pmatrix} \\ &= \lambda E + \begin{pmatrix} 0 & 1 \\ 0 & 0 \end{pmatrix} = \begin{pmatrix} \lambda & 1 \\ 0 & \lambda \end{pmatrix} \end{aligned}$$

This completes the proof of the theorem. With reference to the above proof, the reader should note that in showing that $\mathbf{u} = (A - \lambda E)\mathbf{v}$ is an eigenvector we have actually shown that $(A - \lambda E)^2\mathbf{w} = \mathbf{0}$ for every vector $\mathbf{w}$ in E_2. For this reason we say that $A - \lambda E$ is **nilpotent** on E_2. ▮

We remark that if A is a real matrix and if its eigenvalues are real, then the matrix T constructed in each of the three cases above is real. However, if A is real but has complex eigenvalues (necessarily complex conjugates), the matrix T will not be real and it is of interest to learn the simplest form of $T^{-1}AT$ which can be achieved with a **real** matrix T. The answer lies in the following result.

Theorem 2. *Let A be a real 2×2 matrix with complex conjugate eigenvalues $\alpha \pm i\beta$. Then there exists a real nonsingular matrix T such that*

$$T^{-1}AT = \begin{pmatrix} \alpha & \beta \\ -\beta & \alpha \end{pmatrix}$$

Proof. Let $\mathbf{u} + i\mathbf{v}$ be an eigenvector corresponding to the eigenvalue $\alpha + i\beta$, where $\mathbf{u}$ and $\mathbf{v}$ are real. If $\mathbf{v} = \mathbf{0}$, so that this eigenvector is real,

$$A\mathbf{u} = (\alpha + i\beta)\mathbf{u}$$

the left side of which is real and the right side of which is not real. Thus, $\mathbf{v} \neq \mathbf{0}$, and a similar argument shows that $\mathbf{u} \neq \mathbf{0}$.

We define the matrix T with columns $\mathbf{u}$ and $\mathbf{v}$,

$$T = (\mathbf{u}, \mathbf{v}) = \begin{pmatrix} u_1 & v_1 \\ u_2 & v_2 \end{pmatrix}$$

In order to show that T is nonsingular, we must show that $\mathbf{u}$ and $\mathbf{v}$ are linearly independent. Suppose not; then there exist real constants c_1, c_2 both different from zero such that

$$c_1\mathbf{u} + c_2\mathbf{v} = \mathbf{0}$$

Since $c_1 \neq 0$, and $A(\mathbf{u} + i\mathbf{v}) = (\alpha + i\beta)(\mathbf{u} + i\mathbf{v})$, we have

$$A(\mathbf{u} + i\mathbf{v}) = A\left(-\frac{c_2}{c_1}\mathbf{v} + i\mathbf{v}\right) = (\alpha + i\beta)\left(-\frac{c_2}{c_1} + i\right)\mathbf{v}$$

thus

$$A\mathbf{v} = (\alpha + i\beta)\mathbf{v}$$

Then $\mathbf{v}$ is a real eigenvector, and as remarked above this is impossible. Thus $\mathbf{u}$ and $\mathbf{v}$ are linearly independent and T is nonsingular.

Taking real and imaginary parts in the equation

$$A(\mathbf{u} + i\mathbf{v}) = (\alpha + i\beta)(\mathbf{u} + i\mathbf{v})$$

we obtain

$$A\mathbf{u} = \alpha\mathbf{u} - \beta\mathbf{v} \qquad A\mathbf{v} = \beta\mathbf{u} + \alpha\mathbf{v}$$

Therefore,

$$AT = (A\mathbf{u}, A\mathbf{v}) = (\alpha\mathbf{u} - \beta\mathbf{v}, \beta\mathbf{u} + \alpha\mathbf{v})$$

and

$$T^{-1}AT = \frac{1}{u_1v_2 - u_2v_1}\begin{pmatrix} v_2 & -v_1 \\ -u_2 & u_1 \end{pmatrix}\begin{pmatrix} \alpha u_1 - \beta v_1 & \beta u_1 + \alpha v_1 \\ \alpha u_2 - \beta v_2 & \beta u_2 + \alpha v_2 \end{pmatrix}$$

$$= \begin{pmatrix} \alpha & \beta \\ -\beta & \alpha \end{pmatrix}$$

This completes the proof of the theorem. ▮

Appendix 3 THE LOGARITHM OF A MATRIX

In our discussion of linear systems with periodic coefficients (Section 2.9, p. 96), we made use of the fact that every nonsingular matrix has a logarithm. This appendix is devoted to a proof of that result, and to some remarks concerning the possibility that the logarithm of a real matrix may not be real.

Theorem 1. *Let B be a nonsingular $n \times n$ matrix. Then there exists an $n \times n$ matrix A (called a logarithm of B) such that*

$$e^A = B \tag{1}$$

Proof. If $\hat{B}$ is similar to B, so that $T^{-1}BT = \hat{B}$ for some nonsingular matrix T, and if $\exp \hat{A} = \hat{B}$, then

$$B = T\hat{B}T^{-1} = Te^{\hat{A}}T^{-1} = \exp (T\hat{A}T^{-1})$$

and thus $T\hat{A}T^{-1}$ is a logarithm of B. Therefore, to prove Theorem 1, it is sufficient to prove (1) when B is in a suitable canonical form. Thus if $\lambda_1, \ldots, \lambda_k$ are the distinct eigenvalues of B, with multiplicities $n_1, \ldots, n_k$, respectively, we may assume that

$$B = \begin{bmatrix} B_1 & & & \\ & B_2 & & \\ & & \ddots & \\ & & & B_k \end{bmatrix}$$

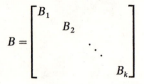

where

$$(B_j - \lambda_j E_{n_j})^{n_j} = 0 \qquad (j = 1, \ldots, k)$$

(See Theorem 2, Appendix 1.) Here, E_{n_j} denotes the $n_j \times n_j$ identity matrix. Thus we may write

$$B_j = \lambda_j\left(E_{n_j} + \frac{1}{\lambda_j} Z_j\right) \qquad Z_j^{n_j} = 0$$

Using the power series expansion

$$\log(1 + x) = \sum_{p=1}^{\infty} \frac{(-1)^{p+1}}{p} x^p \qquad |x| < 1$$

formally, we write

$$\begin{aligned} A_j = \log B_j &= E_{n_j} \log \lambda_j + \log\left(E_{n_j} + \frac{1}{\lambda_j} Z_j\right) \\ &= E_{n_j} \log \lambda_j + \sum_{p=1}^{\infty} \frac{(-1)^{p+1}}{p} \left(\frac{Z_j}{\lambda_j}\right)^p \\ &= E_{n_j} \log \lambda_j + \sum_{p=1}^{n_j-1} \frac{(-1)^{p+1}}{p} \left(\frac{Z_j}{\lambda_j}\right)^p \qquad (j = 1, \ldots, k) \end{aligned} \tag{2}$$

with the power series terminating because $Z_j^{n_j} = 0$. Note, since B is nonsingular, that $\lambda_j \neq 0$, so that $\log \lambda_j$ is defined. If we compute $\exp[\log(1 + x)]$ by expanding the power series for $\log(1 + x)$, substituting into the power series for the exponential function, and then rearranging terms, we obtain simply $(1 + x)$. If we perform the same operations with matrices, we obtain the same terms, and there is no convergence problem since the series (2) for $A_j = \log B_j$ terminates. Thus we have

$$\begin{aligned} \exp(A_j) &= \exp(E_{n_j} \log \lambda_j) \exp\left[\sum_{p=1}^{n_j-1} \frac{(-1)^{p+1}}{p} \left(\frac{Z_j}{\lambda_j}\right)^p\right] \\ &= \lambda_j\left(E_{n_j} + \frac{Z_j}{\lambda_j}\right) = B_j \qquad (j = 1, \ldots, k) \end{aligned}$$

We now write

$$A = \begin{bmatrix} A_1 & & & \\ & A_2 & & \\ & & \ddots & \\ & & & A_k \end{bmatrix}$$

where A_j is defined by (2). Then

$$\exp A = \begin{bmatrix} \exp A_1 & & & \\ & \exp A_2 & & \\ & & \ddots & \\ & & & \exp A_k \end{bmatrix} = \begin{bmatrix} B_1 & & & \\ & B_2 & & \\ & & \ddots & \\ & & & B_k \end{bmatrix} = B$$

and the theorem is proved. ▮

The reader will note that the matrix A in Theorem 1 may not be real even though B is real. Only if all the eigenvalues of B are real and positive will the matrix A given by Theorem 1 be real. It can be shown that if B is a real matrix, then B^2 must have a real logarithm. The proof of this fact depends on the **real canonical form** for matrices, which we have not discussed, and we refer the reader to [7].

We conclude this appendix by giving explicit expressions for the logarithm of a 2×2 real matrix B in canonical form. As pointed out in the proof of Theorem 1, once we have explicit expressions for the logarithm when B is in canonical form, we can easily compute the logarithm in general. As shown in Appendix 2, a nonsingular real 2×2 matrix is similar to one of the following three real matrices:

(i) $B = \begin{pmatrix} \lambda & 0 \\ 0 & \mu \end{pmatrix}$ $\qquad (\lambda \neq 0, \mu \neq 0)$

(ii) $B = \begin{pmatrix} \lambda & 1 \\ 0 & \lambda \end{pmatrix}$ $\qquad (\lambda \neq 0)$

(iii) $B = \begin{pmatrix} \alpha & \beta \\ -\beta & \alpha \end{pmatrix}$ $\qquad (\beta \neq 0)$

We consider each of these cases in turn.

Case (*i*). We take $A = \begin{pmatrix} \log \lambda & 0 \\ 0 & \log \mu \end{pmatrix}$, and then

$$e^A = \begin{pmatrix} \lambda & 0 \\ 0 & \mu \end{pmatrix} = B$$

thus

$$\log B = \begin{pmatrix} \log \lambda & 0 \\ 0 & \log \mu \end{pmatrix}$$

If either λ or μ is negative, A is not real. However

$$B^2 = \begin{pmatrix} \lambda^2 & 0 \\ 0 & \mu^2 \end{pmatrix}$$

and

$$\log (B^2) = \begin{pmatrix} \log \lambda^2 & 0 \\ 0 & \log \mu^2 \end{pmatrix}$$

is real since $\lambda^2 > 0$, $\mu^2 > 0$.

Case (*ii*). We take

$$A = \begin{pmatrix} \log \lambda & 0 \\ 0 & \log \lambda \end{pmatrix} + \begin{pmatrix} 0 & \dfrac{1}{\lambda} \\ 0 & 0 \end{pmatrix}$$

then

$$\exp A = \begin{pmatrix} \lambda & 0 \\ 0 & \lambda \end{pmatrix} \left[\begin{pmatrix} 1 & 0 \\ 0 & 1 \end{pmatrix} + \begin{pmatrix} 0 & \dfrac{1}{\lambda} \\ 0 & 0 \end{pmatrix} \right]$$

$$= \begin{pmatrix} \lambda & 0 \\ 0 & \lambda \end{pmatrix} \begin{pmatrix} 1 & \dfrac{1}{\lambda} \\ 0 & 1 \end{pmatrix} = \begin{pmatrix} \lambda & 1 \\ 0 & \lambda \end{pmatrix} = B$$

Thus

$$\log B = \begin{pmatrix} \log \lambda & 1/\lambda \\ 0 & \log \lambda \end{pmatrix}$$

If $\lambda < 0$, A is not real. However,

$$B^2 = \begin{pmatrix} \lambda^2 & 2\lambda \\ 0 & \lambda^2 \end{pmatrix}$$

and

$$\log (B^2) = \begin{pmatrix} \log \lambda^2 & \dfrac{2}{\lambda} \\ 0 & \log \lambda^2 \end{pmatrix}$$

is real. It is easy to verify that

$$\exp \begin{pmatrix} \log \lambda^2 & \dfrac{2}{\lambda} \\ 0 & \log \lambda^2 \end{pmatrix} = \begin{pmatrix} \lambda^2 & 2\lambda \\ 0 & \lambda^2 \end{pmatrix} = B^2$$

Case (*iii*). We take

$$A = \begin{bmatrix} \log (\alpha^2 + \beta^2)^{1/2} & \tan^{-1} \dfrac{\beta}{\alpha} \\ -\tan^{-1} \dfrac{\beta}{\alpha} & \log (\alpha^2 + \beta^2)^{1/2} \end{bmatrix} = A_1 + A_2$$

where

$$A_1 = \begin{bmatrix} \log (\alpha^2 + \beta^2)^{1/2} & 0 \\ 0 & \log (\alpha^2 + \beta^2)^{1/2} \end{bmatrix}$$

$$A_2 = \begin{bmatrix} 0 & \tan^{-1} \dfrac{\beta}{\alpha} \\ -\tan^{-1} \dfrac{\beta}{\alpha} & 0 \end{bmatrix}$$

Since, as is easily verified, $A_1 A_2 = A_2 A_1$, we have $\exp (A) = \exp (A_1) \exp (A_2)$. But

$$\begin{aligned} \exp(A_1) &= \begin{bmatrix} (\alpha^2 + \beta^2)^{1/2} & 0 \\ 0 & (\alpha^2 + \beta^2)^{1/2} \end{bmatrix} \\ &= (\alpha^2 + \beta^2)^{1/2} E \end{aligned}$$

and as we have shown in Exercise 25, Section 2.5, p. 73,

$$\exp(A_2) = \begin{bmatrix} \cos\left(\tan^{-1}\dfrac{\beta}{\alpha}\right) & \sin\left(\tan^{-1}\dfrac{\beta}{\alpha}\right) \\ -\sin\left(\tan^{-1}\dfrac{\beta}{\alpha}\right) & \cos\left(\tan^{-1}\dfrac{\beta}{\alpha}\right) \end{bmatrix}$$

Since $\cos(\tan^{-1}\beta/\alpha) = \alpha/(\alpha^2+\beta^2)^{1/2}$, $\sin(\tan^{-1}\beta/\alpha) = \beta/(\alpha^2+\beta^2)^{1/2}$, we have

$$\exp(A_2) = \frac{1}{(\alpha^2+\beta^2)^{1/2}}\begin{bmatrix} \alpha & \beta \\ -\beta & \alpha \end{bmatrix}$$

Thus

$$\exp(A) = \exp(A_1)\exp(A_2) = \begin{pmatrix} \alpha & \beta \\ -\beta & \alpha \end{pmatrix} = B$$

and

$$\log B = \begin{pmatrix} \log(\alpha^2+\beta^2)^{1/2} & \tan^{-1}\dfrac{\beta}{\alpha} \\ -\tan^{-1}\dfrac{\beta}{\alpha} & \log(\alpha^2+\beta^2)^{1/2} \end{pmatrix}$$

We note that in this case $\log B$ is real.

Appendix 4 | SOME RESULTS FROM MATRIX THEORY

This appendix contains the proofs of three theorems about matrices that are used in the study of the regulator problem in Chapter 6. Although these theorems are of interest in their own right and have algebraic applications and ramifications, we shall only prove what is needed for the applications in Chapter 6.

I. The first result is concerned with the evaluation of determinants of a particular form.

Theorem 1. *Consider the* $(n + 1) \times (n + 1)$ *matrix.*

$$M = \begin{bmatrix} A & -\mathbf{b} \\ \mathbf{c}^T & -\rho \end{bmatrix}$$

where A *is an* $n \times n$ *matrix,* ρ *is a scalar,* **b** *and* **c** *are* n*-dimensional column vectors,*

$$\mathbf{b} = \begin{bmatrix} b_1 \\ \vdots \\ b_n \end{bmatrix}, \qquad \mathbf{c} = \begin{bmatrix} c_1 \\ \vdots \\ c_n \end{bmatrix}$$

and $\mathbf{c}^T$ *is the transpose of* **c**. *Suppose that* A *is nonsingular. Then*

$$\det M = (-\rho + \mathbf{c}^T A^{-1}\mathbf{b}) \det A \tag{1}$$

Proof. We prove the theorem by induction on n. If $n = 1$, so that $A \neq 0$, b, c, ρ are all scalars, then M is a 2×2 matrix and then (using $A = \det A$, $A^{-1} \det A = 1$)

$$\det M = -A\rho + \mathbf{c}^T\mathbf{b} = (-\rho + c^T A^{-1} b) \det A$$

Thus (1) is true for $n = 1$. Assume that (1) is true for any $(n-1) \times (n-1)$ matrix M, any $(n-1)$-dimensional vectors $\mathbf{b}$, $\mathbf{c}$, and for any scalar ρ. We consider

$$\begin{bmatrix} A^{-1} & \mathbf{0} \\ \mathbf{0} & 1 \end{bmatrix} M = \begin{bmatrix} A^{-1} & \mathbf{0} \\ \mathbf{0} & 1 \end{bmatrix} \begin{bmatrix} A & -\mathbf{b} \\ \mathbf{c}^T & -\rho \end{bmatrix}$$
$$= \begin{bmatrix} E & -A^{-1}\mathbf{b} \\ \mathbf{c}^T & -\rho \end{bmatrix} = \begin{bmatrix} E & \mathbf{r} \\ \mathbf{c}^T & -\rho \end{bmatrix} \tag{2}$$

where $\mathbf{r} = A^{-1}\mathbf{b}$. Expanding the determinant of this matrix by cofactors of the first row, we obtain

$$\det \begin{bmatrix} E & \mathbf{r} \\ c^T & -\rho \end{bmatrix} = \det \begin{bmatrix} 1 & 0 & \cdots & 0 & r_1 \\ 0 & & \ddots & 0 & r_2 \\ \vdots & & \ddots & \vdots & \vdots \\ & & & 0 & \\ 0 & \cdots & 0 & 1 & r_n \\ c_1 & & \cdots & c_n & -\rho \end{bmatrix}$$

$$= \det \begin{bmatrix} 1 & 0 & \cdots & 0 & r_2 \\ 0 & & \ddots & & \vdots \\ \vdots & \ddots & & 0 & \vdots \\ 0 & \cdots & 0 & 1 & r_n \\ c_2 & & & c_n & -\rho \end{bmatrix} + (-1)^n r_1 \det \begin{bmatrix} 0 & 1 & 0 & \cdots & 0 \\ 0 & 0 & & \ddots & \vdots \\ \vdots & \vdots & \ddots & & 0 \\ 0 & 0 & \cdots & 0 & 1 \\ c_1 & c_2 & & & c_n \end{bmatrix} \tag{3}$$

We now apply the induction hypothesis to evaluate the first determinant using $A = E$, the identity matrix; thus its value is $-\rho - \sum_{j=2}^{n} c_j r_j$. The second determinant on the right-hand side of (3) may be evaluated by expanding by cofactors of its first column; its value is $(-1)^{n-1} c_1$. Substituting the results into (3), we obtain

$$\det \begin{bmatrix} E & \mathbf{r} \\ \mathbf{c}^T & -\rho \end{bmatrix} = -\rho - \sum_{j=2}^{n} c_j r_j + (-1)^{2n-1} c_1 r_1$$
$$= -\rho - \sum_{j=1}^{n} c_j r_j = -\rho - \mathbf{c}^T\mathbf{r} = -\rho + \mathbf{c}^T A^{-1}\mathbf{b} \tag{4}$$

Taking determinants of both sides of (2), and using

$$\det \begin{bmatrix} A^{-1} & \mathbf{0} \\ \mathbf{0} & 1 \end{bmatrix} = \det (A^{-1}) = (\det A)^{-1}$$

as well as (4) yields

$$(\det A)^{-1} \det M = -\rho + \mathbf{c}^T A^{-1} \mathbf{b}$$

This proves (1) and completes the proof of Theorem 1.

II. The second result is Sylvester's theorem on positive definite matrices.

Theorem 2. *Let*

$$B = \begin{bmatrix} b_{11} & \cdots & b_{1n} \\ \vdots & & \vdots \\ b_{n1} & \cdots & b_{nn} \end{bmatrix}$$

be a real symmetric matrix. Then B is positive definite if and only if each of the principal minors

$$\det B_j = \det \begin{bmatrix} b_{11} & \cdots & b_{1j} \\ \vdots & & \vdots \\ b_{j1} & \cdots & b_{jj} \end{bmatrix} \qquad (j = 1, \ldots, n)$$

is positive.

Proof. We prove the theorem by induction on n. It is obviously true for $n = 1$. Now suppose that it is true for $n = p$. Consider the $(p + 1) \times (p + 1)$ symmetric matrix

$$B_{p+1} = \begin{bmatrix} B_p & \mathbf{d} \\ \mathbf{d}^T & b_{p+1,p+1} \end{bmatrix} \qquad \mathbf{d} = \begin{bmatrix} b_{1,p+1} \\ \vdots \\ b_{p,p+1} \end{bmatrix} \tag{5}$$

Let $\mathbf{x}$ be a $(p + 1)$-dimensional column vector with components $x_1, x_2, \ldots, x_p, x_{p+1}$, and let $\mathbf{y}$ be the p-dimensional column vector formed from the first p

components of **x**, so that

$$\mathbf{y} = \begin{bmatrix} x_1 \\ \vdots \\ x_p \end{bmatrix} \qquad \mathbf{x} = \begin{bmatrix} \mathbf{y} \\ x_{p+1} \end{bmatrix}$$

A straighttorward calculation shows that

$$\mathbf{x}^T B_{p+1} \mathbf{x} = \mathbf{y}^T B_p \mathbf{y} + x_{p+1} \mathbf{d}^T \mathbf{y} + x_{p+1} \mathbf{y}^T \mathbf{d} + x_{p+1} b_{p+1,p+1} x_{p+1} \tag{6}$$

It is clear from (6), with $x_{p+1} = 0$, that B_p must be positive definite if B_{p+1} is to be positive definite. We may also rewrite (6) in the form

$$\mathbf{x}^T B_{p+1} \mathbf{x} = (\mathbf{y}^T + x_{p+1} \mathbf{d}^T B_p^{-1}) B_p (\mathbf{y} + x_{p+1} B_p^{-1} \mathbf{d}) + (b_{p+1,p+1} - \mathbf{d}^T B_p^{-1} \mathbf{d}) x_{p+1}^2$$

From this it is clear that B_{p+1} is positive definite if and only if B_p is positive definite; that is, if and only if

$$\det B_1 > 0, \det B_2 > 0, \ldots, \det B_p > 0$$

and if and only if

$$b_{p+1,p+1} - \mathbf{d}^T B_p^{-1} \mathbf{d} > 0$$

But by Theorem 1 above,

$$\det B_{p+1} = (b_{p+1,p+1} - \mathbf{d}^T B_p^{-1} \mathbf{d}) \det B_p$$

Thus B_{p+1} is positive definite if and only if

$$\det B_1 > 0, \det B_2 > 0, \ldots, \det B_p > 0, \det B_{p+1} > 0$$

which proves the theorem by induction. ▮

III. The final result needed is Lyapunov's theorem on the solvability of matrix equations.

Theorem 3. *Let A be a given stable matrix and let C be any positive definite symmetric matrix. Then there exists a unique matrix B such that*

$$AB + BA^T = -C \tag{7}$$

Moreover, the matrix B is symmetric and positive definite.

Proof. We begin by solving the matrix differential equation

$$Y' = AY + YA^T \tag{8}$$

(see Exercise 11, p. 105). We let

$$Y(t) = U(t) \exp(A^T t)$$

where U is a matrix to be determined. Then (8) becomes

$$Y' = U' \exp(A^T t) + U \exp(A^T t)A^T = AU \exp(A^T t) + U \exp(A^T t)A^T$$

or

$$U' = AU \tag{9}$$

The solution $U(t)$ of (9) with $U(0) = C$ is

$$U(t) = e^{At}C$$

(Theorem 2.7, p. 57). Thus the solution $Y(t)$ of (8) with $Y(0) = C$ is

$$Y(t) = \exp(At)\, C \exp(A^T t) \tag{10}$$

The next step is to use (10) to solve the matrix equation (7). We define the matrix

$$B = \int_0^\infty \exp(At)C \exp(A^T t)\, dt = \int_0^\infty Y(t)\, dt \tag{11}$$

provided this integral converges; here $Y(t)$ is given by (10). Before making use of the matrix B, we must show that under our hypothesis the integral converges. Since A is assumed to be a stable matrix, all of its eigenvalues have negative real part. The eigenvalues of A^T are the same as the eigenvalues of A. By the Corollary to Theorem 2.10, p. 81, there exist constants $K > 0$, $\sigma > 0$ such that

$$|e^{At}| \leq Ke^{-\sigma t}$$

Thus

$$\begin{aligned}|\exp(At)C\exp(A^Tt)| &\le |\exp(At)|\,|C|\,|\exp(A^Tt)| \\ &\le K^2|C|\exp(-2\sigma t)\end{aligned}$$

This proves the convergence of the integral (11).

We next show that the matrix B provides the solution of the system (7). Thus

$$\begin{aligned}AB + BA^T &= A\int_0^\infty Y(t)\,dt + \int_0^\infty Y(t)\,dt\,A^T \\ &= \int_0^\infty [AY(t) + Y(t)A^T]\,dt = \int_0^\infty Y'(t)\,dt = \lim_{R\to\infty} Y(R) - Y(0)\end{aligned}$$

where we have used the fact that $Y(t)$ defined by (10) satisfies (8) and that $Y(0) = C$. The convergence of the integral (11) implies $\lim_{R\to\infty} Y(R) = 0$, and thus $AB + BA^T = -Y(0) = -C$, and therefore B is a solution of (7).

We may regard (7) as a system of n^2 algebraic equations for the elements of B. Since this system has a solution for every nonhomogeneous term C, the determinant of its coefficients is nonzero, and therefore this solution is unique.

Finally, we must show that the solution B given by (11) is symmetric, and positive definite. Since C is symmetric,

$$\begin{aligned}Y^T(t) &= (\exp(At)C\exp(A^Tt))^T = \exp(At)C^T\exp(A^Tt) \\ &= \exp(At)C\exp(A^Tt) = Y(t)\end{aligned}$$

and $B^T = \int_0^\infty Y^T(t)\,dt = \int_0^\infty Y(t)\,dt = B$. If $\mathbf{x}$ is any nonzero vector,

$$\begin{aligned}\mathbf{x}^T B\mathbf{x} &= \mathbf{x}^T\int_0^\infty \exp(At)C\exp(A^Tt)\,dt\,\mathbf{x} \\ &= \int_0^\infty (\exp(A^Tt)\mathbf{x})^T C(\exp(A^Tt)\mathbf{x})\,dt\end{aligned}$$

Since C is positive definite, $(\exp(A^Tt)\mathbf{x})^T\,C\,(\exp(A^Tt)\mathbf{x}) > 0$, and therefore $\mathbf{x}^T B\mathbf{x} > 0$. This completes the proof of the theorem. ▮

BIBLIOGRAPHY

1. M. A. Aizerman and F. R. Gantmacher, *Absolute Stability of Control Systems* (translation from the Russian) (Holden-Day, San Francisco, California, 1964).
2. F. Brauer and J. A. Nohel, *Ordinary Differential Equations: A First Course* (W. A. Benjamin, New York, 1967).
3. L. Cesari, *Asymptotic Behavior and Stability Problems in Ordinary Differential Equations*, 2nd edition (Springer, Berlin, 1963).
4. E. A. Coddington and N. Levinson, *Theory of Ordinary Differential Equations* (McGraw-Hill, New York, 1955).
5. W. A. Coppel, *Stability and Asymptotic Behavior of Differential Equations* (Heath, Boston, Massachusetts, 1965).
6. P. Franklin, *A Treatise on Advanced Calculus* (Wiley, New York, 1940).
7. F. R. Gantmacher, *Theory of Matrices* (translation from the Russian) (Chelsea, New York, 1957).
8. W. Hahn, *Stability of Motion* (Springer, Berlin, 1967).
9. A. Halanay, *Differential Equations—Stability, Oscillations, Time Lags* (Academic Press, New York, 1966).
10. J. K. Hale, *Oscillations in Nonlinear Systems* (McGraw-Hill, New York, 1963).
11. P. R. Halmos, *Finite Dimensional Vector Spaces*, 2nd edition (Van Nostrand, Princeton, New Jersey, 1958).
12. F. B. Hildebrand, *Advanced Calculus for Applications* (Prentice-Hall, Englewood Cliffs, New Jersey, 1962).

13. W. Hurewicz, *Lectures on Ordinary Differential Equations* (MIT Press, Cambridge, Massachusetts, 1958).
14. J. Korevaar, *Mathematical Methods*, Vol. 1: *Linear Algebra, Normed Spaces, Distributions, Integration* (Academic Press, New York, 1968).
15. N. N. Krasovskii, *Stability of Motion* (translation from the Russian) (Stanford Univ. Press, Stanford, California, 1963).
16. J. P. LaSalle and S. Lefschetz, *Stability by Liapunov's Direct Method with Applications* (Academic Press, New York, 1961).
17. E. B. Lee and L. Markus, *Foundations of Optimal Control Theory* (Wiley, New York, 1967).
18. S. Lefschetz, *Stability of Nonlinear Control Systems* (Academic Press, New York, 1965).
19. W. S. Loud, "Periodic Solutions of $x'' + cx' + g(x) = \varepsilon f(t)$", *Memoir No. 31*, Amer. Math. Soc., Providence, Rhode Island, 1959.
20. I. G. Petrovskii, *Ordinary Differential Equations* (translation from the Russian) (Prentice-Hall, Englewood Cliffs, New Jersey, 1966).
21. G. Sansone and R. Conti, *Non-linear Differential Equations* (translation from the Italian) (Macmillan, New York, 1964).
22. J. J. Stoker, *Nonlinear Vibrations in Mechanical and Electrical Systems* (Wiley [Interscience], New York, 1950).
23. R. A. Struble, *Nonlinear Differential Equations* (McGraw-Hill, New York, 1962).
24. W. R. Wasow, *Asymptotic Expansions for Ordinary Differential Equations* (Wiley, New York, 1966).
25. T. Yoshizawa, *Stability Theory by Liapunov's Second Method* (Math. Soc. Japan, Tokyo, 1966).
26. K. Yosida, *Lectures on Differential and Integral Equations* (Wiley [Interscience], New York, 1960).

INDEX

A CATALOG OF SELECTED

DOVER BOOKS

IN ALL FIELDS OF INTEREST

A CATALOG OF SELECTED DOVER BOOKS IN ALL FIELDS OF INTEREST

HOW TO WRITE, Gertrude Stein. Gertrude Stein claimed anyone could understand her unconventional writing—here are clues to help. Fascinating improvisations, language experiments, explanations illuminate Stein's craft and the art of writing. Total of 414pp. 4⅝ × 6⅝. 23144-5 Pa. $5.95

ADVENTURES AT SEA IN THE GREAT AGE OF SAIL: Five Firsthand Narratives, edited by Elliot Snow. Rare true accounts of exploration, whaling, shipwreck, fierce natives, trade, shipboard life, more. 33 illustrations. Introduction. 353pp. 5⅜ × 8½. 25177-2 Pa. $7.95

THE HERBAL OR GENERAL HISTORY OF PLANTS, John Gerard. Classic descriptions of about 2,850 plants—with over 2,700 illustrations—includes Latin and English names, physical descriptions, varieties, time and place of growth, more. 2,706 illustrations. xlv + 1,678pp. 8½ × 12¼. 23147-X Cloth. $75.00

DOROTHY AND THE WIZARD IN OZ, L. Frank Baum. Dorothy and the Wizard visit the center of the Earth, where people are vegetables, glass houses grow and Oz characters reappear. Classic sequel to *Wizard of Oz*. 256pp. 5⅜ × 8. 24714-7 Pa. $4.95

SONGS OF EXPERIENCE: Facsimile Reproduction with 26 Plates in Full Color, William Blake. This facsimile of Blake's original "Illuminated Book" reproduces 26 full-color plates from a rare 1826 edition. Includes "The Tyger," "London," "Holy Thursday," and other immortal poems. 26 color plates. Printed text of poems. 48pp. 5¼ × 7. 24636-1 Pa. $3.50

SONGS OF INNOCENCE, William Blake. The first and most popular of Blake's famous "Illuminated Books," in a facsimile edition reproducing all 31 brightly colored plates. Additional printed text of each poem. 64pp. 5¼ × 7. 22764-2 Pa. $3.50

PRECIOUS STONES, Max Bauer. Classic, thorough study of diamonds, rubies, emeralds, garnets, etc.: physical character, occurrence, properties, use, similar topics. 20 plates, 8 in color. 94 figures. 659pp. 6⅛ × 9¼. 21910-0, 21911-9 Pa., Two-vol. set $15.90

ENCYCLOPEDIA OF VICTORIAN NEEDLEWORK, S. F. A. Caulfeild and Blanche Saward. Full, precise descriptions of stitches, techniques for dozens of needlecrafts—most exhaustive reference of its kind. Over 800 figures. Total of 679pp. 8⅛ × 11. Two volumes. Vol. 1 22800-2 Pa. $11.95
Vol. 2 22801-0 Pa. $11.95

THE MARVELOUS LAND OF OZ, L. Frank Baum. Second Oz book, the Scarecrow and Tin Woodman are back with hero named Tip, Oz magic. 136 illustrations. 287pp. 5⅜ × 8½. 20692-0 Pa. $5.95

WILD FOWL DECOYS, Joel Barber. Basic book on the subject, by foremost authority and collector. Reveals history of decoy making and rigging, place in American culture, different kinds of decoys, how to make them, and how to use them. 140 plates. 156pp. 7⅞ × 10¾. 20011-6 Pa. $8.95

HISTORY OF LACE, Mrs. Bury Palliser. Definitive, profusely illustrated chronicle of lace from earliest times to late 19th century. Laces of Italy, Greece, England, France, Belgium, etc. Landmark of needlework scholarship. 266 illustrations. 672pp. 6⅛ × 9¼. 24742-2 Pa. $14.95

THE BLUE FAIRY BOOK, Andrew Lang. The first, most famous collection, with many familiar tales: Little Red Riding Hood, Aladdin and the Wonderful Lamp, Puss in Boots, Sleeping Beauty, Hansel and Gretel, Rumpelstiltskin; 37 in all. 138 illustrations. 390pp. 5⅜ × 8½. 21437-0 Pa. $5.95

THE STORY OF THE CHAMPIONS OF THE ROUND TABLE, Howard Pyle. Sir Launcelot, Sir Tristram and Sir Percival in spirited adventures of love and triumph retold in Pyle's inimitable style. 50 drawings, 31 full-page. xviii + 329pp. 6½ × 9¼. 21883-X Pa. $6.95

AUDUBON AND HIS JOURNALS, Maria Audubon. Unmatched two-volume portrait of the great artist, naturalist and author contains his journals, an excellent biography by his granddaughter, expert annotations by the noted ornithologist, Dr. Elliott Coues, and 37 superb illustrations. Total of 1,200pp. 5⅜ × 8.
Vol. I 25143-8 Pa. $8.95
Vol. II 25144-6 Pa. $8.95

GREAT DINOSAUR HUNTERS AND THEIR DISCOVERIES, Edwin H. Colbert. Fascinating, lavishly illustrated chronicle of dinosaur research, 1820's to 1960. Achievements of Cope, Marsh, Brown, Buckland, Mantell, Huxley, many others. 384pp. 5¼ × 8¼. 24701-5 Pa. $6.95

THE TASTEMAKERS, Russell Lynes. Informal, illustrated social history of American taste 1850's–1950's. First popularized categories Highbrow, Lowbrow, Middlebrow. 129 illustrations. New (1979) afterword. 384pp. 6 × 9. 23993-4 Pa. $6.95

DOUBLE CROSS PURPOSES, Ronald A. Knox. A treasure hunt in the Scottish Highlands, an old map, unidentified corpse, surprise discoveries keep reader guessing in this cleverly intricate tale of financial skullduggery. 2 black-and-white maps. 320pp. 5⅜ × 8½. (Available in U.S. only) 25032-6 Pa. $5.95

AUTHENTIC VICTORIAN DECORATION AND ORNAMENTATION IN FULL COLOR: 46 Plates from "Studies in Design," Christopher Dresser. Superb full-color lithographs reproduced from rare original portfolio of a major Victorian designer. 48pp. 9¼ × 12¼. 25083-0 Pa. $7.95

PRIMITIVE ART, Franz Boas. Remains the best text ever prepared on subject, thoroughly discussing Indian, African, Asian, Australian, and, especially, Northern American primitive art. Over 950 illustrations show ceramics, masks, totem poles, weapons, textiles, paintings, much more. 376pp. 5⅜ × 8. 20025-6 Pa. $6.95

SIDELIGHTS ON RELATIVITY, Albert Einstein. Unabridged republication of two lectures delivered by the great physicist in 1920–21. *Ether and Relativity* and *Geometry and Experience*. Elegant ideas in non-mathematical form, accessible to intelligent layman. vi + 56pp. 5⅜ × 8½. 24511-X Pa. $2.95

THE WIT AND HUMOR OF OSCAR WILDE, edited by Alvin Redman. More than 1,000 ripostes, paradoxes, wisecracks: Work is the curse of the drinking classes, I can resist everything except temptation, etc. 258pp. 5⅜ × 8½. 20602-5 Pa. $4.50

ADVENTURES WITH A MICROSCOPE, Richard Headstrom. 59 adventures with clothing fibers, protozoa, ferns and lichens, roots and leaves, much more. 142 illustrations. 232pp. 5⅜ × 8½. 23471-1 Pa. $3.95

PLANTS OF THE BIBLE, Harold N. Moldenke and Alma L. Moldenke. Standard reference to all 230 plants mentioned in Scriptures. Latin name, biblical reference, uses, modern identity, much more. Unsurpassed encyclopedic resource for scholars, botanists, nature lovers, students of Bible. Bibliography. Indexes. 123 black-and-white illustrations. 384pp. 6 × 9. 25069-5 Pa. $8.95

FAMOUS AMERICAN WOMEN: A Biographical Dictionary from Colonial Times to the Present, Robert McHenry, ed. From Pocahontas to Rosa Parks, 1,035 distinguished American women documented in separate biographical entries. Accurate, up-to-date data, numerous categories, spans 400 years. Indices. 493pp. 6½ × 9¼. 24523-3 Pa. $9.95

THE FABULOUS INTERIORS OF THE GREAT OCEAN LINERS IN HISTORIC PHOTOGRAPHS, William H. Miller, Jr. Some 200 superb photographs capture exquisite interiors of world's great "floating palaces"—1890's to 1980's: *Titanic, Ile de France, Queen Elizabeth, United States, Europa,* more. Approx. 200 black-and-white photographs. Captions. Text. Introduction. 160pp. 8⅞ × 11¾. 24756-2 Pa. $9.95

THE GREAT LUXURY LINERS, 1927–1954: A Photographic Record, William H. Miller, Jr. Nostalgic tribute to heyday of ocean liners. 186 photos of Ile de France, Normandie, Leviathan, Queen Elizabeth, United States, many others. Interior and exterior views. Introduction. Captions. 160pp. 9 × 12. 24056-8 Pa. $9.95

A NATURAL HISTORY OF THE DUCKS, John Charles Phillips. Great landmark of ornithology offers complete detailed coverage of nearly 200 species and subspecies of ducks: gadwall, sheldrake, merganser, pintail, many more. 74 full-color plates, 102 black-and-white. Bibliography. Total of 1,920pp. 8⅝ × 11¼. 25141-1, 25142-X Cloth. Two-vol. set $100.00

THE SEAWEED HANDBOOK: An Illustrated Guide to Seaweeds from North Carolina to Canada, Thomas F. Lee. Concise reference covers 78 species. Scientific and common names, habitat, distribution, more. Finding keys for easy identification. 224pp. 5⅜ × 8½. 25215-9 Pa. $5.95

THE TEN BOOKS OF ARCHITECTURE: The 1755 Leoni Edition, Leon Battista Alberti. Rare classic helped introduce the glories of ancient architecture to the Renaissance. 68 black-and-white plates. 336pp. 8⅜ × 11¼. 25239-6 Pa. $14.95

MISS MACKENZIE, Anthony Trollope. Minor masterpieces by Victorian master unmasks many truths about life in 19th-century England. First inexpensive edition in years. 392pp. 5⅜ × 8½. 25201-9 Pa. $7.95

THE RIME OF THE ANCIENT MARINER, Gustave Doré, Samuel Taylor Coleridge. Dramatic engravings considered by many to be his greatest work. The terrifying space of the open sea, the storms and whirlpools of an unknown ocean, the ice of Antarctica, more—all rendered in a powerful, chilling manner. Full text. 38 plates. 77pp. 9¼ × 12. 22305-1 Pa. $4.95

THE EXPEDITIONS OF ZEBULON MONTGOMERY PIKE, Zebulon Montgomery Pike. Fascinating first-hand accounts (1805–6) of exploration of Mississippi River, Indian wars, capture by Spanish dragoons, much more. 1,088pp. 5⅜ × 8½. 25254-X, 25255-8 Pa. Two-vol. set $23.90

SIR HARRY HOTSPUR OF HUMBLETHWAITE, Anthony Trollope. Incisive, unconventional psychological study of a conflict between a wealthy baronet, his idealistic daughter, and their scapegrace cousin. The 1870 novel in its first inexpensive edition in years. 250pp. 5⅜ × 8½. 24953-0 Pa. $5.95

LASERS AND HOLOGRAPHY, Winston E. Kock. Sound introduction to burgeoning field, expanded (1981) for second edition. Wave patterns, coherence, lasers, diffraction, zone plates, properties of holograms, recent advances. 84 illustrations. 160pp. 5⅜ × 8¼. (Except in United Kingdom) 24041-X Pa. $3.50

INTRODUCTION TO ARTIFICIAL INTELLIGENCE: SECOND, ENLARGED EDITION, Philip C. Jackson, Jr. Comprehensive survey of artificial intelligence—the study of how machines (computers) can be made to act intelligently. Includes introductory and advanced material. Extensive notes updating the main text. 132 black-and-white illustrations. 512pp. 5⅜ × 8½. 24864-X Pa. $8.95

HISTORY OF INDIAN AND INDONESIAN ART, Ananda K. Coomaraswamy. Over 400 illustrations illuminate classic study of Indian art from earliest Harappa finds to early 20th century. Provides philosophical, religious and social insights. 304pp. 6⅝ × 9⅝. 25005-9 Pa. $8.95

THE GOLEM, Gustav Meyrink. Most famous supernatural novel in modern European literature, set in Ghetto of Old Prague around 1890. Compelling story of mystical experiences, strange transformations, profound terror. 13 black-and-white illustrations. 224pp. 5⅜ × 8½. (Available in U.S. only) 25025-3 Pa. $5.95

ARMADALE, Wilkie Collins. Third great mystery novel by the author of *The Woman in White* and *The Moonstone*. Original magazine version with 40 illustrations. 597pp. 5⅜ × 8½. 23429-0 Pa. $9.95

PICTORIAL ENCYCLOPEDIA OF HISTORIC ARCHITECTURAL PLANS, DETAILS AND ELEMENTS: With 1,880 Line Drawings of Arches, Domes, Doorways, Facades, Gables, Windows, etc., John Theodore Haneman. Sourcebook of inspiration for architects, designers, others. Bibliography. Captions. 141pp. 9 × 12. 24605-1 Pa. $6.95

BENCHLEY LOST AND FOUND, Robert Benchley. Finest humor from early 30's, about pet peeves, child psychologists, post office and others. Mostly unavailable elsewhere. 73 illustrations by Peter Arno and others. 183pp. 5⅜ × 8½. 22410-4 Pa. $3.95

ERTÉ GRAPHICS, Erté. Collection of striking color graphics: *Seasons, Alphabet, Numerals, Aces* and *Precious Stones*. 50 plates, including 4 on covers. 48pp. 9⅜ × 12¼. 23580-7 Pa. $6.95

THE JOURNAL OF HENRY D. THOREAU, edited by Bradford Torrey, F. H. Allen. Complete reprinting of 14 volumes, 1837-61, over two million words; the sourcebooks for *Walden*, etc. Definitive. All original sketches, plus 75 photographs. 1,804pp. 8½ × 12¼. 20312-3, 20313-1 Cloth., Two-vol. set $80.00

CASTLES: THEIR CONSTRUCTION AND HISTORY, Sidney Toy. Traces castle development from ancient roots. Nearly 200 photographs and drawings illustrate moats, keeps, baileys, many other features. Caernarvon, Dover Castles, Hadrian's Wall, Tower of London, dozens more. 256pp. 5⅜ × 8¼. 24898-4 Pa. $5.95